The Use of the
Scanning Electron Microscope

The Use of the Scanning Electron Microscope

BY

J. W. S. HEARLE

J. T. SPARROW

AND

P. M. CROSS

PERGAMON PRESS

OXFORD · NEW YORK · TORONTO

SYDNEY · BRAUNSCHWEIG

Pergamon Press Ltd., Headington Hill Hall, Oxford
Pergamon Press Inc., Maxwell House, Fairview Park, Elmsford, New York 10523
Pergamon of Canada Ltd., 207 Queen's Quay West, Toronto 1
Pergamon Press (Aust.) Pty. Ltd., 19a Boundary Street, Rushcutters Bay, N.S.W. 2011, Australia
Vieweg & Sohn GmbH, Burgplatz 1, Braunschweig

First edition 1972

Library of Congress Catalog Card No. 72–79072

Printed in Great Britain by A. Wheaton & Co., Exeter

08 016246 0

Contents

List of Contributors

DR. J. E. CASTLE, Department of Metallurgy, University of Surrey, Guildford, Surrey.

MRS. P. M. CROSS, R.R.2., Preston, Ontario, Canada.

DR. P. ECHLIN, The Department of Botany, The University of Cambridge, Downing Street, Cambridge CB2 3EA.

DR. J. W. S. HEARLE, Department of Textile Technology, University of Manchester Institute of Science and Technology, P.O. Box No. 88, Sackville Street, Manchester 1.

MR. G. S. LANE, Engineering Research Division, Railway Technical Centre, London Road, Derby.

DR. D. C. NORTHROP, Electrical Engineering Dept., University of Manchester Institute of Science and Technology, P.O. Box No. 88, Sackville Street, Manchester 1.

MR. A. J. SHERRIN, Textile Research Dept., Dunlop Research Centre, Kingsbury Road, Birmingham 24.

MR. J. T. SPARROW, Department of Textile Technology, University of Manchester Institute of Science and Technology, P.O. Box No. 88, Sackville Street, Manchester 1.

CHAPTER 1

Introduction to Scanning Electron Microscopy

J. W. S. HEARLE

1.1. THE BACKGROUND

1.1.1. History of microscopy

The first microscopes were made in the seventeenth century. The most effective were probably those made by Leeuwenhoek—tiny glass beads mounted in a metal plate, held close to the eye, and capable of showing detail down to 1 μm on a carefully sited and illuminated object. Other types of microscopes of the same period gave a poor image, and it was not until the nineteenth century that the compound microscope achieved technical superiority over the simple-lens microscope: it was much easier to use, and by advances in design technology, the resolution was improved and aberrations were reduced. In 1876 Abbé showed by his theory of image formation that there was a limit of about 0·2 μm, set by the wavelength of light. The instrument was near its peak, and since 1900 the major advances have been mainly in techniques of use, in methods of illumination and in ways of promoting contrast.

By 1900 another possibility was available, electrons could be guided in curved paths and so could be used to form magnified images. Time was needed for the development of the technology, but in 1932 the first electron microscope was made. Contrary to the usual myth, this was based on "particle optics" and owed little to de Broglie's presentation in 1924 of a successful theory of the wave nature of the electron. Only later was this used in conjunction with Abbé's theory to calculate the limit of resolution of the electron microscope, far beyond that of the light microscope because of the much smaller wavelength of electrons. Until the 1950s development of electron microscopy was concentrated on technical improvements in the instrument; but since then advances have come more from developments in methods of use based on an understanding of the interaction between electrons and the specimen, and the methods of image formation.

It is interesting to see in Table 1.1 the similar sequence of stages in the development of each type of microscope, for the same stages are apparent in scanning electron microscopy, though some have been reduced to years rather than decades.

By 1960 the combination of light and electron microscopes covered the whole range of magnification needed to study the super-atomic world; but there were still important limitations in their use, notably the limited depth of focus of the optical microscope and the fact that the ordinary electron microscope, which was not much use in reflection, was almost entirely restricted to the examination of very thin specimens, requiring special preparation. Scanning electron microscopy has filled these gaps, and also contributed some other new possibilities.

TABLE 1.1. STAGES IN THE DEVELOPING TECHNOLOGY OF A MICROSCOPE

Stage	Significant dates (approximate)		
	Optical microscope	Direct electron microscope	Scanning electron microscope
1. The ideal "geometrical idea" of the microscope, with no aberrations and no limit to magnification; realization limited by available technology	1600	1900	1935
2. Early design improvements	1800–1870	1932–1950	1945–1955
3. Basic understanding	1875	1935	1955
4. Commercial availability	1800	1940	1965
5. Further design improvements	1875–1900 1950–1960	continuing	continuing
6. Better techniques of use	1930–1960	continuing	continuing

1.1.2. Direct and indirect images

In an ordinary optical microscope the light passes directly into the eye and no real magnified image is formed: the image on the retina is such that the eye "sees" an enlarged virtual image. Using electrons this is not, of course, possible; and the electrons have to be focused on a screen where they stimulate light emission and so give a real, direct, enlarged image of the specimen. The same principle is used in the projection light microscope. Figure 1.1 is a schematic diagram of either type of microscope.

The best term for an electron microscope of this sort is probably a *direct electron microscope* (referring to the direct image formation) but more commonly used adjectives have been *conventional* or *traditional* (an historical distinction which will soon lose significance) or *transmission* (unsatisfactory because scanning electron microscopes are now being used in transmission).

Instead of direct image formation, it is possible, with light (as in the flying-spot microscope) or electrons, to form a magnified image indirectly by a scanning system. One way—the way in which an ordinary television system works—is to illuminate the object generally, collect the radiation, form an image and scan across it so as to pick up separately the response from each part. A corresponding signal is then transmitted on a spot scanning across a screen in register with the original scan. This method is illustrated in Fig. 1.2 (a). The magnification, or reduction, in image size is controlled by the relative areas of scan.

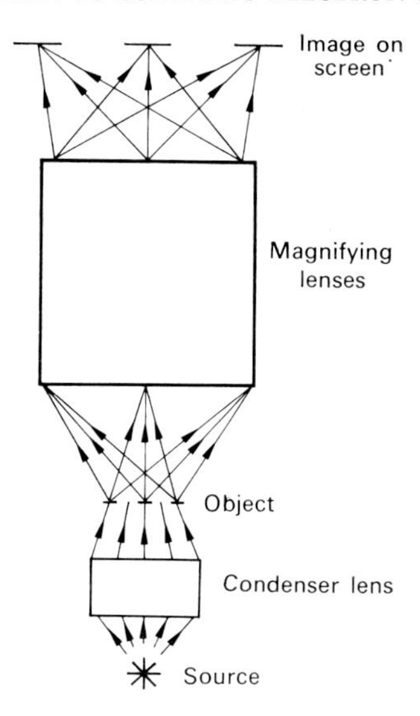

FIG. 1.1. Schematic diagram of light microscope or direct electron microscope.

For microscopy, an alternative scanning system is preferable. There is no need for general illumination. The light or electrons can be concentrated on one spot, which is then traversed across the specimen as indicated in Fig. 1.2 (b), and the complete response at each instant used to modulate the signal governing the image on the screen.

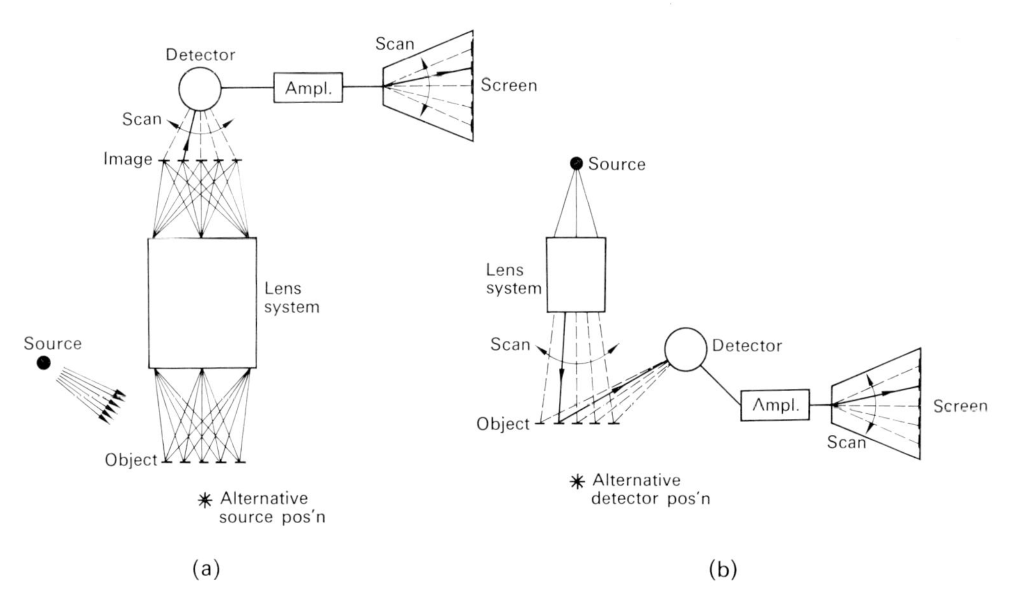

FIG. 1.2. (a) Scanning system applied to radiation which has come from specimen. (b) Scanning system applied to radiation incident on specimen. Asterisks (*) show position of source or detector for viewing in transmission.

1.2. THE SCANNING ELECTRON MICROSCOPE

1.2.1. The instrument

The idea of making an electron microscope on this principle was suggested by M. Knoll in 1935, and as a result M. von Ardenne constructed a scanning electron microscope in 1938. This, and other early instruments, seemed of little practical value: the resolution was little better than an optical microscope and recording times were long. Great improvements in design were made by C. W. Oatley and his group at Cambridge, and resolution of the order of 250 Å, similar to that of the first commercial

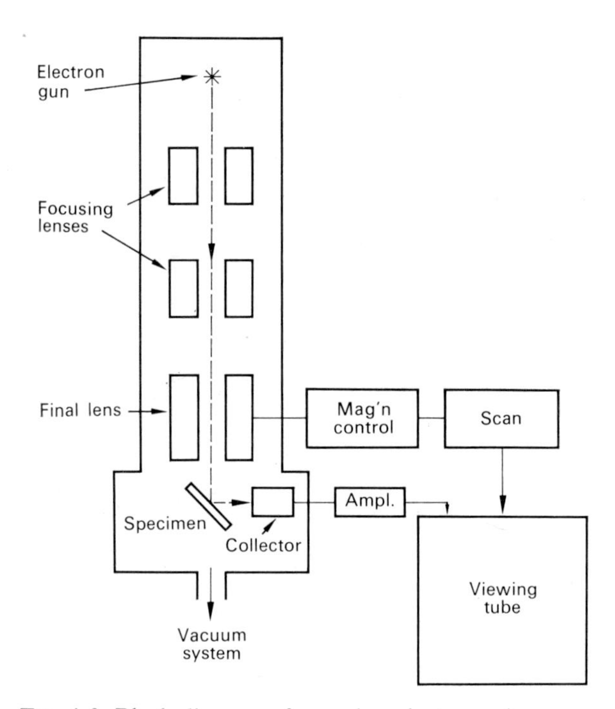

FIG. 1.3. Block diagram of scanning electron microscope.

instruments, had been achieved by 1955. This was still much worse in resolution than could be achieved with direct electron microscopes. Consequently with all the thrust in electron microscopy directed towards magnification and high resolution, the fact that most electron microscope work does not make use of the available resolution,* and that the humdrum virtues of depth of focus and convenience in use are very important, was not appreciated. Widespread exploitation of the scanning electron microscope was delayed; but, once started, it grew at an enormous rate. In 1955 there were only two or three instruments; in 1965, when the first commercial model was introduced, perhaps ten; in 1970 there were over 500 in use.

* A recent survey indicated that 60% of conventional electron micrographs were taken at less than 40,000 times magnification.

The essential features of a scanning electron microscope, indicated in Fig. 1.3, are:

1. an electron source,
2. a means of focusing a tiny spot of electrons from the source on the specimen,
3. a means of scanning the spot across the specimen,
4. a means of detecting the response from the specimen,
5. a display system, capable of being scanned in register with the incident scan,
6. a means of transmitting the response from the specimen to the display system.

Figure 1.3 indicates the commonest mode of use in which it is the electrons scattered by the specimen which are collected as the response of the specimen.

1.2.2. The practical advantages

Table 1.2 compares the three important instruments—optical, direct electron, and scanning electron microscopes. They should not be regarded as competitive, but as complementary; and, indeed, some of the most useful research results come from a combination of the use of more than one type of microscope, or from a combination with other instruments.

TABLE 1.2. COMPARISON OF MICROSCOPES

	Optical	Scanning electron	Direct electron
Resolution—easy	*5 μm*	0·2 μm	**100 Å (10 nm)**
—skilled	*0·2 μm*	100 Å (10 nm)	**10 Å (1 nm)**
—special	*0·1 μm*	5 Å (0·5 nm)	**2 Å (0·2 nm)**
Depth of focus	*poor*	**high**	moderate
Mode—transmission	yes	yes	yes
—reflection	yes	yes	*not satisfactory*
—diffraction	yes	yes	yes
—other	some	**many**	*no*
Specimen—preparation	usually easy	**easy**	*skilled, liable to artefacts*
—range and type	versatile real or replica	versatile real or replica	*only thin, or replica*
—maximum thickness for transmission	**thick**	medium	*very thin*
—environment	**versatile**	usually vacuum but can be modified	*vacuum*
—available space	small	**large**	small
Field of view	large enough	large enough	*limited*
Signal	only as image	**available for processing**	only as image
Cost	**low**	high	high

Advantages over others are indicated in bold type; disadvantages in italics.

The obvious advantages of the scanning electron microscope are:

(a) great depth of focus, as illustrated in Fig. 1.4 in comparison with the use of an optical microscope at comparatively low magnification—this is a much greater asset than might at first be imagined;
(b) the possibility of direct observation of the external form of real objects, such as complex fracture surfaces, at high magnification—thus avoiding the necessity to make thin replicas for use in direct transmission electron microscopy;
(c) the ability to switch over a wide range of magnification, so as to zoom down to fine detail on some part identified in position on the whole object;
(d) the ease of operation, and the large space available for dynamic experiments on the specimen.

There are two other advantages which are more specialized. Firstly, an image can be formed as a result of any response of the specimen stimulated differentially by the electron spot which impinges on the specimen. The usual mode of operation is to pick up secondary electrons knocked out of atoms on the surface of the specimen, but scattered primary electrons, light emission, X-ray emission, current in the specimen, and many other responses described in the next chapter can also be used to generate

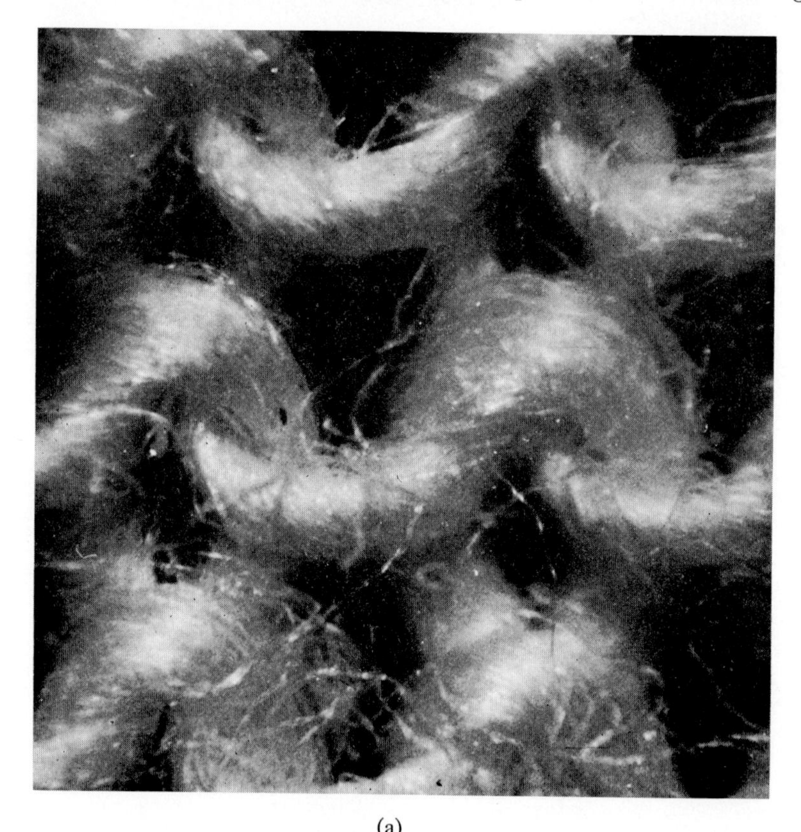

(a)

FIG. 1.4. Comparison of depth of focus in optical and scanning electron microscope. (a) Optical micrograph of knitted cotton fabric.

an image and obtain useful information about the specimen. Secondly, since the information is available in the form of an electric signal, it can be processed in various ways to present images in different forms. For example, two different responses may be added or subtracted; a contour map of image intensity may be presented; the image may be tilted or rotated; or quantitative measurements made.

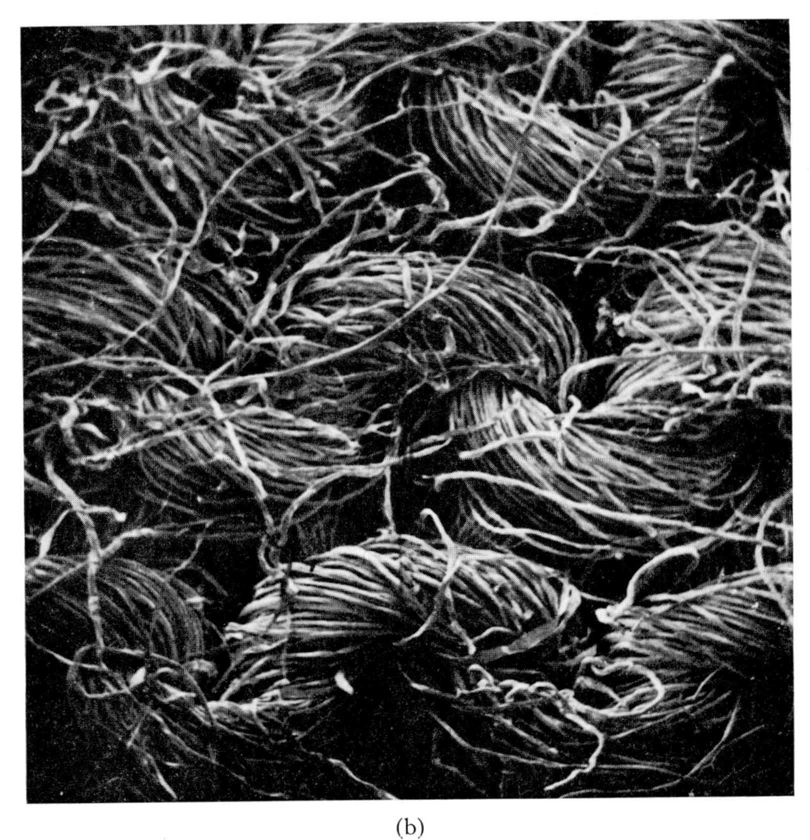

(b)

FIG. 1.4 (*cont.*). (b) SEM micrograph of the same fabric.

Apart from its cost, the obvious limitations of a scanning electron microscope are:

(a) lack of the highest resolution;
(b) the vacuum environment of the specimen;
(c) the inability to show up internal detail, visible in an optical microscope;
(d) the lack of colour response, which gives a means of contrast in light microscopy, additional to intensity differences (though some analagous contrast mechanisms may be involved in scanning electron microscopy).

Some of the factors which limit the quality of a scanning electron microscope image in practice are listed in Table 1.3. Much of this book will be concerned with ways of minimizing these faults.

TABLE 1.3. FACTORS LIMITING QUALITY OF A SCANNING ELECTRON MICROSCOPE IMAGE

Spot size + penetration and spread in specimen.
Spherical aberration.
*Distortion, anisotropic distortion.
*Astigmatism, anisotropic astigmatism.
Chromatic aberration, rotational chromatic aberration.
Space-charge distortion.
Diffraction.
*Departure from symmetry.
*Distortion by external fields.
*Scan faults.
*Other design or operational faults.
Distortion by internal fields, e.g. charging on specimen.
Uncontrolled emission due to charging of specimen.
Noise.
Vibration.
Specimen damage.

* Avoided by good design and maintenance.

1.3. THE LOGIC OF SCANNING ELECTRON MICROSCOPY

1.3.1. Image formation

Image formation can be approached in various ways, and, at first sight, the method indicated in Fig. 1.2 (b) or 1.3 looks very different from what happens in other microscopes. Indeed, it was apparently a source of surprise that the image formed on the screen looked just like the object! And what is more, like the object viewed from the source of electrons.

While, in the absence of ordinary familiarity with the viewing of objects, the relations between incident radiation, object, emitted radiation and image might most logically be dealt with in other ways, it turns out that the analogy with a conventional microscope is surprisingly close, and extremely useful.

1.3.2. The analogy with the reflected light microscope

For the commonest usage of the scanning electron microscope, with an image formed by scattered electrons in order to show up topographical detail, the analogy is with a projection light microscope used to examine a specimen in reflection. Figure 1.5 is a diagram which applies to either type of microscope, with the following differences:

	Direct optical	*Scanning electron*
Radiation	light	electrons
Direction of radiation	$C \rightarrow B \rightarrow A$	$A \rightarrow B \rightarrow C$
Focusing action *or* object–image relation	$PBQ \equiv LAM$ (image formed by separation in space)	$A \equiv P$ at time t_1 $A \equiv B$ at time t_2 $A \equiv Q$ at time t_3 (image formed by separation in time)

The focused relation between points A and B is a fundamental feature of both types of microscope, and shows why the image in the scanning electron microscope appears as if the object were viewed down the tube. Indeed the effect of the scan in the image tube can be regarded as swinging point A (regarded as an image point) from L to M.

Any distortion of the path between B and A will appear as distortion of the image. Thus if a stray electric field is present, it may deflect the beam, giving rise to the

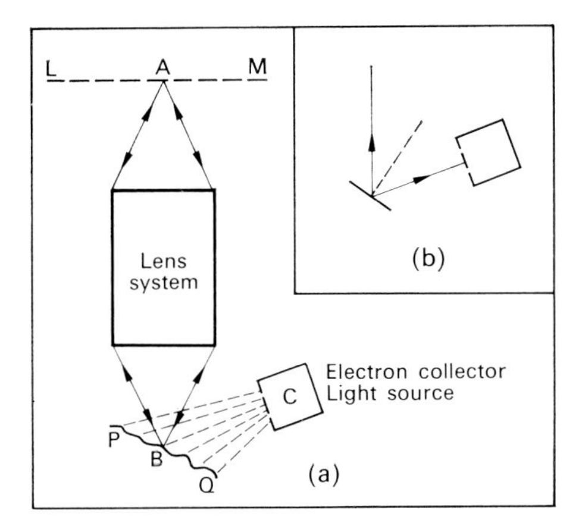

FIG. 1.5. (a) Analogy of reflected light microscope and scanning electron microscope. (b) Condition for formation of image of collector.

distortion illustrated in Fig. 1.6. It is also possible, when a specimen charges up, to reflect the beam to the collector, as indicated in Fig. 1.5 (b), and thus to see an image of the collector, as in Fig. 1.7. The specimen is acting as a mirror.

As an alternative to merely adopting an explanation by analogy, the position can be stated in the following way. At a particular instant, the electron beam is intended to be aimed at B and, provided there are no operational faults, *the image beam is certainly aimed at the point corresponding to B on the image tube.* If there is no distortion, then the state of the object at B will determine the image intensity at that point. But if the beam is deflected to some other point, then this will appear on the image as if it were at B—while B may appear elsewhere or will be missing altogether if the beam never reaches B.

The comparison between the microscopes is shown in another way in Table 1.4, which also brings out the analogy in path geometry between the scanning electron microscope and a conventional microscope in reverse. However, besides considering the geometric paths relating object to image, we must also consider what causes contrast.

In a reflected light microscope, the most important factor is that parts of the specimen facing the source of light are better illuminated than those which face away or are in shadow. Similarly, in a scanning electron microscope, electrons coming from parts of a specimen which face towards the collector are more easily picked up than those which come from parts which face away from the collector. This is illustrated in

FIG. 1.6. Distortion of beam by charged object. The charged fibre has distorted the image of the circular stub lying below.

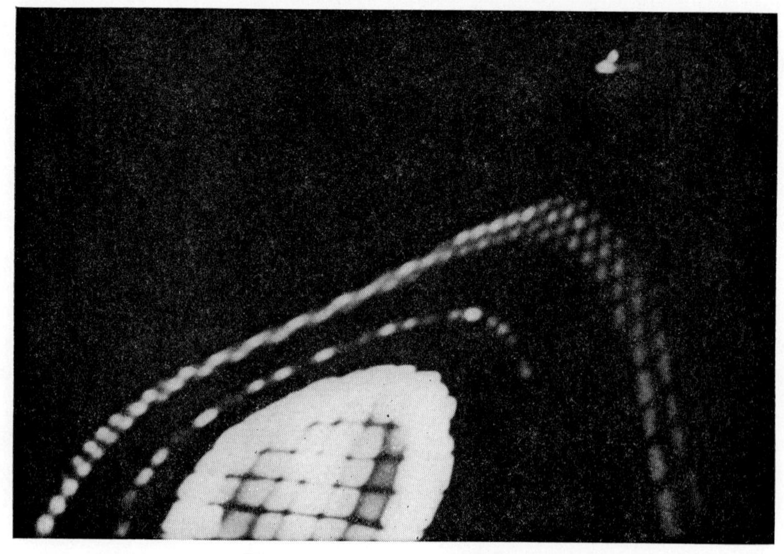

FIG. 1.7. Image of collector.

TABLE 1.4. IMAGE FORMATION RELATIONS IN MICROSCOPES

OM = light optical microscope
EM = direct electron microscope
SEM = scanning electron microscope

	(collected radiation) =	(function of specimen) ×	(source radiation)
Path geometry			
OM, EM	FOCUSED (space)	all irradiated all time	general (may be condensed)
SEM	general (biased by collection system)	point irradiated variable in time	FOCUSED (time)
Intensity			
OM, EM	variable in space	CONTRAST	constant
SEM	variable in time		constant
Influence of angle to specimen on contrast			
OM, EM	weak		strong
SEM	strong		weak
Projected image			
OM, EM	direct	all space	all time
SEM	indirect	point	variable in time

Fig. 1.8: the effect is more marked with high-energy electrons which travel in straight lines (and are thus analagous to the use of point illumination) than with low-energy secondary electrons which can follow curved paths (and are thus analagous to diffuse illumination). Figure 1.9 shows the changing contrast as a ridge-shaped specimen is

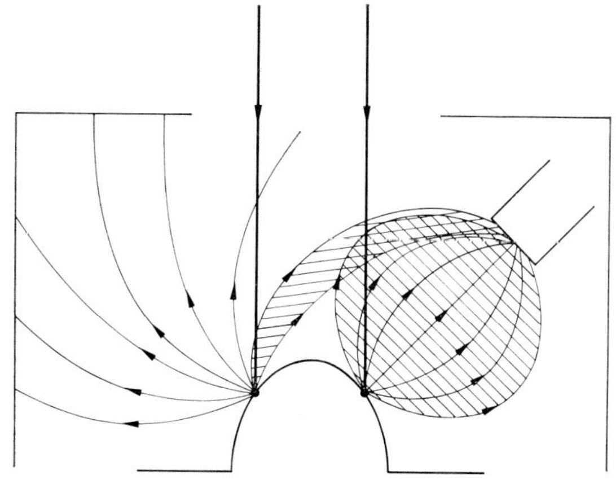

FIG. 1.8. Collection of electron in SEM, illustrating difference in efficiency of collection depending on which side faces collector.

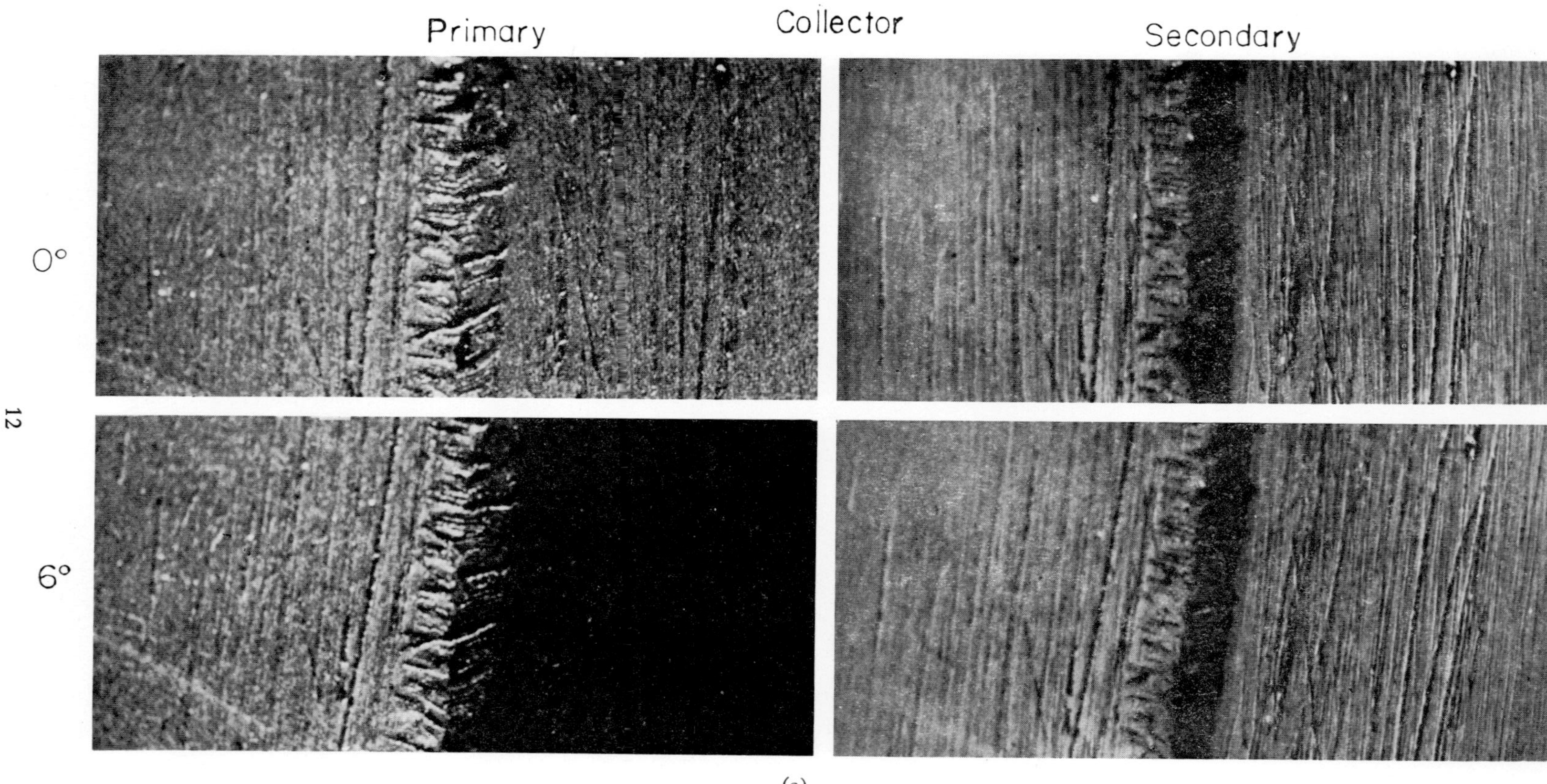

(a)

Fig. 1.9 (a)–(c). Changing contrast in ridge-shaped specimen. The angles indicated are between the line from specimen to collector and the line of the ridge.

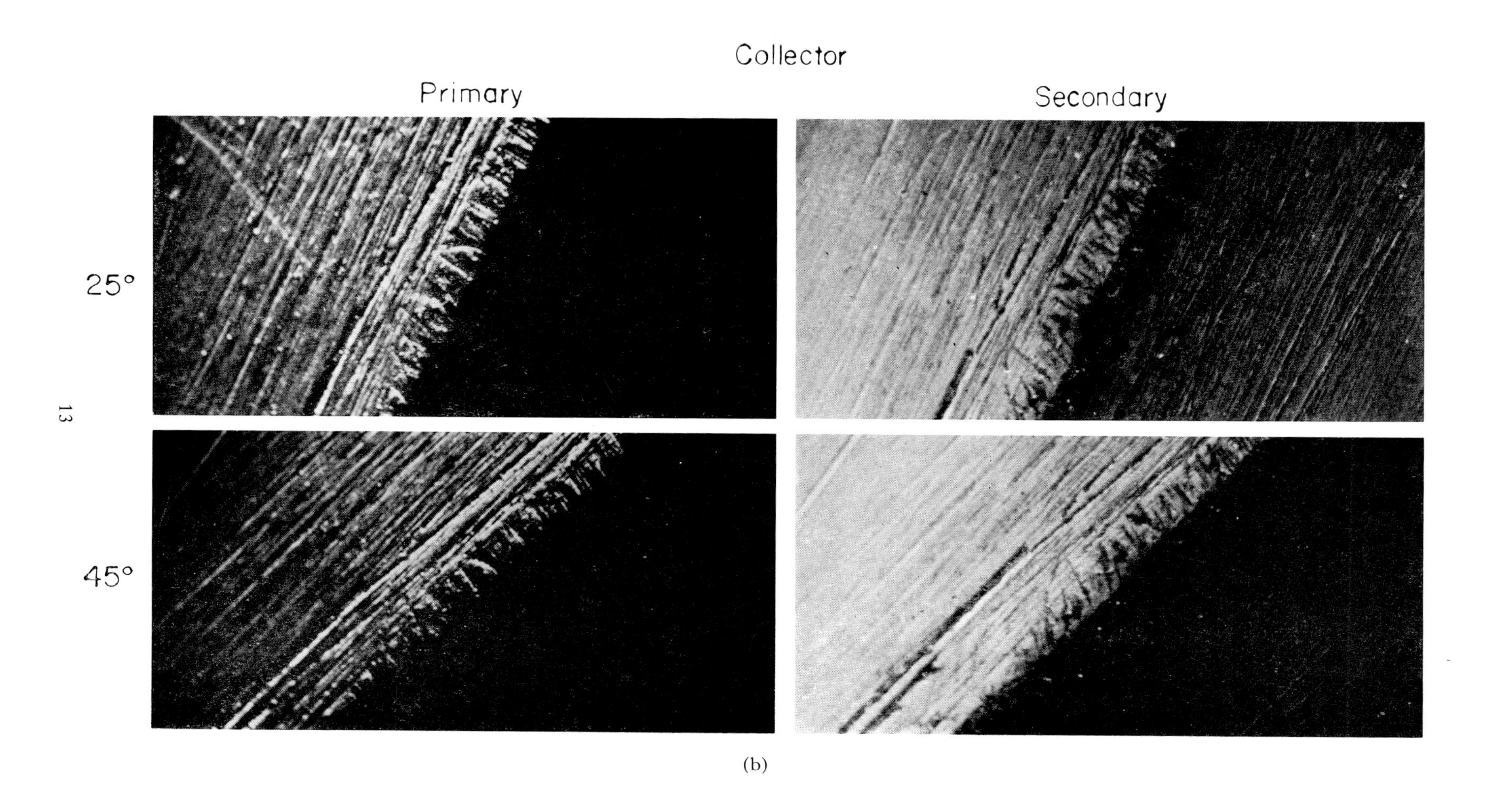

(b)

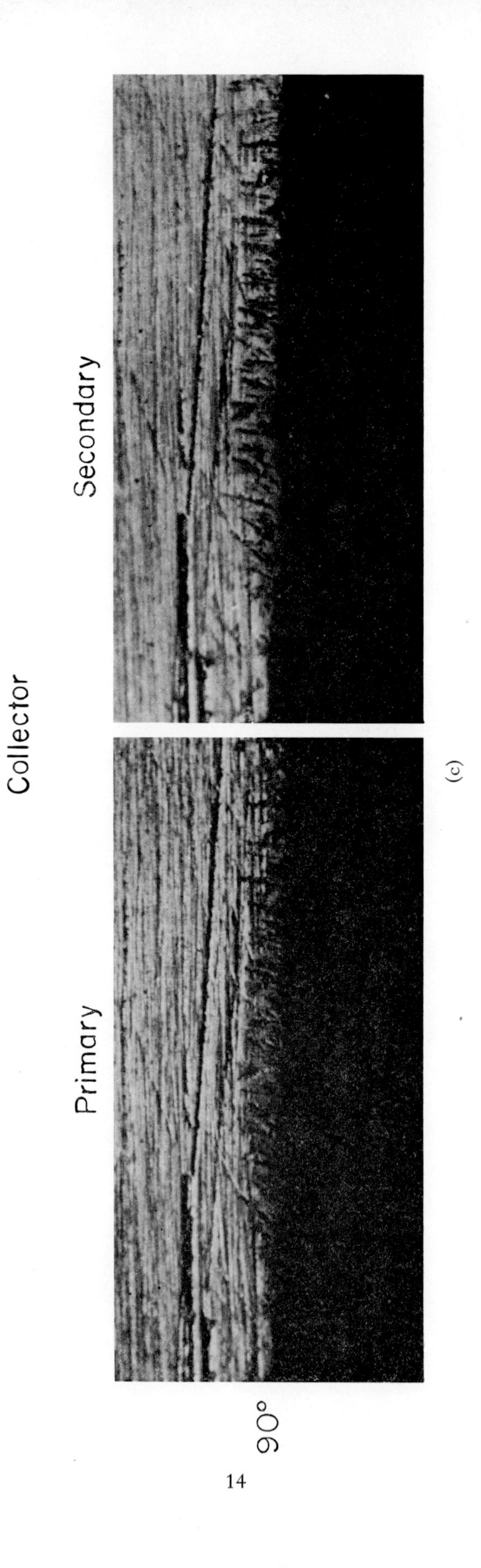

90°

(c)

turned away from the collector, while Fig. 1.10 shows the effect of changing the collector position in highlighting different facets of a fracture surface. The use of several collectors picking up primary electrons, equivalent to "portrait illumination" with several spotlights, is illustrated in Fig. 1.11 (a), (b).

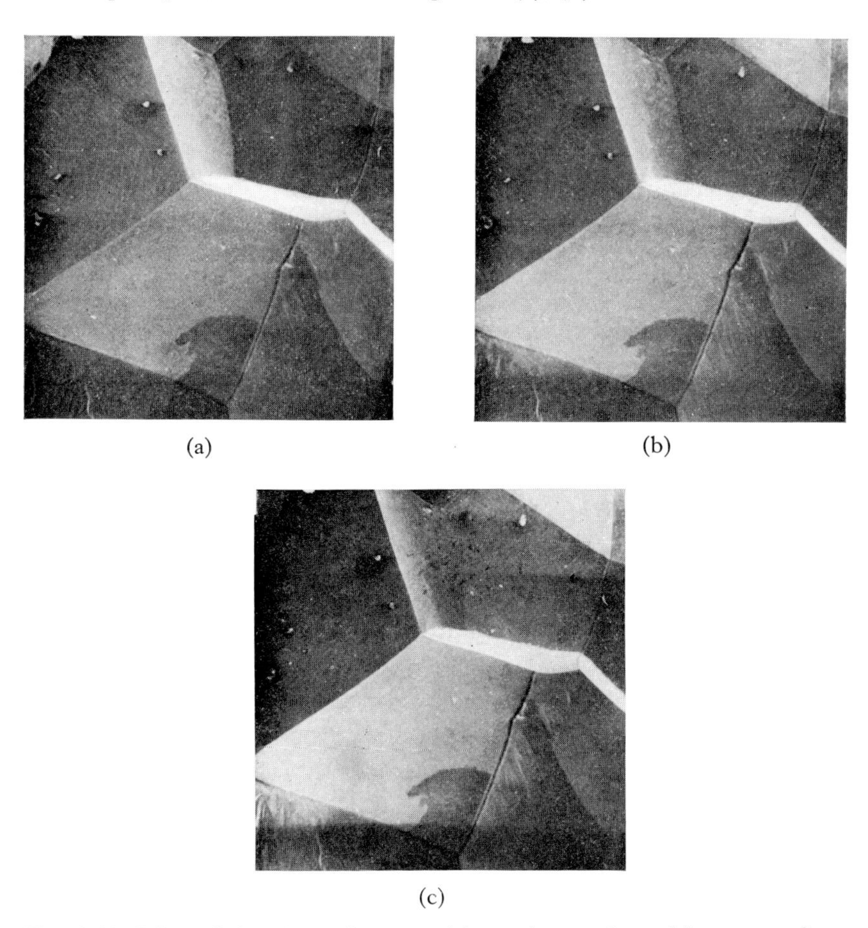

FIG. 1.10. Effect of changing collector position on image of metal fracture surface. (a) Collector to left. (b) Collector central. (c) Collector to right. Note changing contrast of faces.

More formally one can write the equation defining contrast as:

$$I_R = I_S \Sigma (R_{so} \times E \times R_{or}) + N \tag{1.1}$$

where the received intensity I_R for any image point depends on the incident intensity I_S, the geometrical relation between source and object R_{so}, the emission characteristic E of the object point, the relation between object and receiver R_{or}, and the unwanted noise N. The summation sign is included as a reminder that more than one object point may contribute to the image. For example, the primary beam may pass through the specimen and generate some emission from the under-surface.

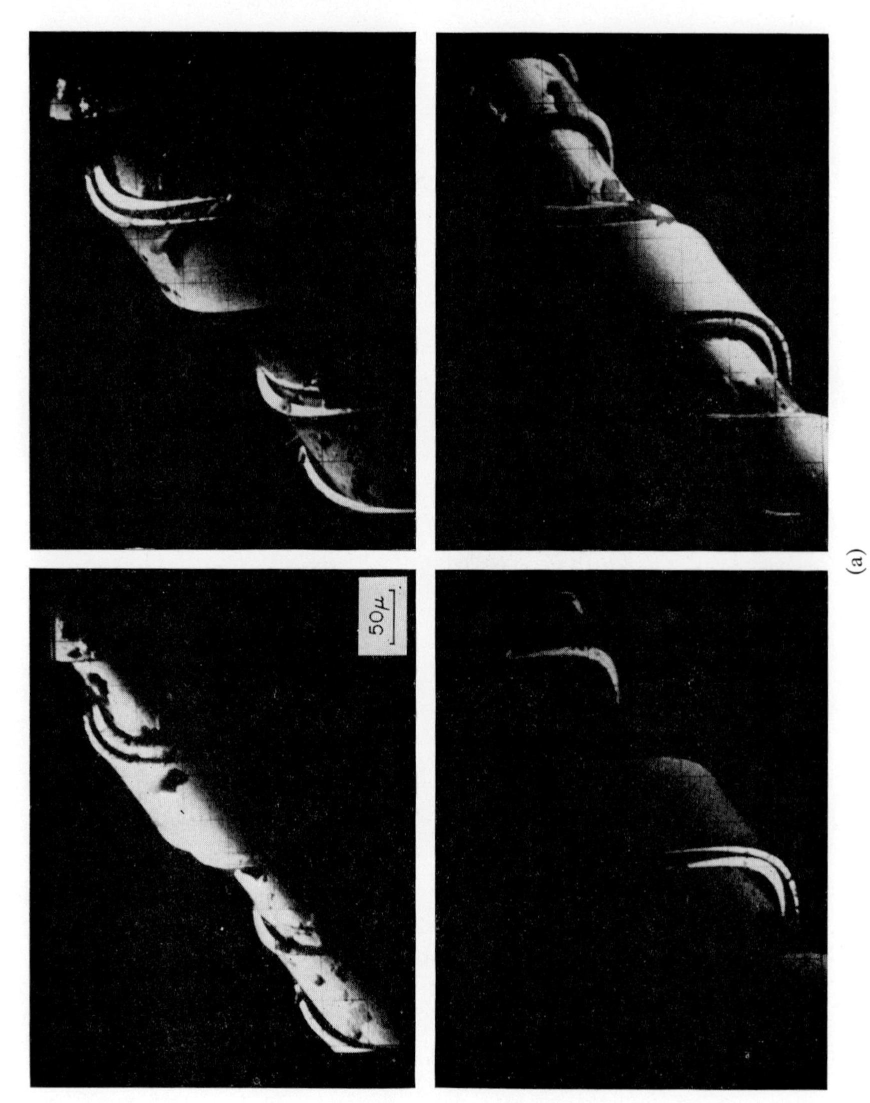

(a)

FIG. 1.11. "Portrait illumination." (a) Object viewed with single collector in different positions.

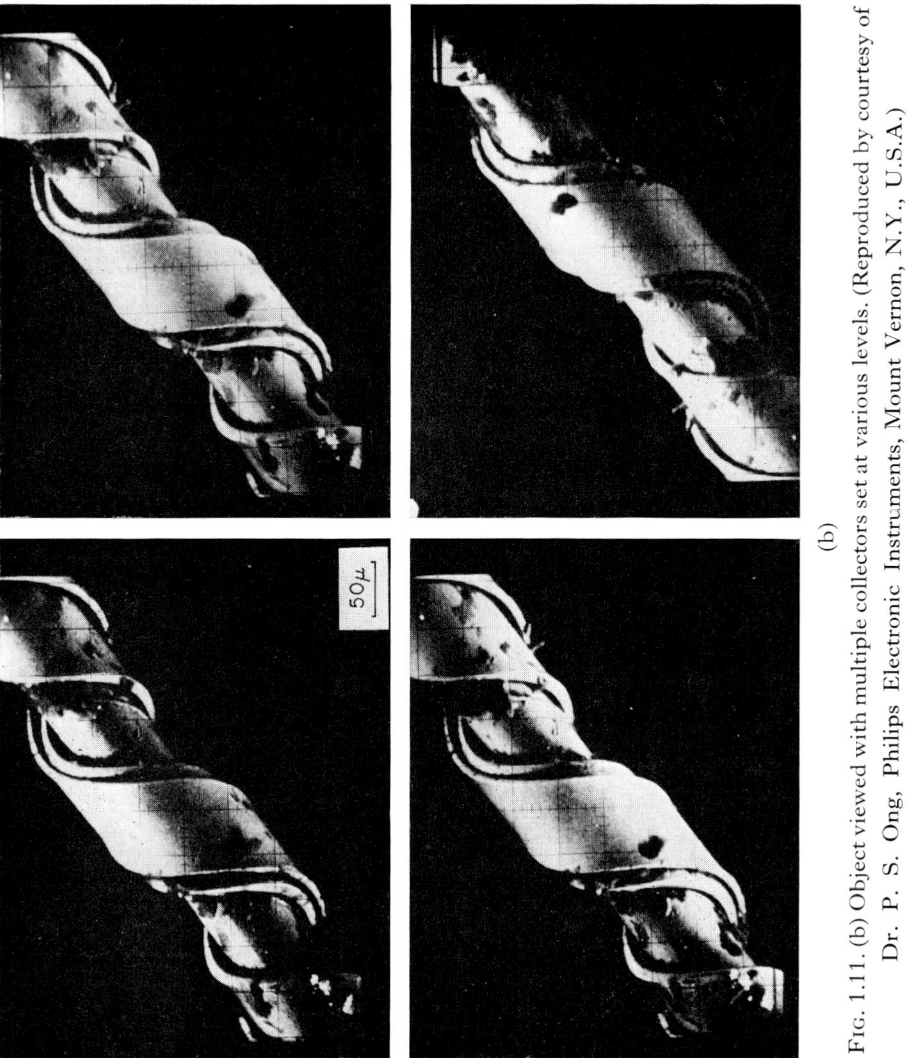

(b)

FIG. 1.11. (b) Object viewed with multiple collectors set at various levels. (Reproduced by courtesy of Dr. P. S. Ong, Philips Electronic Instruments, Mount Vernon, N.Y., U.S.A.)

As has been indicated, the major factor causing topographical contrast is R_{or}, the angle to the collector, which is closely analogous to the showing up of contrast by oblique illumination in optical microscopy. The reverse analogy again applies, because this would be the incident angle in optical microscopy; in scanning electron microscopy the incident angle has a lesser effect, analogous to the effect of viewing angle in optical microscopy, but does have some as shown in the next chapter. The factor E is mainly dependent on the atomic numbers of the elements on the surface, and will thus give rise to element contrast: it is also, of course, of dominant importance in some other modes of use of the scanning electron microscope, such as X-ray emission or the conduction mode.

In the absence of noise, the relative contrast C_0 between two points will be given by:

$$\frac{I'_R}{I_R} = \left(\frac{R'_{so}}{R_{so}}\right)\left(\frac{E'}{E}\right)\left(\frac{R'_{or}}{R_{or}}\right) = C_0. \tag{1.2}$$

The important factor is thus the relative values of the terms in equation (1.1). The presence of noise will reduce contrast, according to the equation:

$$C = \frac{C_0 + N/I_E}{1 + N/I_E}$$

where $I_E = (I_R - N)$ is the wanted part of the collected signal.

1.3.3. Detail, contrast and resolution

Detail can be obscured in an image for various reasons of which the most obvious are loss of resolution and loss of contrast; but, since the situation in scanning electron microscopy is not quite so simple, it is worth examining the fundamentals of the matter.

Suppose that we have an object whose real form is as shown in Fig. 1.12 (a). The idealized intensity variation along a line AB across an image of the object should then be as in Fig. 1.12 (b). If resolution is inadequate, then sharpness at the edges is lost as in Fig. 1.12 (c), or, as resolution becomes worse, the strips merge as in Fig. 1.12 (d), and finally are completely lost as in Fig. 1.12 (e). It is to be noted that Fig. 1.12 (c) to (e) can refer either to a reduction in instrument performance with a given object or to a reduction in dimensions of the object and increase of magnification in a given instrument. In a scanning electron microscope, as indicated in Fig. 1.13 (a), there must be a blurring of the image over distances corresponding to the spot size. Aberrations may further reduce resolution.

The effect of reduction in contrast, leading to an inability to distinguish any detail, is shown in Fig. 1.12 (f) to (h). A rather similar result comes from increased noise, whether due to stray electrons in the collection chamber or to effects in the electric circuits; this reduces the relative contrast as shown in Fig. 1.12 (i) to (k). The effect shown is that of a uniform noise; but there will be worse obscuring of the true image if the noise is variable, as in Fig. 1.12 (l). Similarly the true image may be obscured by an unwanted stray image giving a "double exposure", Fig. 1.12 (m).

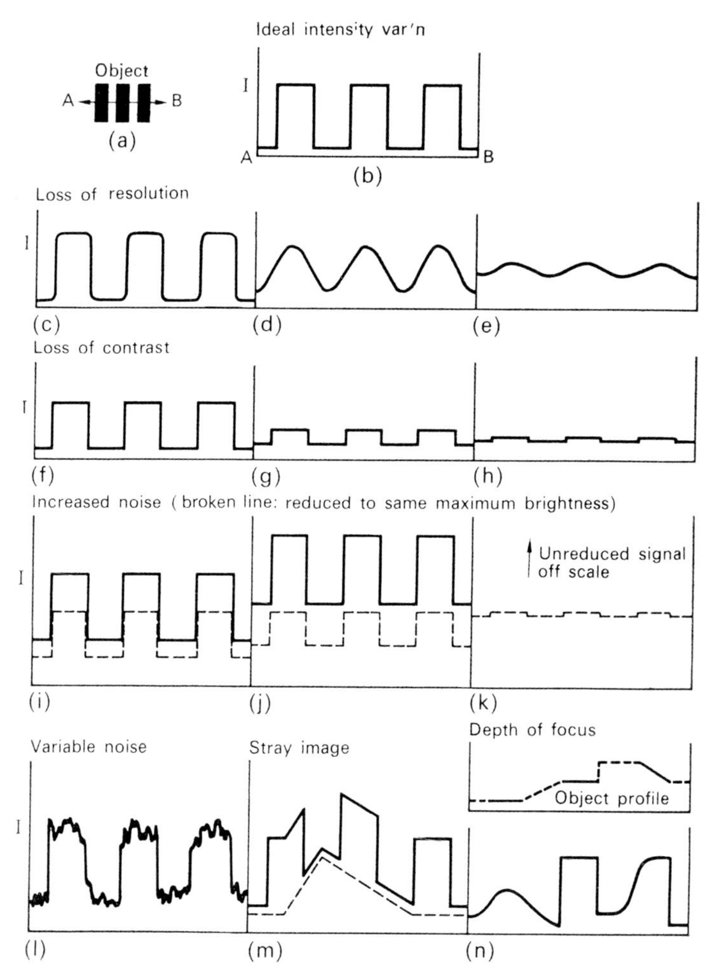

FIG. 1.12. General causes of loss of detail.

So far we have assumed that the object is in one plane perpendicular to the direction of viewing, but if it has a three-dimensional surface as indicated by the section shown on the upper part of Fig. 1.12 (n), then only one level will be in perfect focus. Depending on the depth of focus, detail at other levels will be obscured as indicated in the lower part of Fig. 1.12 (n).

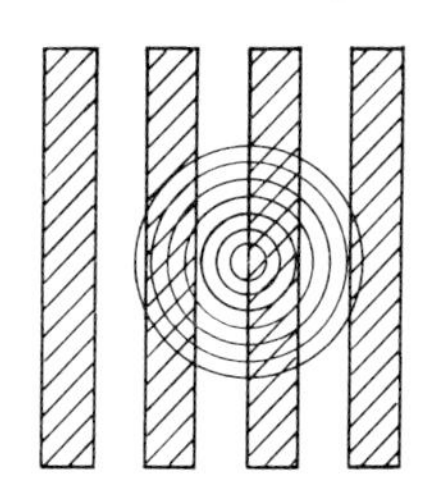

FIG. 1.13. Spot size in relation to object.

Finally, we come to a factor which is of special importance in scanning electron microscopy. If the image is formed solely by electrons from the surface of the object, then, as shown in Fig. 1.13, spot size will be the major reason for loss of resolution. But, particularly in low-density specimens, penetration can occur into the specimen and then secondary electrons are emitted from a volume below the surface. The zone affected spreads out laterally and is roughly pear-shaped. If we have a particularly simple form of object, in which the material giving differential contrast is in vertical bands as in Fig. 1.14 (a) to (c), then increased penetration will lead to the loss of detail

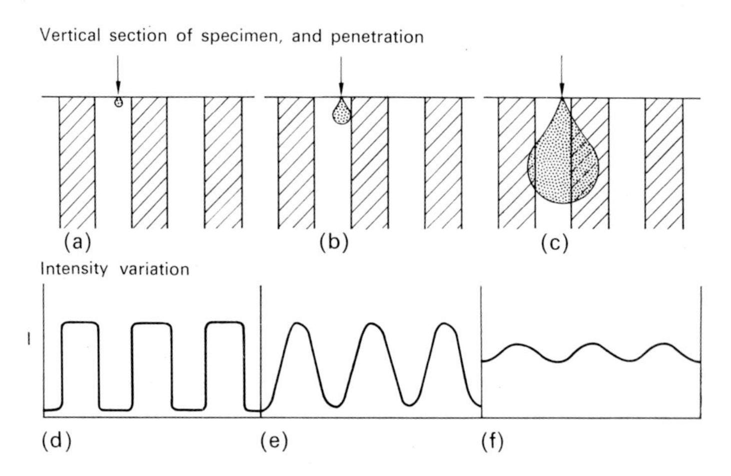

FIG. 1.14. Effect of penetration on resolution.

as indicated in Fig. 1.14(d) to (f). For simplicity it has been assumed that the spot size on the surface is small, so that there is no effective loss of resolution due to this cause. With this type of object, the effect of increased penetration is clearly to cause a loss of resolution, and, as above, the diagrams would correspond either to increased penetration in a given object or to reduced dimensions of object with a given penetration.

When using the scanning electron microscope in the most usual way, it is not, of course, the penetration of the primary beam which is important, but the size of the zone from which secondary electrons can escape, and so reach the collector. With other modes of use, the resolution will be determined by the zone over which the particular stimulus being used to activate a signal is effective.

The type of object indicated in Fig. 1.14 is, however, exceptional. A more probable situation is shown in Fig. 1.15(a), where increased penetration will effectively increase noise or the accumulation of multiple images, which will be increasingly out of focus, and so will lead to an obscuring of detail through loss of contrast due to this increased noise. It does, however, retain one important characteristic of resolution, since the loss of contrast is dependent on the relative sizes of the zone and the features of the object. Another related effect is that continued penetration of the primary beam may lead, in relatively thin specimens, to other zones of secondary emission on the lower surface or on some other object, as indicated in Fig. 1.15 (b): this gives double or triple images, whose sharpness will depend on the depth of focus and the extent to which the spot size is increased by scattering of the primary beam.

It is interesting to note that the consequences of penetration include all the factors shown in Fig. 1.12 as causing loss of detail. This has been a source of some confusion in terminology in the past.

There are also other factors which, while not necessarily obscuring detail on the image, can lead to a loss of correct detail. High emission can lead to glare at the edge of specimens. Electrostatic charging causes false contrast, due to its influence on emission, and to image distortion. More seriously, the specimen may be damaged by the electron bombardment, either directly or through the effect of heating, and it may move due to the impact of the electron beam or to external vibration.

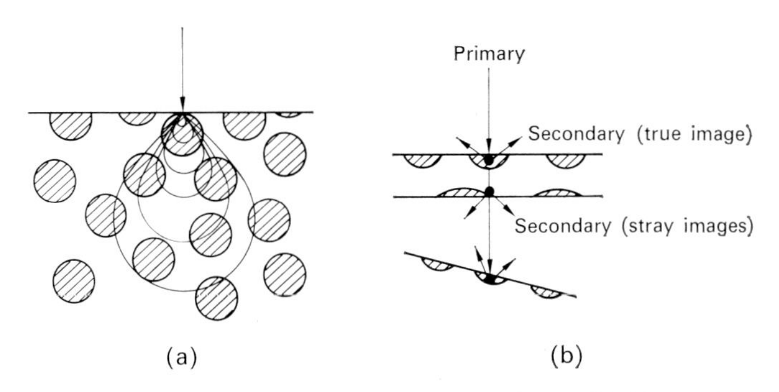

FIG. 1.15. (a) Penetration into a more typical specimen. (b) Formation of stray images from lower surface or second object.

While there are a few factors, such as dirt in the column, scanning faults, external vibration and astigmatism, which should always be minimized to secure the best results in scanning electron microscopy, the other important operating variables are so interrelated that no universally applicable rules can be laid down. Success in operation with difficult specimens or at the highest magnification comes from achieving the best compromise for the particular specimen and the particular purpose of examination. Thus there is little point in trying to increase resolution by increasing beam voltage, so as to reduce spot size if this gives an undue increase in penetration; or to make an adjustment which leads to increased resolution if this leads to loss of contrast because of an unduly low signal-to-noise ratio in the instrument, though this may then be compensated in some other way, which in turn may affect the image in yet another respect. The compromise will involve not only resolution and contrast, but also other factors such as depth of focus, exposure time, beam damage and so on, which will vary in importance with the circumstances.

1.3.4. Some numbers

It is worth pointing out that some features of the instrument are virtually self-designing. For convenience of operation, we need a screen at a convenient viewing distance of about 25 cm. At this distance the field of view of the eye is about 10 cm so there is little to be gained by having a screen larger than 10 cm^2: in order to use a larger screen, one would have to move one's eyes in order to look around it. The resolution of the eye at 25 cm is about 0·20 mm. Hence the number of lines needed, in order to avoid their showing up, is about 500: in practice 1000 are usually used.

It is usually convenient to express magnifications in relation to instrument screen size, which is adapted to the distance of easiest viewing. Table 1.5 gives comparative figures for resolution and magnification, and also indicates the depth of focus and the range of application of different types of microscope.

TABLE 1.5. COMPARATIVE FIGURES

Based on a 10-cm^2 screen viewed at 25 cm

	Magnification	Resolution	Field	Depth of focus OM	Depth of focus SEM
OM {	1×	0·2 mm	100 mm		
SEM {	10×	0·02 mm	10 mm	0·1 mm	10 mm
	100×	2 μm	1 mm	1 μm	1 mm
EM {	1000×	0·2 μm	0·1 mm		10 μm
	10,000×	20 nm (200 Å)	10 μm		1 μm
	100,000×	2 nm (20 Å)	1 μm		
	10^6×	0·2 nm (2 Å)	0·1 μm		

(Brackets: OM spans 1× to 1000×; SEM spans 10× to 100,000×; EM spans 1000× to 10^6×.)

1.3.5. The transmission scanning electron microscope: reciprocity

Figure 1.16 shows a diagram which can apply to transmission microscopes (electron or optical). Once again the scanning system has the electrons going in the opposite direction. Since diffraction paths are not changed by reversing the direction of radiation, it follows that there is a principle of reciprocity between the two instruments, and, in principle, what can be done with one ought to be possible with the other. This does apply in some circumstances, and then a direct electron microscope is usually better. However, there are other circumstances where the scanning system has advantages. One reason is that the scattered electrons are of widely different energy and so are difficult to focus together: there will be appreciable error due to chromatic aberration. But this difficulty does not arise in the scanning instrument since only the incident electrons have to be focused. Consequently it is possible to examine thicker specimens, to process the information, or in a high-voltage instrument to put the specimen outside the vacuum chamber in the atmosphere.

1.3.6. The formal advantages of the scanning system

Despite the extent to which the principle of reciprocity applies, and hence tends to indicate that there is no fundamental difference between scanning and direct electron microscopes, there are three basic features of difference, which, though they have already been mentioned, should be stressed.

The first is that, in a direct system, electrons which have interacted with the specimen have to be focused, whereas in a scanning system it is the more homogeneous incident electrons which have to be focused. As a rider to this it follows that any reaction stimulated by the incident electrons can be used to obtain an image, but this reaction does not have to consist of focusable radiation; conversely, in a direct system, the emitted radiation must be focusable but the incident stimulus need not be.

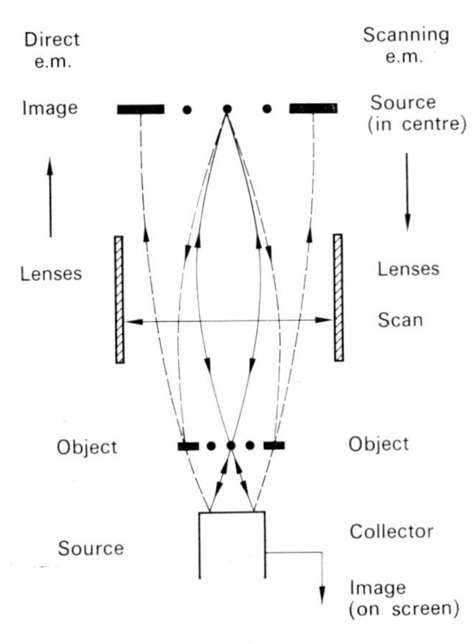

FIG. 1.16. Basic diagram of transmission microscope.

The second feature is that it is only necessary to focus on one spot of the specimen at any one time: in direct microscopes, all points must be in focus all the time. Lens design is thus likely to be inherently simpler in a scanning system and aberrations easier to reduce.

The greater depth of focus is a result of these differences. It is possible, in a scanning system, to concentrate the incident beam in a cone of very small angle, so that the circle of confusion remains small over a long distance, whereas in a direct system it is necessary to collect over a wider angle to secure sufficient brightness. The scanning of a small spot over the specimen also serves to minimize damage to the specimen.

The third feature is that the image information in a scanning electron microscope is carried on an electric signal which can be processed in various ways to give useful knowledge of the specimen. Thus the signals may be fed to a computer for numerical analysis, or by combining different signals the effects of particular contrast mechanisms may be isolated.

CHAPTER 2

The Interaction of Electrons with Solids

D. C. NORTHROP

2.1. INTRODUCTION

In this chapter we shall be concerned with a small part of a very large subject, and we shall be forced to treat it in a qualitative and empirical manner. Our difficulty is that the interaction of a high-energy electron with the assembly of electrons in a solid is complex, depending on the residual energy of the primary electron, and it can be dealt with only on a statistical basis because each primary electron will interact differently, depending on the aspect of its first collision and on the distribution of all the other electrons in the solid. Another kind of complexity which is very important in scanning electron microscopy arises when we consider the physical effects induced by the interaction of an electron beam with a solid specimen. These include secondary electron emission, X-ray emission, cathodoluminescence and induced currents and voltages in the specimen, as well as the reflection and transmission of elastically scattered primary electrons. Each of these is a transient effect whose magnitude depends on the constitution of the specimen, so that each can in principle be used to give a "picture" of the specimen if its magnitude is displayed on a scan synchronized with the scan of the primary electron beam.

Figure 2.1 shows schematically the most commonly used effects (contrast mechanisms), but it must be recognized that not all materials will give large enough effects for all the mechanisms to be available. For example, rather few materials have large enough cathodoluminescence efficiencies to be useful and only very thin specimens will show appreciable transmitted primary electron currents. On the other hand, virtually all materials will exhibit usefully large secondary electron currents and X-ray emissions.

Our aim must be to understand the interaction of the primary electrons with the specimen sufficiently well to be able to use the electron microscope effectively, using the best contrast mechanism for each specimen, and using the best beam voltage and current to obtain sensitivity and resolution from the chosen mechanism.

Different contrast mechanisms will, of course, yield information of different kinds about the specimen, and the range of possibilities is virtually unlimited, but it may be helpful at this stage to give a very brief guide to the commonest kinds of information sought from each mechanism. These are listed in Table 2.1. Two important points

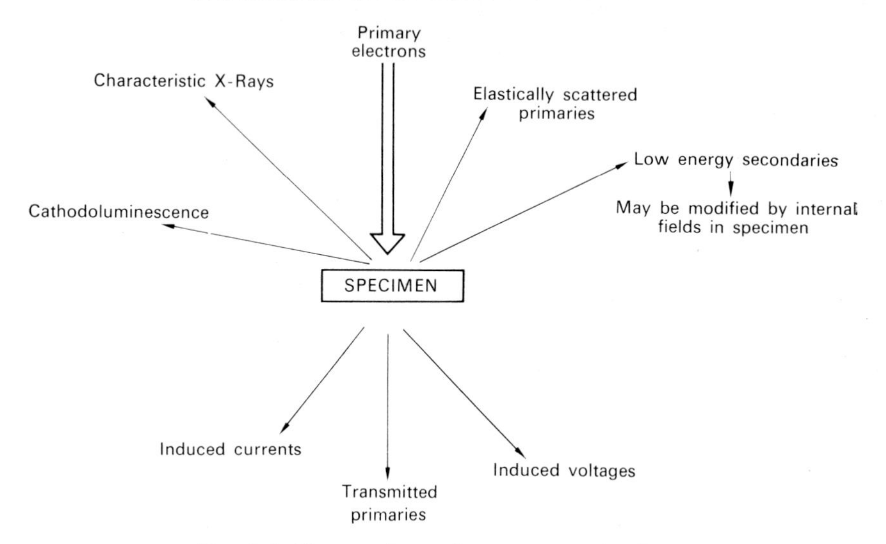

FIG. 2.1. Commonly used contrast mechanisms.

TABLE 2.1. INFORMATION FROM SCANNING ELECTRON MICROSCOPY

Property	Contrast mechanism
Surface topography	Principally low-energy secondaries but also elastically scattered primaries and cathodoluminescence.
Material analysis	Principally X-ray emission (primary energy $\sim$ 30 keV) but also elastically scattered primaries for specimens of low atomic number and sometimes cathodoluminescence.
Specimen thickness	Transmitted primaries.
Electric fields in specimen	Voltage contrast modulation of low energy secondaries.
Electrical properties of specimen (especially in semiconductor devices)	Internal currents and voltages. Voltage contrast modulation.

need to be made at this stage. The first is that most users of scanning electron microscopes have expert knowledge of the materials being investigated and therefore are able to make the correct decision about contrast mechanisms they should in some cases be able to invent new mechanisms. The second general point is that two pictures of a specimen obtained by different contrast mechanisms frequently yield more information than either would do alone.

2.2. ELECTRON RANGE

All the electron-scattering processes which take place are "large-angle" ones, that is the two electrons follow trajectories after the collision which may make large angles with the trajectory of the incident electron. Calculation of the electron range therefore consists of following the fortunes of an individual electron to determine first its total path length, which depends on the energy lost in each collision, and secondly its range from the point at which it entered the solid, taking into account the scattering angles. A typical track could look like the one drawn schematically in Fig. 2.2.

In describing the operation of the SEM we are interested in the volume of the solid target which contains all possible electron paths, and in the energy distribution of the electrons within this volume. The volume of the target which is sampled will depend

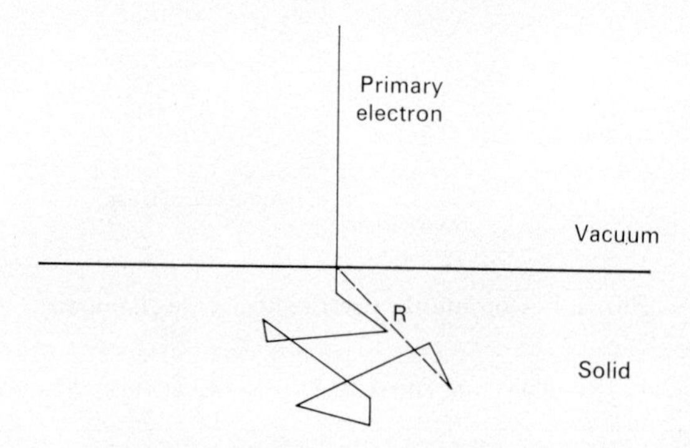

FIG. 2.2. Schematic diagram showing the path of a single primary electron and its range R. (Note that the two electrons taking part in any one collision are not distinguishable. In determining the range we follow the highest energy electron leaving each collision.)

on both these factors and on the contrast mode being used; for example, electrons of less than a certain energy cannot cause secondary emission, but there will in general be a different threshold energy for the generation of X-rays, or the production of cathodoluminescence. The effective volume of the target interacting with the electron beam for any particular contrast mode will be one of the factors determining the spatial resolution of the microscope, and it can in some circumstances be the most important factor.

Although the physical considerations governing electron scattering and electron mean range are useful in gaining an insight into the operation of the SEM, calculations based on these ideas have not been taken far enough to give us precise values of range and interaction volume. For numerical data to guide us in our choice of operating conditions for the microscope we have to rely on a more empirical approach. Experimentally we may determine $N(z)$, the number of electrons which have penetrated a distance z into the material in the direction of the primary beam. It is found that the extrapolated range, as defined in Fig. 2.3, may be expressed as

$$R_z = k\, E_0^n$$

where k and n are constants and E_0 is the primary electron energy. This law is obeyed to within $\pm 10\%$ for a large number of metallic absorbers and for values of E_0 lying between 2 keV and 50 keV. For a more accurate fit to the experimental data it is

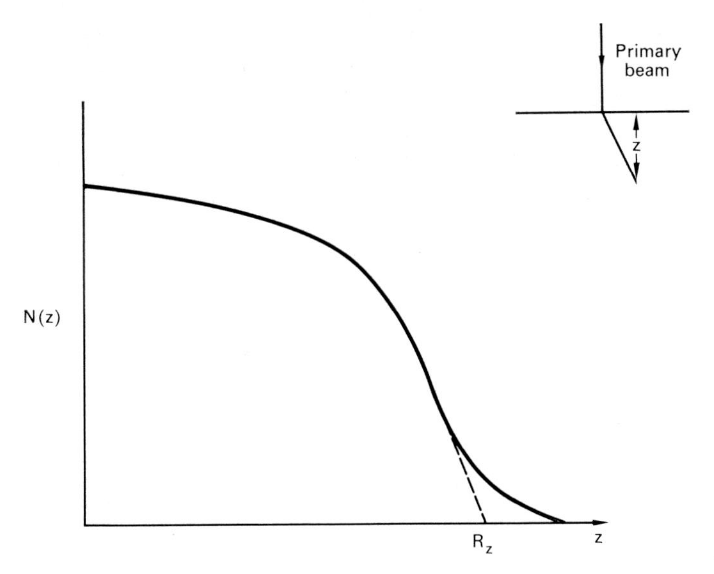

FIG. 2.3. The penetration of electrons into a solid, defining the extrapolated range R_z.

necessary to allow n to be a slowly varying function of E_0. The value of k is large for weakly absorbing materials and small for strongly absorbing ones, and since the important interactions are electron–electron collisions it is expected that a small value of k will correspond to a large electron density. Furthermore, because the volume of a solid occupied by a single atom does not vary greatly with atomic number, Z (the number of orbital electrons per atom), it is expected that k will vary in a systematic way with atomic number. Measurements on metals confirm these expectations. Table 2.2 shows the values of n and k for 15 keV electrons in four metals of widely varying atomic number.

It is clear then that $E_0^{1\cdot 5}$ is a good approximation to the experimental results in this energy range, although at about 25 keV the situation begins to change, and the better known Whiddington law $R_z \propto E_0^2$ becomes a better approximation.

A more useful concept in many cases is that of the range–density product. It is expected that range depends on the electron density in the absorber, and since the

TABLE 2.2

	Atomic number	k	n	R_z (15 keV)
Aluminium	13	27	1·5	1·50 microns
Copper	29	10	1·5	0·58 microns
Silver	47	8·2	1·5	0·48 microns
Gold	79	4·5	1·5	0·26 microns

The units of k are such that when E_0 is in keV, R_z is in millimicrons.

weight of an atom is approximately proportional to its atomic number, we can further expect that the density of a material will be proportional to its electron density. This in turn leads us to expect that the product ρR_z will be roughly constant if the range R_z of electrons of a particular energy is measured in a number of different materials of density ρ. Experimental results for electrons of up to 20 keV energy and for a variety of metallic absorbers bear out this expectation, as Table 2.3 shows.

TABLE 2.3. MEASURED RANGE–DENSITY PRODUCTS OF ELECTRONS IN METALS ($\mu g/cm^2$)

Metal	E_0			
	2·5 keV	5 keV	10 keV	15 keV
Aluminium	36	90	230	—
Copper	43	102	305	530
Silver	42	96	280	510
Gold	40	97	280	510

It is not expected that organic solids and molecular crystals will have range–density products in such good agreement as those for the metals in Table 2.3. They are sufficiently close, however, for the range–density product to be a satisfactory qualitative guide with a possible error of $\pm 30\%$ in the range. Thus if we assign a range–density product in units of $\mu g/cm^2$ to electrons of each energy we can determine their range in any particular material by dividing by the density of that material. When the density is in g/cm^3 this gives the range in units of 10^{-6} cm.

The importance of the electron range for scanning electron microscopy is in the determining effect it can have on the spatial resolution of the instrument. We can think in a qualitative way of the electron beam interacting at any instant with a volume of the target material which is approximately hemispherical and has a radius equal to the extrapolated range. The volume of the specimen which is "sampled" by the beam at any instant, and which defines the limiting resolution of the instrument, will in general be smaller than this interaction volume for two reasons.

(a) The current density falls as $1/r^2$ from the point at which the electron beam enters the specimen, and the magnitude of the observed effect of the beam (secondary electron emission, cathodoluminescence, specimen photocurrent, etc.) will fall off at least as fast as this.

(b) The energy of individual electrons also falls monotonically as r increases, and it will eventually fall below the threshold necessary for the production of the effect being measured. For example, the energy which a secondary electron must have if it is to escape increases with depth below the specimen surface; this causes the "sampled volume" to have a quite different shape from the interaction volume, making it smaller and flatter.

Clearly the "sampled volume" will depend on the contrast mode in which the microscope is used. For this reason the resolution of the instrument may be different for different modes when the beam voltage and current are held constant. Another way of saying the same thing is that the optimum conditions for observing a specimen in one mode may be quite different from those for a different mode. Each contrast mechanism must be considered separately.

2.3. SECONDARY ELECTRON EMISSION

2.3.1. The energy distribution of emitted secondaries

It is possible to give a qualitative description of the energy distribution of secondary electrons which applies to all target materials. Such a description is shown in Fig. 2.4, which allows us to distinguish three clearly defined regions.

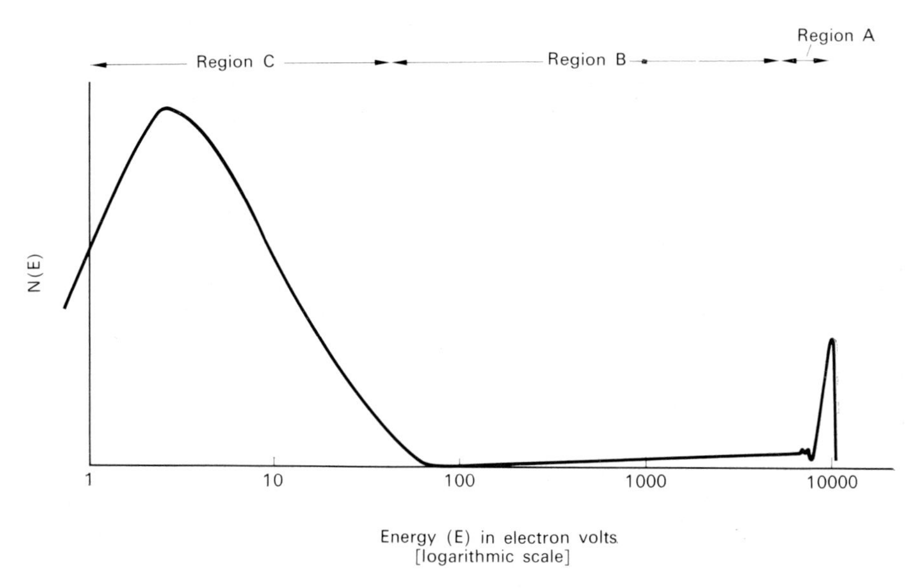

FIG. 2.4. Energy distribution of secondary electrons induced by 10 keV primaries (schematic only).

Region A. Energies very near the primary energy

There is a very pronounced peak in the distribution due to primary electrons which have undergone one or two large angle elastic collisions and have then left the specimen without appreciable loss of energy. Often there is some fine structure on the low-energy side of this peak due to the excitation of plasma oscillations in conducting specimens.

Region B. Intermediate energies

Between the primary energy and about 50 eV there is a background of electrons which have been back-scattered from the specimen after a number of inelastic collisions. Although this is a low intensity distribution it contains many more electrons than Region A of the spectrum.

Region C. Low energies

The largest peak of the electron current leaving the specimen surface consists of electrons with energies between zero and about 50 eV. These are the true secondary electrons, ejected from the specimen by the primary electrons. Many materials emit more low-energy secondaries than the number of bombarding primaries.

The shape of this part of the spectrum varies surprisingly little from one material to another, and we can think qualitatively in terms of a model in which the energy spectrum of the electrons excited *within* the material is always the same, and where the work function of the material decides the electrons which are able to escape. The most probable energy of the secondaries lies between 2 eV and 3 eV, the higher energies being found for the lower work function materials.

2.3.2. Selection of secondaries in the SEM

In the SEM an electron collector collects the electrons emitted from the specimen surface, and the collected current is amplified and used as the signal in the picture display. The kind of information contained in the picture depends on the part of the secondary electron spectrum used. Microscopes normally allow the potential of the collector to be varied continuously between +250 V and −50 V. At the first extreme most, but by no means all, of the electrons are collected. As the potential is varied through zero to negative values the low-energy secondaries are progressively excluded, beginning with those of lowest energies and those whose direction of motion is away from the electron collector. This means, of course, that the signal is reduced, but the character of the picture will also change: exclusion of some secondaries may improve the resolution of particular details; use of the high-energy electrons may give information about the composition of the specimen, especially when it consists of a mixture of materials. (The use of high-energy electrons will be explained further in the discussion of secondary electron yield.) The use of velocity analysers placed between the specimen and the electron collector will be explained in Chapter 10.

2.3.3. Secondary electron yield

The total secondary electron current may be expressed formally in terms of the three energy regions defined above as follows

$$i_t = i_s + \eta i_p + r i_p$$

where i_s is the true secondary electron current (less than 50 eV), i_p is the primary current, and η and r are factors which express the currents in the energy regions B and A of Fig. 2.4 as fractions of the primary current. η and r are in general functions of primary energy and the atomic number Z of the specimen.

The ratio $\delta = i_s/i_p$ is a function of primary energy of the form shown in Fig. 2.5. The curve is basically the same shape for all materials, having a broad flat maximum between 400 and 800 eV primary energy. The maximum value of δ may or may not exceed unity, depending mainly on the work function of the material; it is necessary, however, to qualify this by noting that roughened surfaces have lower emission than

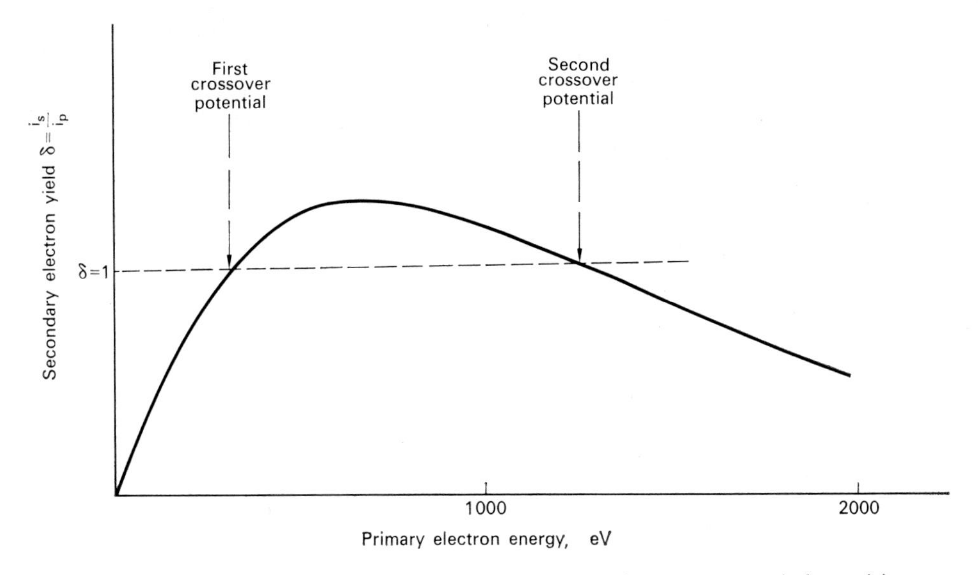

FIG. 2.5. Secondary electron yield versus primary electron energy (schematic).

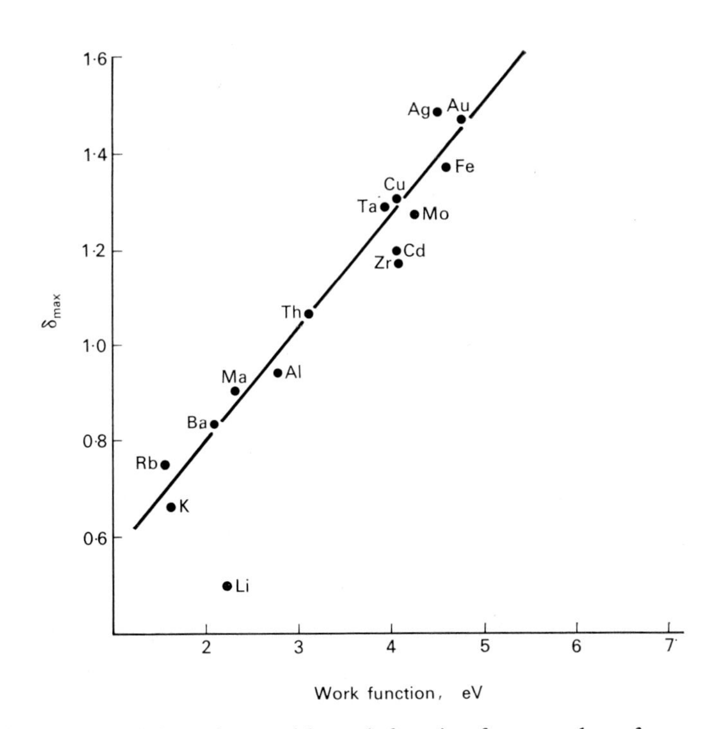

FIG. 2.6. The variation of δ_{max} with work function for a number of pure metals.

smooth ones of the same substance. The variation of δ_{max} with work function is shown in Fig. 2.6. From the data it is clear that high work function materials will give a higher secondary emission current than low work function materials, and other things being equal this will mean a higher signal: noise ratio and a clearer picture. (This is worth bearing in mind when specimens have to be metallized before examination.) It is also evident that materials of different work functions will have different "brightness" in the scanning electron microscope, and this is an important source of contrast.

The factor η, which expresses the probability that a primary electron will cause the emission of an electron of energy greater than 50 eV, but less than the primary

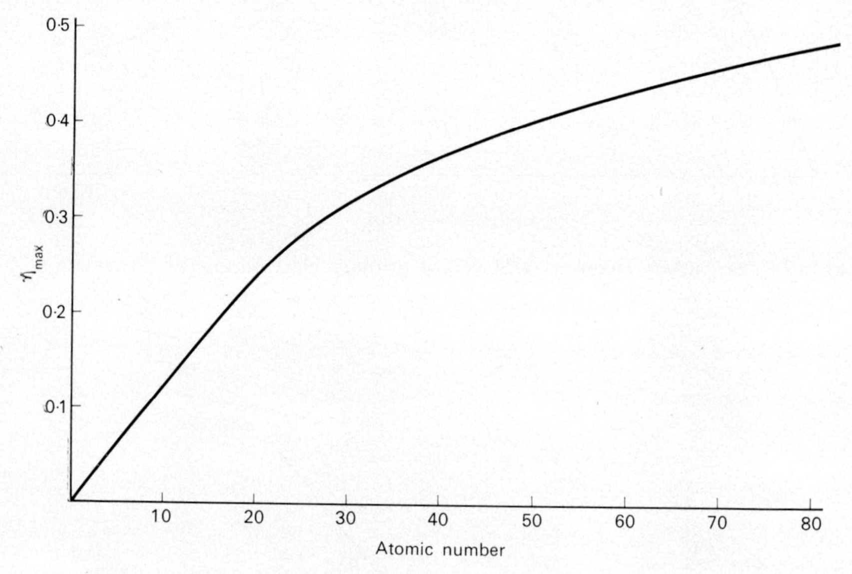

FIG. 2.7. The variation of η_{max} with atomic number for metals.

energy, is normally less than unity. It varies rather slowly with primary energy and usually exhibits a broad flat maximum at about 10 keV. The most important parameter controlling η is the atomic number Z of the target. A graph of η_{max} versus atomic number is shown in Fig. 2.7, where η_{max} is the maximum value of η, measured at the appropriate primary energy for each material. The significant feature here is that η varies rather rapidly with atomic number for the lighter elements up to about $Z = 45$, but much more slowly for the higher atomic numbers. Thus using the electron microscope with the electron collector biased negatively to exclude the low-energy secondaries may give better contrast between adjacent features on the specimen if their atomic numbers are low and the work functions nearly equal.

2.3.4. The effect of angle of incidence on secondary electron emission

It has been shown that the angular distribution of secondary electrons follows approximately a cosine law. Results for the low-energy secondaries are shown in Fig. 2.8, and those for the higher energies are closely similar. No fine structure is evident,

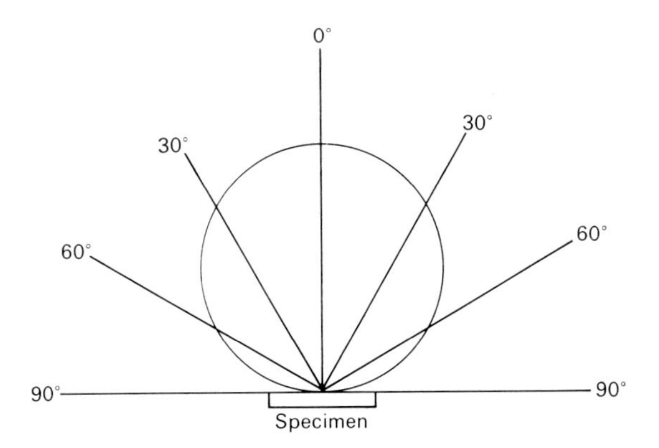

FIG. 2.8. The angular distribution of low-energy secondaries (after Jonker (1957) *Philips Research Reports* **12,** 249).

and the distribution is found to be independent of the crystalline structure of the specimen. We see from this curve that the nearer the electron collector lies to the normal to the specimen the larger the secondary electron current will be.

Usually the angle between the primary electron beam and the electron collector is fixed at about 90°, or slightly more, and changing the specimen tilt changes both the angle of incidence of the primaries and the angle of collection of the secondaries, so that we must consider the effect of these two angles on the current to the electron collector.

The effect of angle of incidence on yield is also known. At primary voltages less than 1 kV, and for rough surfaces (rough that is over distances smaller than the beam diameter), angle of incidence has no effect, but for smooth surfaces and primary energies greater than 1 kV

$$\text{Yield} \propto \frac{1}{\cos\theta}$$

where θ is the angle between the primary beam and the normal to the specimen surface (see Fig. 2.9). The physical model usually used to explain this is to say that the primary electron has more interactions near the surface with a greater chance of producing secondaries.

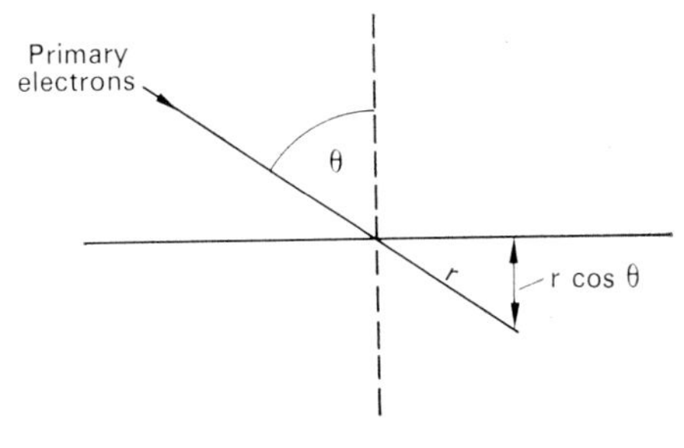

FIG. 2.9. The effect of angle of incidence on secondary electron emission.

The fact that at normal incidence the secondaries are distributed according to a cosine law suggests very strongly that the potential secondary electrons are isotropically distributed in the specimen, and this being so the angular distribution of secondaries is not expected to be strongly dependent on angle of incidence of primaries.

Putting all these factors together it follows that the collected secondary electron current is lowest when the specimen surface is normal to the primary electron beam, and that it increases monotonically as the specimen is tilted through 90° to face the

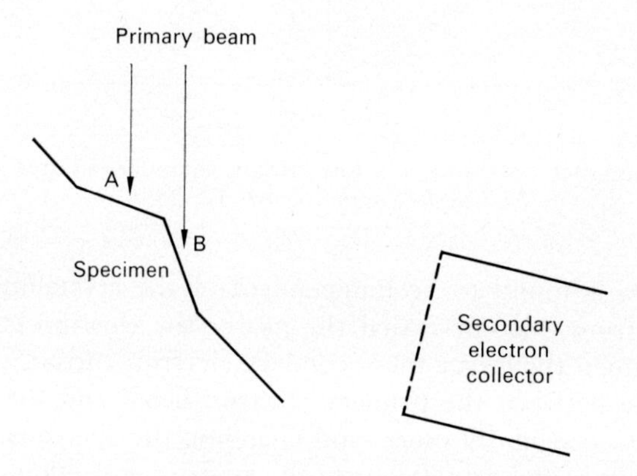

FIG. 2.10. Contrast due to surface contours.

electron collector. However, the magnitude of the current is not the only important factor; at large angles of incidence there is considerable foreshortening of the picture, the whole of the image cannot be in sharp focus, and there may be considerable shadowing by the surface contours of the specimen resulting in there being no emission at all from some parts of the surface. A suitable working compromise has to be found for each kind of specimen, but usually the best angle of incidence will lie near 45°. It should also be noted that the effect of the surface contours in preventing some secondaries from reaching the collector is a most important factor in introducing topographical detail into the picture. Figure 2.10 shows schematically how this comes about. Surface *A* produces secondaries whose angular distribution is unfavourable for collection, whereas surface *B* produces more electrons (because of the near glancing angle of incidence) all of which are favourably directed to reach the collector. Thus the effects of surface contours are complicated, but we can see that the last mentioned shadowing of the secondaries from the collector tends to be lost as the specimen is tilted to face the collector.

2.3.5. Specimen charging effects

It is clear from the secondary electron yield curve of Fig. 2.5 that in general the number of secondary electrons leaving the specimen surface is different from the number striking it in the primary beam, i.e. there is a net current either into or out of the surface. If the specimen surface is to remain an equipotential this current must also pass through the specimen to the earth plate. When the specimen is an insulator this cannot happen and the specimen surface will charge, either positively or negatively depending on whether the secondary electron yield is greater than or less than unity.

If the specimen were a perfect insulator it would eventually reach a potential (relative to the cathode of the electron gun) equal to the second crossover potential, when the net electron current to the surface would be zero, and no further change of potential would occur. This argument applies for any initial beam voltage greater than the first crossover potential, which for most materials is a few hundred volts. The specimen surface will charge positively when the secondary emission ratio is greater than unity, and negatively when the ratio is less than unity. In the first case this increases the potential difference between cathode and specimen and increases the energy of the primary electrons towards the second crossover. In the second case the energy of the primaries is reduced towards the second crossover.

For insulators with a secondary emission ratio which never rises above unity the specimen surface always charges negatively and eventually reaches cathode potential, which is a stable situation. Insulators with secondary emission ratios greater than unity also charge to cathode potential when the primary electron energy is initially lower than the first crossover potential.

Thus we see that for perfect insulators there are two stable values of surface potential. One is the cathode potential of the electron gun, which is not a useful condition because no primary electrons reach the specimen and no signal is available to produce a picture. The other potential, equal to the second crossover potential, which lies between 1500 and 3000 V for most materials may be useful in some cases. In this zero current situation, in which primary and secondary currents are equal, there is a useful signal current available to form a picture. In such a case it is possible to obtain acceptable pictures by setting the primary beam energy equal to the second crossover potential of the specimen (if this is known) so that the specimen surface remains uncharged. If it is necessary to charge that part of the surface being scanned in order to bring the primary electron energy to second crossover, then there will be an electric field parallel to the specimen surface which will be strongest at the edge of the scanned area and which will cause distortion of the scan and large variations in picture brightness.

The situation described above applies to perfect insulators, but in practice most specimens will have sufficient conductivity for the charge on the surface to leak away through the bulk to the earthed specimen stub. For really good conductors the conductivity is so high that the net specimen current (primary current minus secondary current) can be conducted away for very small potential difference between top and bottom surface of the specimen. But where this is not so picture distortion will arise, as illustrated in Fig. 2.11, because the point where the primary beam strikes the surface at any instant is at a slightly different potential from adjacent points. The resulting lateral field has two effects: it spreads out the primary beam, spoiling the resolution of the picture; it also deflects the secondary electrons as they leave the specimen surface in a way which gives non-uniform picture brightness when the specimen has non-uniform resistivity or secondary emission ratio.

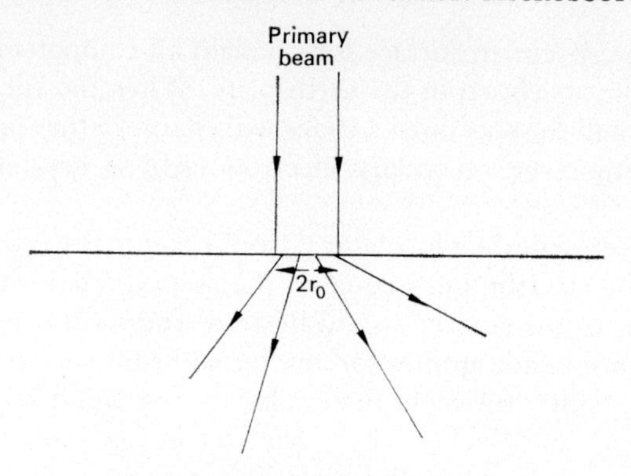

(a) Schematic diagram of hemispherical distribution of current from the primary electron beam

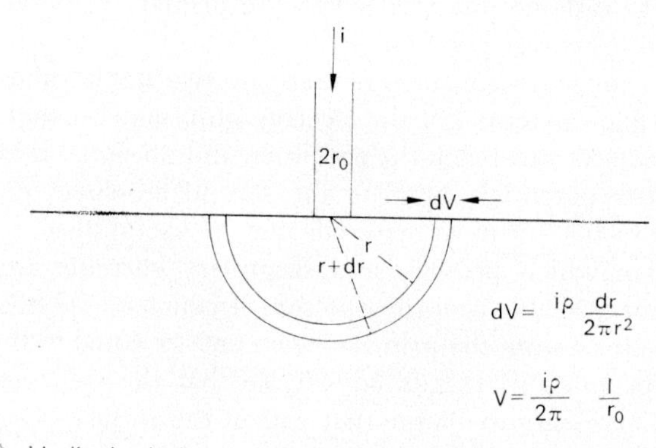

(b) idealised calculation of current flow in a semi-infinite specimen

FIG. 2.11. Specimen charging with conducting materials.

It can be shown that a contact of circular cross-section and area πr_0^2 on a semi-infinite homogeneous conductor has a resistance

$$R = \frac{\rho}{2\pi} \cdot \frac{1}{r_0}.$$

So that the potential drop across the specimen is obtained by multiplying by the current, which is of the order of 10^{-10} A. The potential will be highest for materials of high resistivity and for small beam radius. If the specimen potential change produced by the beam is to be limited to $V = 0{\cdot}1$ V (which would give a transverse field of up to 10^5 V/cm) then we can place a limitation on the specimen resistivity which depends on beam diameter. For a beam diameter of 500 Å and a specimen current of 10^{-10} A the resistivity must be less than about 10 ohm cm. Most bulk specimens will be very thick compared with the beam diameter, but where this is not the case, for example where insulating specimens are coated to avoid the problems of charging, then the potential drop is higher than calculated here, and the resistivity of the coated material must be correspondingly lower.

2.3.6. Voltage and field contrast

It is comparatively easy to observe the phenomena of voltage and field contrast in suitable specimens, but very much harder to measure fields and voltages. These two related effects arise because of modifications in the secondary electron image caused by the application of a potential difference to the specimen.

The voltage effect can be most easily explained by reference to a biased *p–n* junction, as illustrated in Fig. 2.12. If a substantial voltage, perhaps 50 V, is applied to the *p–n* junction in the reverse direction this results in a rather small current flow; most of the *p*-type region is at +50 V with respect to earth and most of the *n*-type region is near earth potential. The potential difference produces a rather narrow high field region around the interface between *n*-type and *p*-type material. When the primary beam falls partially on the *p*-type and partially on the *n*-type sides of the junction, as illustrated, it will produce identical secondary electron currents from the two sides, but they will experience different fields drawing them towards the collector, due to the potential difference of 50 V between the two sides of the junction. Provided that the collector voltage is not too high this arrangement will result in a difference in charge collection efficiency and a different picture brightness.

This effect can be seen for any specimen in which there are potential discontinuities large enough to affect charge collection, but it is hard to say with certainty what "large enough" means in any particular case. It is clearly important to arrange that the electron collection is not too efficient, otherwise the modulation produced by the

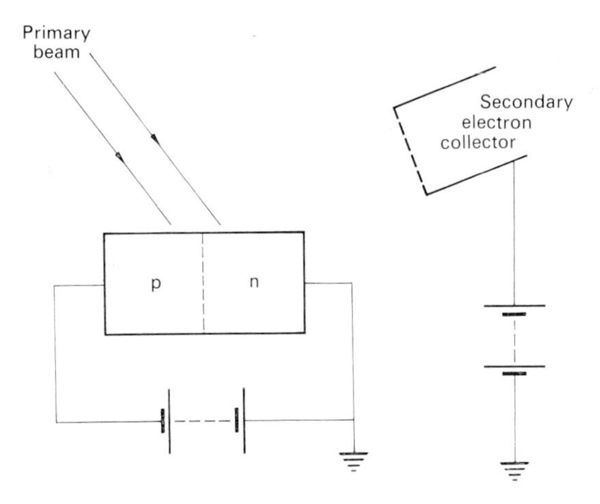

FIG. 2.12. Schematic diagram of voltage contrast.

specimen voltage will be negligible. On the other hand, if the collection is too inefficient we shall collect only the elastically scattered primaries which again will not be modulated by the specimen voltage. Another important factor which affects the sensitivity limit of the voltage contrast is the amount of topographical detail on the specimen. If this is too large it will mask the brightness variations due to the voltage. For most specimens the limit of sensitivity is about a volt between adjacent areas, but modifications to the collector system to incorporate a measure of velocity selection can improve this, perhaps to a few tenths of a volt.

Where there is a potential difference there must be a field, and to this extent the voltage and field effects are related, but the mechanisms whereby they affect the secondary electron image in the scanning electron microscope are quite different. The effect of the electric field, which could be seen if the electron beam in Fig. 2.12 were confined to the high field charge depletion region between *p*- and *n*-type semiconductor, is

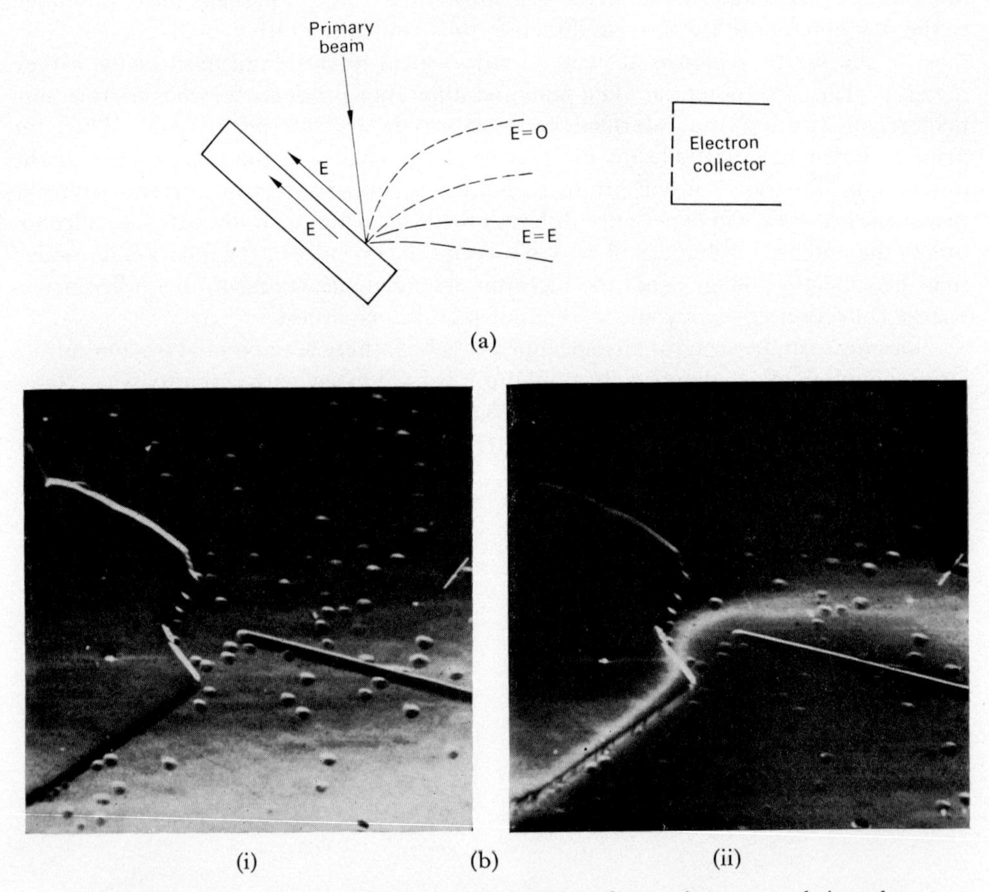

FIG. 2.13. The surface field effect. (a) The field in the specimen extends into the vacuum space outside the specimen and affects the trajectories of some secondary electrons, as shown. (b) The high field at a twin boundary in semi-insulating gallium arsenide (magnification ×150). (i) No applied field. (ii) Average field 20 V cm.

believed to be associated with its effect on the trajectories of the secondary electrons. If the secondary electron collection is not fully efficient, as is usually the case, then not all the secondary electrons reach the collector. A field parallel to the surface of the specimen can affect the collector in a number of ways, viz. it may "bend" the trajectories which would have escaped collection so that they are in fact collected; it may bend the trajectories of some low-velocity electrons so that they are not collected, and so on. Figure 2.13 (a) shows in an entirely schematic way the kind of thing that may happen. A high field region may therefore result in a dark or a bright band on the secondary electron picture, depending on the direction of the field and the orientation of the specimen relative to the electron collector. Figure 2.13(b) shows a semiconductor specimen with and without a current flowing through it. A high field streak appears in the second image where the low-angle grain boundary indicated by the dislocation etch pits reveals that the resistivity is higher here than elsewhere. In favourable circumstances it appears that a field of a few volts/cm can be detected.

It must be emphasized very strongly that these two phenomena are not well understood and that there is no general method of optimizing the SEM for their observation. However, they have obvious importance and it should be possible eventually to use them for quantitative measurements.

2.4. SPECIMEN CURRENTS AND VOLTAGES INDUCED BY THE PRIMARY ELECTRON BEAM

The current flowing to earth through the specimen is

$$i = i_p - i_s,$$

i.e. the difference between the primary and secondary currents. This gives a current of the same order of magnitude as i_p in most cases (i.e. about 10^{-10} amp) which contains information about the variations from point to point on the surface of the secondary emission coefficient and the reflection coefficient for elastically scattered primaries. It can therefore be used as a display signal which emphasizes the variations of composition of the surface and minimizes the effects of surface topography. The signal is likely to be noisy in many cases and will require considerable amplification, which probably explains why it has been comparatively little used.

Another fact worth mentioning is that where this specimen-to-earth current can be measured it is possible to adjust it to zero by reducing the primary electron energy. This procedure determines the second crossover potential for secondary emission and eliminates any possibility of specimen charging effects. A second and more commonly used specimen current is that associated with bombardment-induced conductivity. When an electron of energy E_0 is absorbed by a material whose mean ionization energy for impact ionization is E_i it produces an average of E_0/E_i pairs of charges. In most solids E_i will be tens of electron volts, but in semiconductors only a few electron volts (typically 3·5 eV for silicon). Thus a 20-keV electron will produce between 600 and 6000 extra free electrons, and an equal number of ions or holes. A small fraction of these leave the solid as secondary electron emission, and another small fraction comprise the specimen-to-earth current, but the vast majority eventually recombine after a period of time which may vary from 10^{-12} second in metals and very impure semiconductors to a few seconds in ionic crystals and semiconductors like cadmium sulphide.

If this beam-induced component of the free carrier density is comparable with the equilibrium carrier density it is possible to measure it, either by applying electrodes to the specimen and connecting it to an external battery, or by observing the voltages which may develop in the specimen when the free carriers diffuse from their points of origin towards regions of lower carrier concentration. These phenomena are closely analogous to photocurrents and photovoltages produced by the absorption of light, and as a general rule specimens which exhibit such photoeffects when excited by light will behave similarly when excited by an electron beam. The scanning electron beam allows us to investigate variations from point to point in the specimen of those parameters which determine the magnitude of photoelectric effects and to compare this variation with those of specimen composition and surface topography. The resolution is better than can be obtained using a flying-light-spot scanner, but the measurement is similar in principle and yields the same kind of information. It is not possible here to give a complete catalogue of the properties of photosensitive materials, and it is probably not necessary since most experimenters are already familiar with the basic properties of the materials being investigated before they bring them to the SEM. Examples of the use of specimen currents and voltages will be given in the applications chapters of this book.

A final word is necessary about the resolution available for this mechanism and the relationship between resolution and response time. The design of the SEM is based on the secondary emission mode of operation, where the delay between the arrival of a primary electron and the emission of secondary electrons is so short that it can be neglected. The secondary electron emission current can be measured in a simple collector and ascribed entirely to the point on the specimen where the primary beam is at that instant. This may not be so for the internal current mode in materials where the response time is long. If, for example, the line scan time is one second and the photoconductive response time is also one second then there will still be a measurable photoconductive signal due to primaries incident at point *A* on the specimen when the beam is a whole picture width away at point *B* in Fig. 2.14. A long response time therefore leads to a blurring of the picture unless the time that the primary beam dwells on a single resolved picture element exceeds the decay time of the effect being measured. In extreme cases this time will be too long for the measurement to be practicable.

The resolution is also expected to be worse than that in the emissive mode because ionization near the end of the range of the primary electrons is as effective in inducing currents and voltages as that near the surface (possibly even more effective). Thus the resolution is of the order of the electron range and will depend on the beam voltage and the specimen atomic number. This resolution is worse than that for the emissive mode where secondary electrons come from a small volume near the beginning of the primary tracks.

2.5. CATHODOLUMINESCENCE

Some classes of materials emit light when bombarded by high-energy electrons. This is the phenomenon of cathodoluminescence, by which the specimen under bombardment relaxes back to its equilibrium condition by emitting some of the energy absorbed from the primary electron beam as light. The most efficient cathodoluminescent phosphors may emit up to 10% of the absorbed energy in this way, and these include organic crystals such as anthracene and stilbene, and inorganic crystals like zinc sulphide. Many materials show much weaker emission, for example many plastics

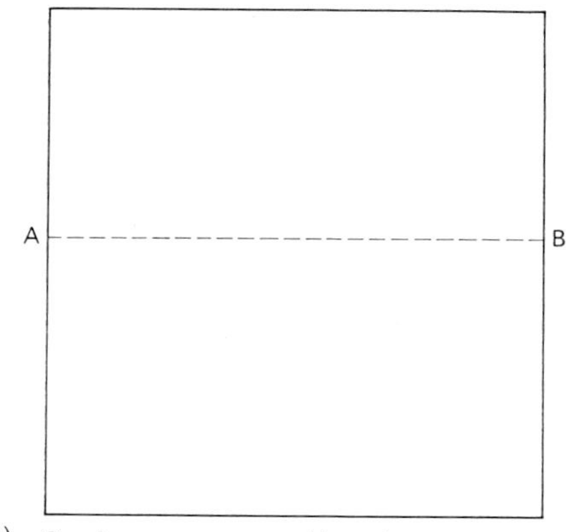

(a) Specimen area scanned by primary beam, showing a single line A.B.

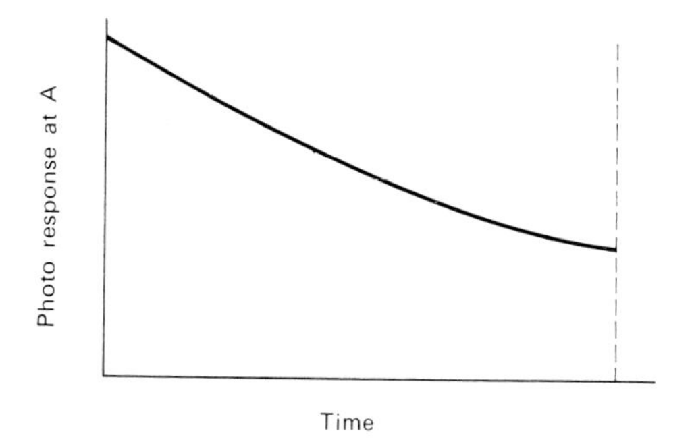

(b) Decay of photo response at A whilst beam scans to B

FIG. 2.14. Effect of response time on resolution.

and glasses. Most materials show no measurable cathodoluminescence at all. The luminescence decay time can lie anywhere between 10^{-10} second and 100 seconds depending on the material and its impurity content.

There are several quite different physical processes which lead to light emission in different classes of solid. Again the worker conducting a SEM study is likely to be familiar with the mechanism which is dominant in his particular specimen. The important point to make is that most luminescent materials are extremely sensitive to impurities; in many cases the luminescence spectrum is determined by the species of impurity, and the brightness by the impurity concentration. Thus the cathodoluminescent mode of operation can be used for analysis, often sensitive to a few parts per million of impurity.

The same considerations about response time and resolution apply as in the case of specimen current and voltage measurements. The "dwell-time" of the primary beam on a single picture element must exceed the luminescent decay time of the phosphor if the resolution of the picture is not to be blurred. Sometimes this will be impossible, and sometimes it will require the long decay time components of the luminescence to be removed with optical filters, leaving only the short decay components to be measured.

The volume from which the luminescence emission comes is as large as the interaction volume of the primary electrons for specimens which are transparent to their own cathodoluminescence emission, so that the best resolution is again comparable to the primary electron range.

2.6. X-RAY EMISSION

Characteristic X-rays are emitted when an inner shell electron is excited to leave an atom or to go into a higher unoccupied level. This electron is then replaced by one of the outer shell electrons conserving energy by emitting an X-ray photon. Figure 2.15 shows schematically the X-ray levels of an atom (the energies apply to the case of molybdenum) together with the spectroscopic and X-ray terminology for the levels.

When an electron is removed from the K-shell ($1s$) it can be replaced from the $2p$ or $3p$, or any higher occupied state from which transitions are allowed. The energy released in these transitions is emitted as X-radiation, and the emission lines are labelled K_α, K_β, etc. When an electron is removed from the L-shell this leads to the characteristic L-emission lines.

This X-ray spectrum is unique to each element, and therefore an analysis of the X-ray emission for any specimen can give an analysis of the constituent elements in the specimen: this is the importance of X-ray emission in the scanning electron microscope and in the closely related electron probe microanalyser. The K_α and K_β lines contain the highest energy (shortest wavelength) X-ray photons from each atom, and because this energy is almost equal to the binding energy of the 1s electron, which in turn is proportional to the atomic number squared (Z^2) we expect the photon energy, $h\nu$, of the K emission lines to be related to the atomic number by

$$[h\nu]^{\frac{1}{2}} = AZ$$

where A is a constant. This is a statement of Moseley's law, which is accurately borne out in practice for each line of the characteristic X-ray spectra of atoms. Figure 2.16 shows a Moseley diagram for the K_α lines of the elements with $[h\nu]$ in units of electron volts. For each element the primary electron energy needed to excite this line is at least as great as the X-ray photon energy, so that uranium requires a primary energy greater than 110 keV.

To carry out analysis by means of the characteristic X-rays emitted in the scanning electron microscope it is necessary to have a primary beam voltage of at least 30 kV and preferably 50 kV, to give a high intensity of K-radiation from atoms up to $Z = 30$. Even 50 kV electrons cannot excite K-radiation from atoms of $Z > 65$, but for the larger atoms with the M and outer shells filled it is possible to use the L-radiation to decide the composition of a sample. [Atoms lower than sodium ($Z = 11$) do not have L-radiation because in the ground state they do not have any L-shell electrons, and for $Z < 45$ the L-radiation is too low in energy to be dispersed by the available X-ray gratings.]

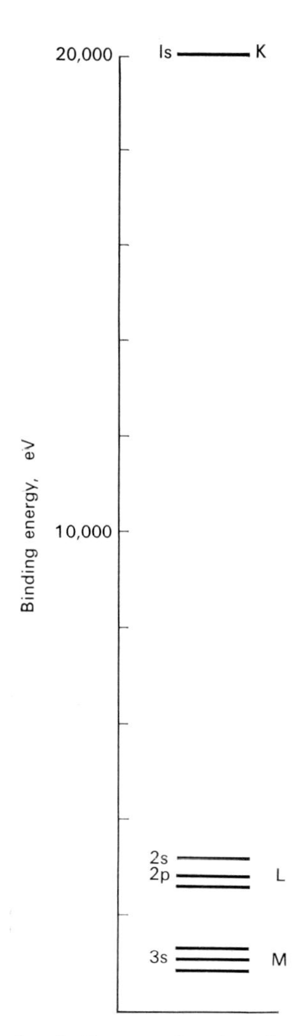

FIG. 2.15. X-ray energy level diagram showing the spectroscopic terminology on the left and the X-ray terminology on the right.

Analysis of specimens in the scanning electron microscope thus requires primary beam voltages of 30–50 keV and, for good sensitivity, specimen currents higher than 10^{-8} A. The interaction volume and the beam diameter are therefore greater than is the case in conventional scanning electron microscopy. Typically the resolution claimed for such measurements is of the order of 1 micron (rather better for low Z elements where a comparatively low beam voltage is satisfactory for the excitation of K-radiation). The sensitivity is about 1 part in 10^4 for the heavier elements, provided that there is no reabsorption of the X-radiation by other elements in the specimen, but low atomic number atoms are difficult to measure because their radiation is not adequately dispersed by X-ray gratings, and this necessitates the use of a proportional counter and pulse height analyser, with consequent loss of sensitivity.

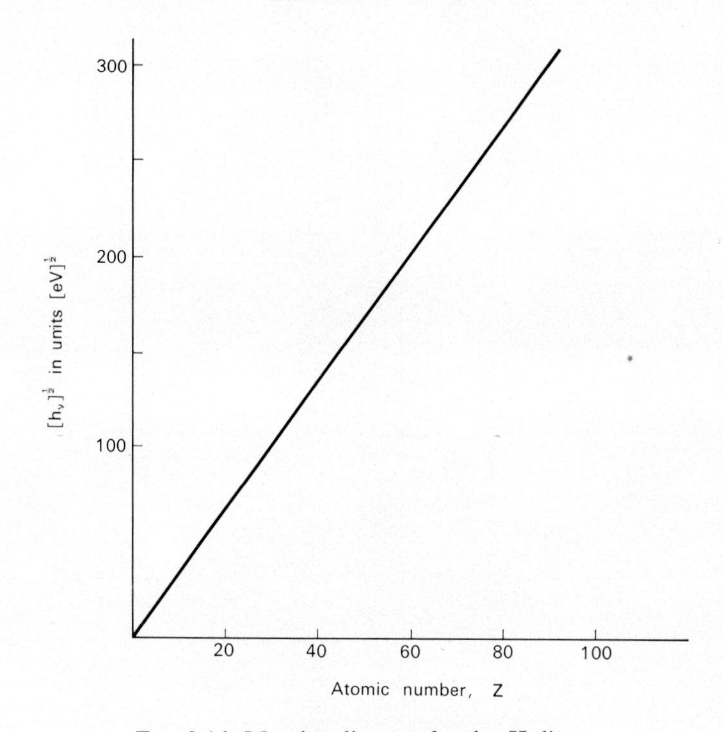

FIG. 2.16. Moseley diagram for the K_α lines.

It is, of course, necessary to have an X-ray spectrometer and detection system inside the specimen chamber of the SEM in order to carry out X-ray microanalysis.

2.7. AUGER ELECTRON EMISSION

Auger electron spectroscopy provides a sensitive method of analysis of surfaces and surface contamination with such high sensitivity that it is possible to identify less than a monolayer of foreign atoms on a surface. The process is closely related to X-ray emission, since it begins with the ejection of an electron from an inner atomic shell, and it ends with the emission of an electron with an energy which is related to the allowed electron energies in the atom. Figure 2.17 describes the Auger emission process in a simple diagrammatic manner. A primary electron of energy E_0 ejects an electron from the K-shell, requiring energy E_K, and is itself inelastically scattered. It may undergo further scattering in the specimen or be reflected with energy E_0–E_K.

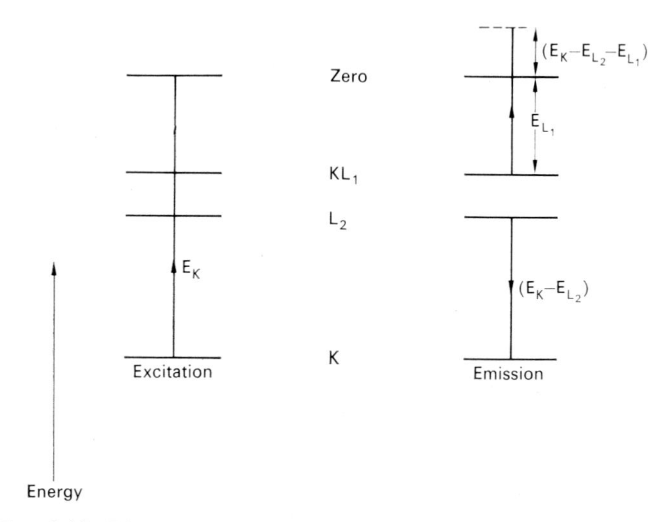

FIG. 2.17. Diagrammatic description of Auger electron emission.

As already explained in the previous section this could lead to X-ray emission, but it is more likely to give rise to Auger emission in the following way. First an electron falls into the K-shell, from, say, the L_2-shell, making available an amount of energy E_K–E_{L2}, this energy is then used to cause the emission of an electron from, say, the L

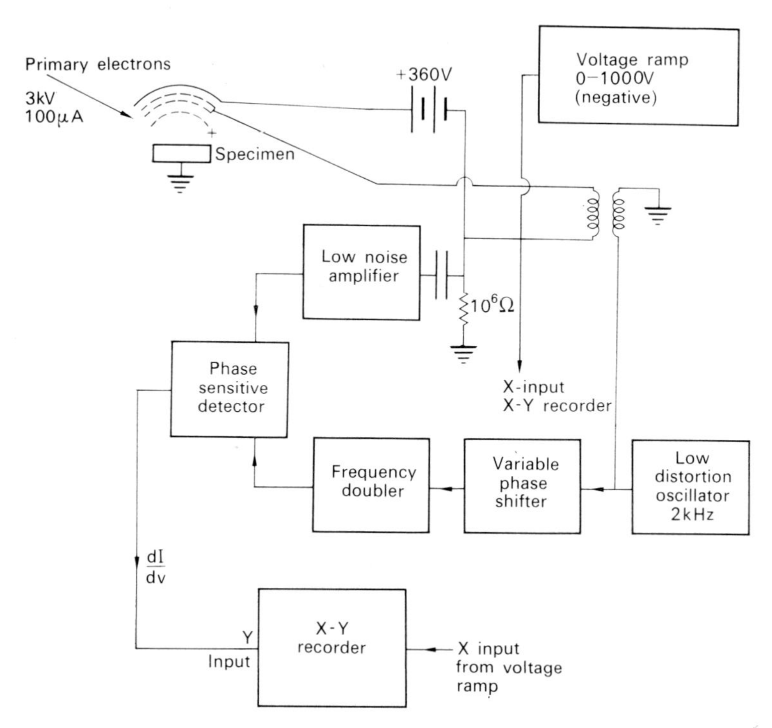

FIG. 2.18. (a) Schematic diagram of Auger electron spectroscopy equipment.

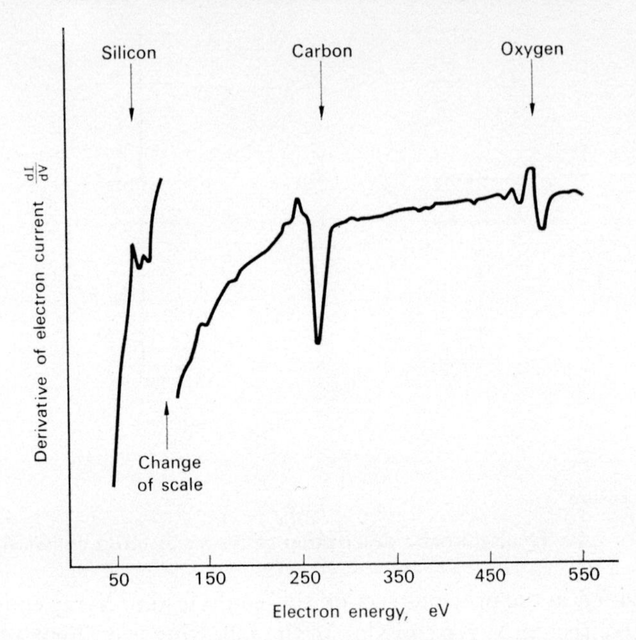

FIG. 2.18. (b) Auger electron spectrum of contaminated silicon surface.

shell, which will leave the atom with energy (E_K–E_{L2}–E_{L1}) as shown on the diagram. In general there will be several such combinations of allowed transitions for most atoms, giving a spectrum of Auger emission energies characteristic of the particular atom, and a careful inspection of spectroscopic data allows the Auger energies to be predicted. Emission energies lie mainly in the range 5–1000 eV and appear as fine structure on the secondary emission spectrum. This is usually emphasized by differentiation of the energy distribution using a double-modulation technique.

The excitation cross-section for Auger electrons is greatest for primary electrons of about 3 times the energy of ionization of the atom, but it falls off rather slowly for energies above this value. Thus the 20 keV electrons of SEM could be used to excite Auger electrons, and a modified electron collector with velocity analysis could be used to give a differentiated secondary electron energy spectrum. Figure 2.18 (a) shows schematically an electron collector system which has been used successfully for this purpose, and (b) a typical Auger electron spectrum.

2.8. SUMMARY AND CONCLUSIONS

Electron interactions in solids are important in scanning electron microscopy in two basic ways:

(i) they initiate a number of physical phenomena, any one of which can be used to give a picture when displayed in synchronism with the scanning beam;

(ii) they play a part in determining the spatial resolution of the instrument, although there are other factors which can be more important, e.g. noise, beam diameter, etc.

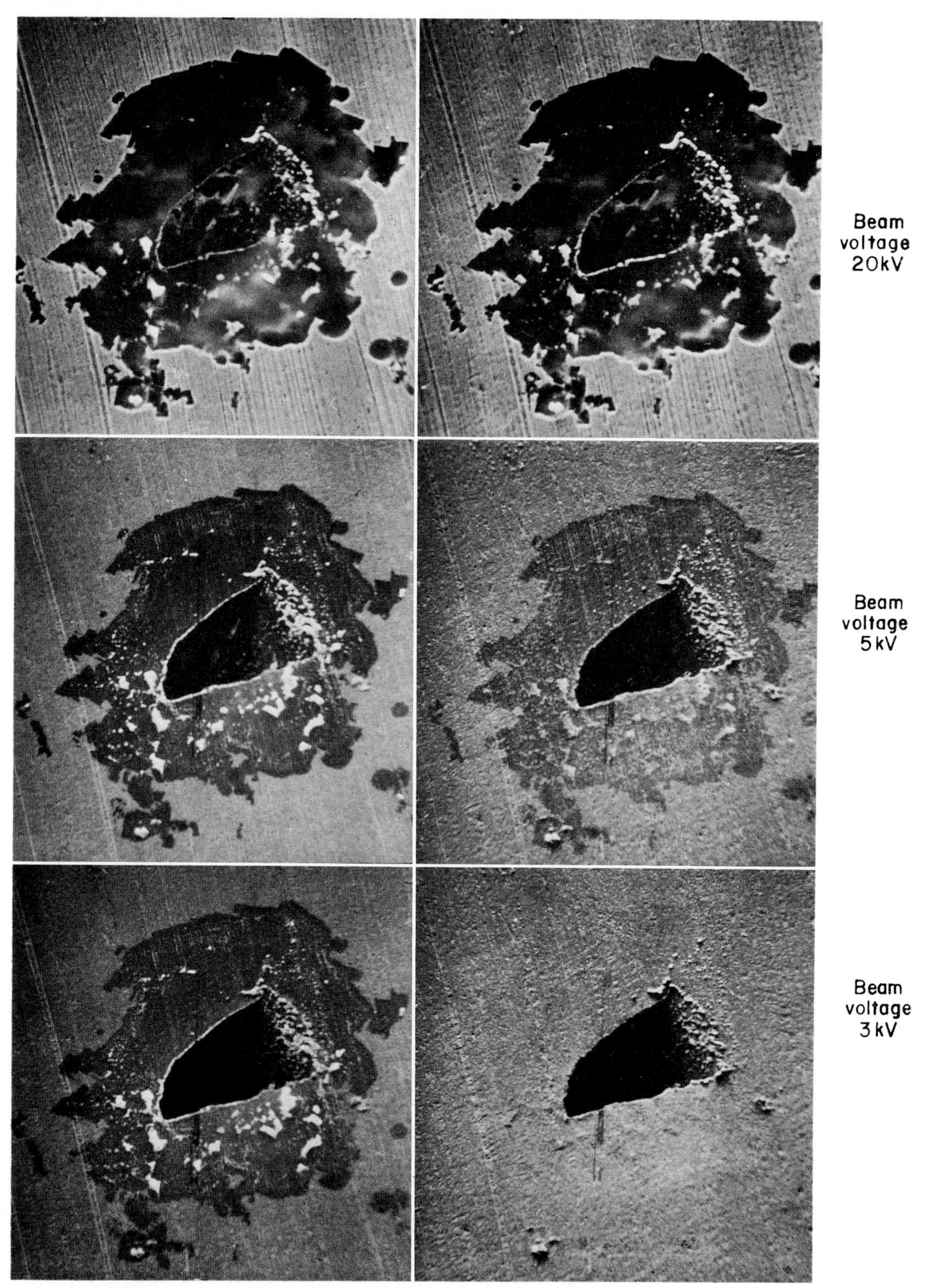

FIG. 2.19. Aluminium specimen examined under different conditions, and showing in bottom right view the surface of the oxide film with a break in the film at the centre. The other views show, to varying extents, a flaw and surface features in the underlying metal. (By courtesy of J. T. Sparrow, U.M.I.S.T.)

The purpose of this chapter has been to explain the nature of electron interactions from these two points of view so that SEM users will have guidance about the kind of information which is available to them and the quality of the pictures that can be obtained. In obtaining the best picture the operator has a number of variables at his disposal, and Table 2.4 lists the criteria by which these variables may be optimized.

TABLE 2.4. SUMMARY TABLE

Microscope control	Factors affected
Specimen tilt	Magnitude of secondary electron emission Efficiency of collecting secondary electrons, X-rays, etc. Foreshortening of picture
Primary beam voltage	Magnitude of effects produced, and thickness of specimen sampled Resolution obtained (important in some cases only) Specimen charging effects
Secondary electron collector voltage	Energy of secondaries used (true secondaries or elastically scattered primaries) [Secondary electron velocity analyser is better, and is vital for Auger electron spectroscopy]
Lens currents and aperture	Increasing the lens currents and reducing the aperture size give smaller beam diameter and smaller current. This may improve the resolution obtained if the beam diameter is a limiting factor but not where speed of response or amplifier noise are the main limitations

More detailed guidance will be given in later chapters in connection with specific kinds of measurements, but as an illustration of the importance of selecting the optimum value of beam voltage the reader should study Fig. 2.19, which shows a specimen of oxidized aluminium examined using a number of different primary beam voltages. Increasing the beam voltage causes secondary electron emission from deeper in the specimen and reveals a fault in the aluminium which is not apparent in the lower voltage pictures where the secondaries come only from the aluminium oxide.

REFERENCES

O. HACHENBURG and W. BRAUER, Secondary electron emission from solids, *Advances in Electronics and Electron Physics* **11,** 413 (1959).
C. C. KLICK and J. H. SCHULMAN, Luminescence of solids, *Solid State Physics* **5,** 97 (1957).
K. G. MCKAY, Secondary electron emission, *Advances in Electronics* **1,** 66 (1948).
C. W. OATLEY, W. C. NIXON and R. F. W. PEASE, Scanning electron microscopy, *Advances in Electronics and Electron Physics* **21,** 181 (1965).
G. SHINODA, Behaviour of electrons in a specimen, *Advances in Electronics and Electron Physics Supplement* **6,** 15 (1969).
P. R. THORNTON, *Scanning Electron Microscopy*, Chapman & Hall, 1968.

CHAPTER 3

The Design and Use of the Scanning Electron Microscope

J. E. CASTLE

3.1. INTRODUCTION

We have seen that when a beam of electrons strikes a solid it loses its energy in many ways but that each type of interaction signals the state of its surface, or of its substructure. In this chapter we shall explore the way in which modern instruments have been designed to accept these signals and display them sequentially so as to image the surface. We shall consider the factors relating to noise, resolution and contrast in the scanning electron microscope, so as to provide a background for Chapter 5 on the Procedures for Using a Scanning Electron Microscope, and to indicate what is attainable by scanning electron microscopy. In the present chapter also we shall consider the type of facilities which are built into commercial instruments or which are available as options. We cannot, however, supplant what is available in the handbooks appropriate to individual instruments.

The scanning electron microscope belongs to the family of scanning instruments, which also includes radar, sonar and the microprobe analyser. In all these systems the probing beam scans the surface repeatedly, either along a single line or in a sequence of lines forming a rectangular raster of the television type. A portion of the signal or reflected beam is picked up at a receiver, amplified and used to modulate the brightness generated by an electron beam scanning the face of a cathode-ray tube in synchronism with the movements of the probing beam. The image, therefore, is no more than a display of the received signals in their correct relative positions and may look quite different from an ordinary optical image of the surface. In practice, we soon learn from examination of familiar objects in the SEM that the display usually depicts the surface as we expect it to look, particularly when the collected secondary electrons are used as the signal. This being so, it is easy to forget that the micrograph is sequential in character, that at any instant of time there is no image, and that the usual rules governing resolution and contrast in optical systems do not apply.

The physical form of the SEM is now well known: the "Stereoscan" of the Cambridge Instrument Company is shown in Fig. 3.1, but all commercial instruments represent different approaches to the arrangement of the blocks in Fig. 3.2. The design of the instrument is, however, determined to a large extent by the requirements for resolution and contrast in a raster and this will be our main topic. Before going on to this, however, we must consider the basic electron optics required for the formation of an electron beam.

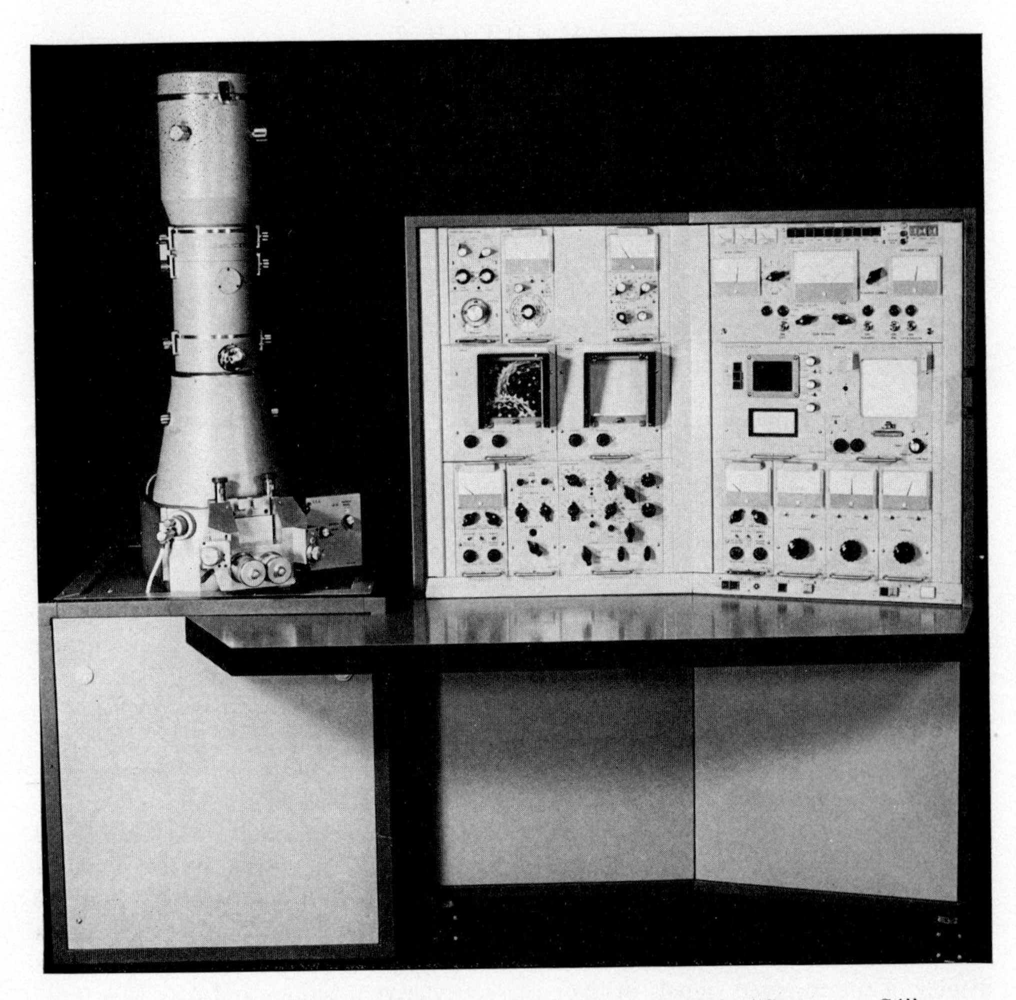

FIG. 3.1. A high resolution scanning electron microscope: the "Stereoscan S4". (Courtesy of the Cambridge Instrument Co. Ltd.)

3.2. THE FORMATION OF AN ELECTRON BEAM

The electron beam is produced in the electron gun at one end of the vacuum column and is brought to a focus on the specimen mounted at the far end. Focusing and movement of the beam is achieved using electron lenses and deflector coils mounted outside the vacuum tube.

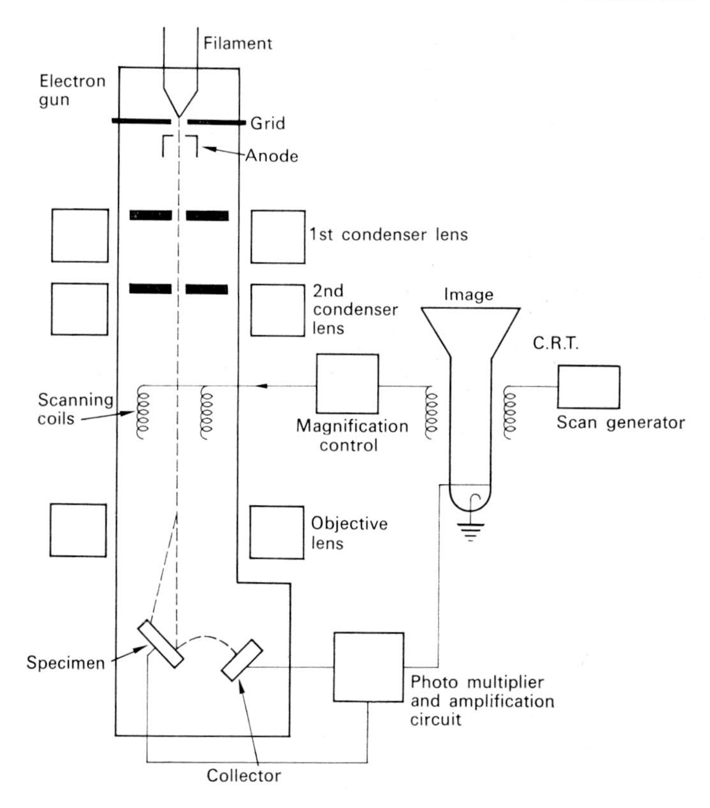

FIG. 3.2. Block diagram of a typical SEM.

3.2.1. The electron gun

A method by which an electron beam can be formed is shown in Fig. 3.3. Here a triode gun, as used in most scanning electron microscopes, is combined with a single lens. Formally, this triode may be treated as an electrostatic immersion lens (Cosslett, 1950). The accelerating voltage (or e.h.t.) is applied between the cathode and anode and the field at the cathode (filament) surface is moderated by the bias potential applied to the grid. This is normally cylindrical (Wehnelt cylinder) with only a single aperture which is large enough for the anode field to penetrate to the cathode surface. Electrons are thus emitted from that part of the filament bounded by the zero equipotential (Fig. 3.3 (b)*) with only thermal energies. They are accelerated through the grid aperture and into the field of the anode which imparts a strong converging action. Hence, as the ray diagram (Fig. 3.3 (c)) shows, a first focus or "cross-over" is formed and the beam is strongly diverging as it leaves the vicinity of the gun. The role of the lens is to bring this diverging beam to a focus in the specimen plane. This focus is an image of the emission area of the filament whose size depends on the value of the bias potential. Increasing the bias causes the equipotentials to move away from the filament

* In the case of most scanning electron microscopes the zero equipotential is equal to the negative e.h.t. applied to the filament: the anode is at earth potential.

until eventually emission ceases. Thus to draw a satisfactory beam current the area must be rather large. The size of the image produced at the first cross-over depends on the focal length of the grid/anode combination acting as a lens and is approximately V_g/V_A where V_g–V_A are their respective potentials. By suitable choice of filament–grid and grid–anode distances the image at the first cross-over can be made smaller than

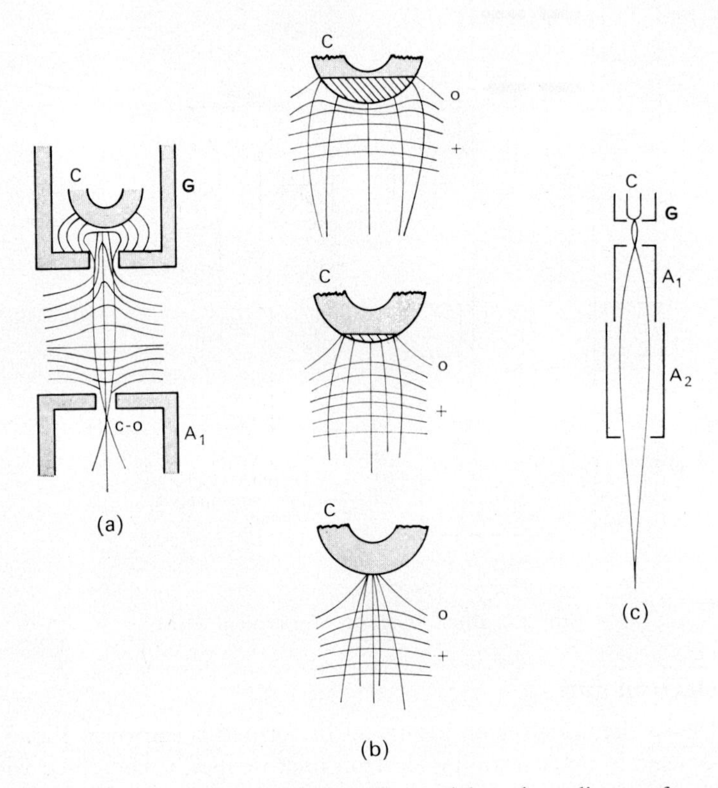

FIG. 3.3. The electron gun. (a) Typical equipotentials and ray diagram for a triode electron gun (after Cosslett, 1950). C, cathode; G, grid, A, anode; C-O, 1st cross-over. (b) Equipotentials showing the influence of grid bias on the emission area of the filament (after Elion, 1966). (c) Ray diagram showing the focusing action of a second anode A_2 at a potential greater than A_1 (after Cosslett, 1950).

the emitting area and thus assists in the production of a fine beam. Because of the finite solid angle of the electrons emitted from any point on the filament, and because of their mutual repulsion, the size of the image at the cross-over is always finite no matter how well the gun is designed (see for example, Moss, 1968). This point will be returned to in a later section.

3.2.2. The cathode

From the discussion above it can be seen that the emission area must be kept as small as is consistent with a usable current. For this reason and because they are unsuitable unless the vacuum is permanently maintained, it is not possible to use the indirectly heated, coated, cathodes found in electron tubes and cathode-ray tubes. A "hairpin" filament of tungsten is normally used and this reaches a space-charge limited emission current of about 2 A/cm^2 at 2700°C. At this rating a beam current of 200 μA can be drawn from an area of 100 μm square. Most of this current falls on the grid and acts as a useful monitor of the filament performance (i.e. the beam current monitor); only a small proportion passes through the grid aperture. The alignment of the filament both with respect to the aperture and the grid/anode distance is important, as we saw above, and facilities should be available to move the filament whilst it is in action. It is all the more important because the life of the filament is only tens of hours and therefore it cannot be set permanently.

Very much greater beam currents can be drawn from the tip of a field emission cathode. This type of gun has been developed by Crewe (1966). A drawback to the use of such a cathode in regular instruments is the need for an ultra high vacuum system which is not compatible, at reasonable cost, with the mechanical feed-throughs necessary in a working microscope.

3.2.3. The lens system

The simplest form of lens is the second anode as shown in Fig. 3.3(c). This will converge the beam to a focus if its potential is much greater than that on the first anode. Electrostatic lenses were used in the early instruments built by Professor Oatley and his co-workers but electromagnetic lenses (usually simply referred to as magnetic lenses) are now universally adopted. The main reasons for this are their relative freedom from contamination problems, the saving on the dimension of the vacuum column because they can be externally mounted and their smaller aberration coefficients (Oatley *et al.*, 1965).

The positioning of the scanning coils is also important. If they are mounted on the image side of the final lens the working space is seriously reduced. (The image distance cannot be increased without limit since once it equals the object distance the lens loses its demagnifying function.) The scanning coils are therefore mounted either on the object side of the lens when a double deflection must be used so that the beam passes through the centre of the lens (Oatley *et al.*, 1965) or within the lens itself (Fig. 3.10). As we shall see in the next section the total demagnification required to produce the electron "spot" is so large that at least two and preferably three lenses are required in addition to the electron gun. The first or condenser lenses are made strongly demagnifying whilst the final lens is designed to close limits in order to reduce spherical aberration (Elion, 1966). The condenser lenses are fitted with spray apertures to remove stray and aberrated electrons and the objective lens with interchangeable apertures to define the angle of the converging cone of electrons forming the final image.

3.3. RESOLUTION AND CONTRAST IN THE SCANNING ELECTRON MICROSCOPE

The discussion of the electron optics cannot proceed until we have some figures for the final image size required to achieve a useful magnification together with figures for the final beam current which will enable us to form a display. We shall thus turn to the problem of resolution, contrast and noise in a raster and only then return to the remaining instrumental controls which are so apparent in Fig. 3.1. Each of the components of the SEM in Fig. 3.2 has its own characteristics of resolution and noise and the whole network must be optimized by the designer—with enough leeway for the user to accommodate the needs of his particular object. We will deal separately with the three main components of the network: the display system, the electron optics, and the specimen stage and signal receiver.

3.3.1. The display system

It is with the display that the instrumental resolution becomes related to that attainable with the human eye and is here that we can relate the probe size to the magnification attainable. The display system consists of the cathode-ray tubes—one of high resolution for photographic recording—with the usual contrast and brilliance controls found on a television receiver. All movements of the electron beam within the cathode-ray tube are, however, controlled by way of the "scan generator" which is also directly linked to the scanning coils in the microscope column. In this way each adjustment which the operator makes to the display, whether it be height or area of the raster, the position of a line scan or a stationary spot, is faithfully reproduced by the movement of the probing beam across the face of the specimen.

In the normal operating mode the waveform fed to the scanning coils in the cathode-ray tube gives a constant amplitude to the beam deflection so that the raster always exactly fills the screen. The waveform fed to the scanning coils in the column is modified by the magnification control, so the amplitude of the movement there is $1/m$ of that observed in the cathode-ray tube, where m is the magnification. It is usually arranged that the attenuation of the waveform current necessary to achieve this is switched in stages which have been precalibrated to read in magnification directly.

The other link, in the main network, between display and the specimen, is by way of the signal receiver which relates the brightness of the object as perceived in the electron illumination to the brightness of the equivalent point in the display.

The human eye can resolve some 50 lines per centimetre and therefore, since the display CRT will normally have a screen of about 10 cm in edge length, some 500 lines of information are required to form the image. Conversely the resolution of the screen itself does not need to be better than 500 lines. In general we can write:

$$N' = \frac{l}{\delta} \tag{3.1}$$

where N' is the number of lines resolved, l is the edge length of the display and δ is the resolving power of the eye. The scan generator enables the operator to select the line rate, i.e. the time for the beam to sweep across one line of the raster, and the frame rate, i.e. the time to complete one frame. The ratio of these two times

$$N = T_{\text{frame}}/T_{\text{line}} \tag{3.2}$$

clearly gives the number of lines of information on the screen. When the specimen is stationary, that is correctly aligned in the field of view and in focus, T_{frame} can be increased until it is limited by the persistence of the screen phosphor. This enables

T_{line} to be increased, which as we shall see improves the signal-to-noise ratio in the display. Whilst the specimen is being aligned, T_{frame} needs to be as short as possible, so that no significant movement of the specimen is lost between frames.

Since also, as we shall see in the next section, it is best to have $N = N'$

$$T_{frame} = T_{line} \times \frac{l}{\delta} \tag{3.3}$$

we can reduce T_{frame}, whilst keeping the resolving power constant, by reducing l. The operator will often make use of a facility which enables him to reduce l during the setting-up period. This reduction in field of view is carried out without altering the attenuating factor between the scanning coils in the display and in the microscope so that the magnification remains unchanged.

When an image is being recorded on a photographic emulsion the limitation on the frame rate almost disappears. The image is built up line by line in a single-frame exposure and as there is no need to use a tube with a persistent phosphor the limiting feature is the stability of the instrument and the operator's patience. When recording photographically the number of lines of information in the raster will probably be doubled to give a typical value of $N = 1000$ in order to allow for subsequent photographic enlargement on the micrograph. The resolution of the raster can be defined in terms of the number of lines N; alternatively, since in practice there is no observable difference between a micrograph produced from a "vertical" or a "horizontal" raster, we may define it as containing a number of picture elements equal to N^2. This latter definition will be useful in discussing the requirements of the electron probe.

3.3.2. The electron probe

We are now in a position to appreciate what the designer of the electron gun and lens assembly has to achieve in order to obtain a given useful magnification in the final display. We shall assume for the moment that the specimen is a passive reflector of the portion of the signal in which we are interested and that no spreading of the electron "probe" occurs within or at the specimen surface. We have seen that the separation of raster lines or of any picture elements on the specimen surface is $1/N$ of that in the print. Thus the spacing of the raster lines is

$$D = \frac{1}{mN}. \tag{3.4}$$

Taking the print size of 10 cm and a line density of 50 lines per centimetre, the spacing of raster lines on the specimen will be $D = 1000$ Å for a magnification of 2000×, and 100 Å for a magnification of 20,000 diameters respectively. If the effective size of the electron probe is equal to the required raster spacing at a given magnification, then each point in the field of view will be swept once only and none will be missed. We can therefore define the required effective probe diameter as

$$d = D = \frac{1}{mN}. \tag{3.5}$$

The relationship between the "gaussian" diameter of the probe to the circle of least confusion and to the aberrated images is given in Fig. 3.4. It is usual to consider the effective diameter of the probe be that given by the quadratic sum of the gaussian probe diameter, d_0, and the diameters of the aberration circles (Grivet, 1965).

In a conventional gun the image of the cathode at the first cross-over has a radius of about 100 μm: thus to achieve a useful magnification of 20,000 diameters the cross-section of the beam must be reduced by a factor of about 10^4 during its passage through

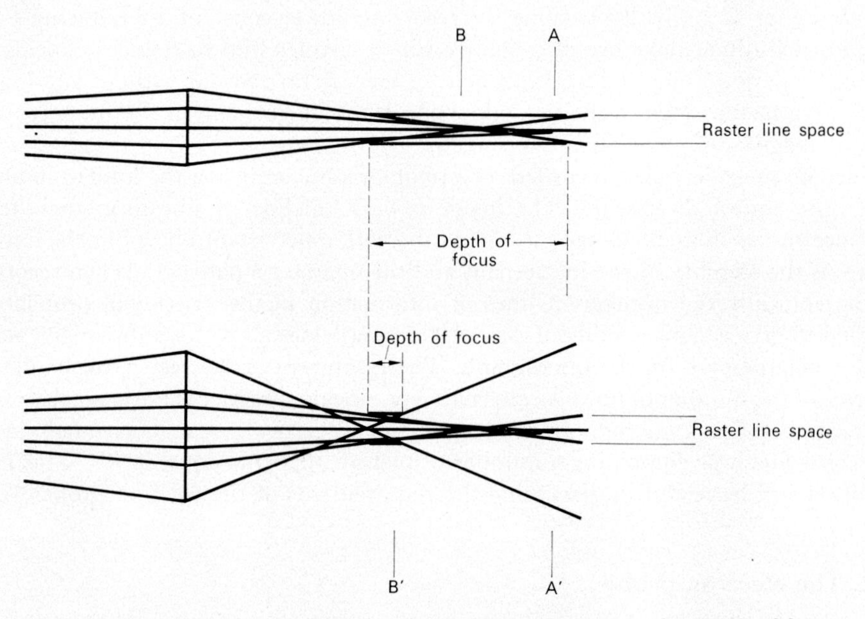

FIG. 3.4. Ray diagrams showing the effect of lens aperture on the depth of focus. Section AA' represents the gaussian focal plane and BB' the position of the circle of least confusion.

the column. The philosophy of the Engineering Department at Cambridge University, used in designing their series of research instruments, was that three lenses, two condenser and one objective, are necessary to achieve this degree of demagnification (Oatley *et al.*, 1965). This lead has been followed by Cambridge Instruments in their high resolution "Stereoscan" series of scanning microscopes. Other manufacturers, notably Japan Electron Optics Co. Ltd., have adopted a two-lens arrangement with success.

The resolution and contrast limits which may be reached with an electron-beam instrument have been explored by Oatley, Nixon and their co-workers at the Engineering Laboratories at Cambridge and it is their treatment of the subject that is adopted here.

Pease and Nixon (1965) showed that the relationship between the minimum probe diameter and the spherical aberration of the final lens, when the beam current is vanishingly small, is given by

$$d_{min} = 1{\cdot}22\, C_s^{\frac{1}{4}}\, \lambda^{\frac{3}{4}}. \qquad (3.6)$$

The wavelength, λ, of the electrons is given by

$$\lambda = 12{\cdot}4\ V^{-\frac{1}{2}}\ \text{Å} \tag{3.7}$$

and C_s is the coefficient of spherical aberration of the lens. In practice, of course, the beam must contain a substantial current in order to elicit a signal from the specimen. The magnitude of the necessary current may be obtained from the relationship derived by Smith and Oatley (1955),

$$C_{Lt} = 10\left(\frac{eP}{it}\right)^{\frac{1}{2}} \tag{3.8}$$

where i is the probe current, P is the number of picture elements in the raster, e is the electron charge, t is the recording time in seconds and C_{Lt} is the limiting contrast, i.e. the contrast which just enables adjacent picture elements to be discerned. The factor 10 is introduced to satisfy the empirical criterion of Rose (1948) for the relationship between the signal-to-noise ratio and C_{Lt} and is modified to allow for the assessment by Oatley *et al.* (1965) for the increase in noise during amplification and display of the signal.

The implication of this relationship becomes clearer when some values are inserted. For example it is reasonable to argue (Oatley *et al.*, 1965) that the minimum discernible difference in signal strength is equal to 5% of the total signal, i.e. $C_{Lt} = 0{\cdot}05$. This lower limit of performance is reached, for a raster of 400 lines ($p = 1{\cdot}6 \times 10^3$ points per picture) covered in 5 minutes, when the electron count in the signal drops to 10^4 per picture point. Since the lines in this raster would be just at the limit of resolution of the eye and a 5-minute exposure is getting inconveniently long for routine work we can reasonably say that this is the minimum number of electrons to make a picture. At low magnifications, with consequently large spot sizes, this limit poses no problems. For example, the probe current may reach 10^{-9} A (10^{10} electrons/sec) which would enable a 1000-line raster to be accomplished in 1 second. As the spot size is reduced, however, the intensity of the electron beam does not rise *pro rata*. If we examine the Langmuir formula,

$$J_A = J_c\left(\frac{eV}{kT} + 1\right)\sin^2\alpha, \tag{3.9}$$

where J_A is the current *density* in the gaussian image and J_c the emission current density at the cathode, temperature T°K; we see that the intensity is not related to the area of the probe but only to $\sin^2\alpha$, where α is the semi-angle of the electron beam converging to the image. If the aperture of the final lens is 0·1 mm and the working distance 10 mm, $\alpha = 0{\cdot}005$ rad. The effect on this value of any reduction in spot size is negligible. Hence the current density will remain constant and the actual signal strength will decrease according to the square of the spot size. On the other hand, a reduction in the working distance to 1 mm would increase the brightness by a factor of 100. The Langmuir formula can, of course, be used to relate the exposure time to probe size, d_0, instead of the signal-per-picture point as is done in (3.8). We can simplify the procedure by first noting that for any useful accelerating voltage,

$eV/kT \gg 1$ and $\sin \alpha = r/x$ where r is the radius of the final aperture and x the working distance. Thus, from (3.9),

$$\mathcal{J}_A = \mathcal{J}_c \frac{eV}{kT} \frac{r^2}{x^2} \tag{3.10}$$

also

$$\mathcal{J}_A = \frac{4i}{\pi} \cdot d_0^2 \tag{3.11}$$

where, as before, i is the current in the probe.

Thus

$$i = \frac{\mathcal{J}_c}{4} \frac{eV}{kT} \cdot \frac{\pi d_0^2 r^2}{x^2} \tag{3.12}$$

or by combination with (3.8)

$$C_{Lt} = \frac{20x}{d_0 r} \left[\frac{kT\,P}{\pi V \mathcal{J}_c t}\right]^{\frac{1}{2}}. \tag{3.13}$$

When all other parameters are maintained constant we see that the exposure time is inversely proportional to the square of the probe size that is directly proportional to the square of the magnification which is attainable. The term contained under the root sign in (3.13) is fixed by the emission properties of tungsten, by the resolution of the human eye and by human patience in making exposures. This means that, eventually, as we reduce the spot size d_0, we can only compensate for loss of signal by increasing α, i.e. r/x. Beyond a point further reduction in working distance must reduce signal strength because of spatial hindrance by the final lens. On the other hand, increases in the size of the final aperture are eventually limited by the lens aberrations which increase the effective diameter of the probe. This limit, which represents the optimization of the requirements for contract separation of picture elements with those for spatial separation of picture elements, has been obtained by Oatley *et al.* (1965). They related the effective probe diameter to the root of the square on the gaussian probe size plus those on the circles of least confusion due to spherical and chromatic aberration, respectively:

$$d^2 = d_0^2 + d_c^2 + d_s^2. \tag{3.14}$$

As we have seen

$$d_0^2 \propto \alpha^{-2} \tag{3.15}$$

whilst

$$d_s \propto \alpha^3 \tag{3.16}$$

and

$$d_c \propto \alpha. \tag{3.17}$$

Thus

$$d^2 = A\alpha^{-2} + B\alpha^2 + C\alpha^6 \tag{3.18}$$

where A, B, C are constants (Oatley *et al.*, 1965). This equation was solved to yield the optimum aperture size at which d is a minimum and hence to predict the best performance attainable using a tungsten filament as the electron source. Curves derived from their parameters are given in Fig. 3.5. For an exposure time of 15 minutes, a spot diameter of 45 Å is predicted and this has been obtained in practice. As we shall see in the next section the loss of resolution due to spreading of the beam within the

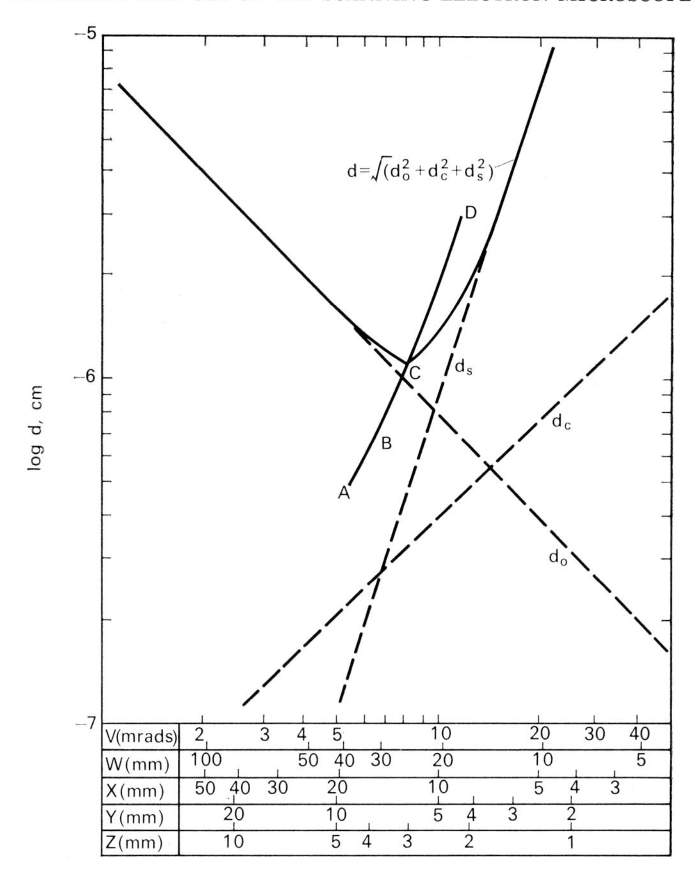

FIG. 3.5. The effective spot size as a function of the semi-solid angle of the aperture at the specimen (α): d_0 is the size of the gaussian image determined by the requirements of signal strength; d_s and d_c the diameter of the circles due to spherical and chromatic aberrations respectively. The line *ABCD* is the locus of the minima in the curve as a function of counting time: $A = 0{\cdot}4$ lines/sec; $B = 1$ line/sec; $C = 10$ lines/sec; $D = 100$ lines/sec. The x-axis designations are as follows: V, α in mrads. W–Z, working distance (mm) for aperture sizes 400 μm, 200 μm, 100 μm and 50 μm, respectively.

specimen itself will ensure that except in exceptional cases, picture elements separated by distances as small as this will not be resolved. The relationships are useful, however, since they give a feel for the dependence of the exposure time on the probe diameter. By the use of large values of d_0 the current i can be increased to the point where the image can be presented at normal television scanning speeds. For example, Nixon (1969) has suggested that at a frame time of 0·5 seconds a spot size of 300 Å should still be attainable. Scan times of this speed have obvious advantages if kinetic changes in the specimen are to be observed. Moreover, this figure does not take account of the large gain in signal/noise ratio by the time averaging property of both the phosphor in the display screen and of the eye. The static specimens can certainly be examined at 10,000 diameters on the 625-line display of a conventional television monitor.

In fixing α at α_{opt} as represented by the line *ABCD* in Fig. 3.5, we also fix the depth of focus which is proportional to $1/\alpha$ as can be seen from Fig. 3.4. This not only fixes the height of objects which will be in focus along the axis of the lens system but also fixes the field, within the raster, which will be in focus. Thus for a given counting time we can use the solution of equation (3.18) to plot the depth of focus as a function of the magnification. This is given in Fig. 3.6 for the Stereoscan. In many cases it is

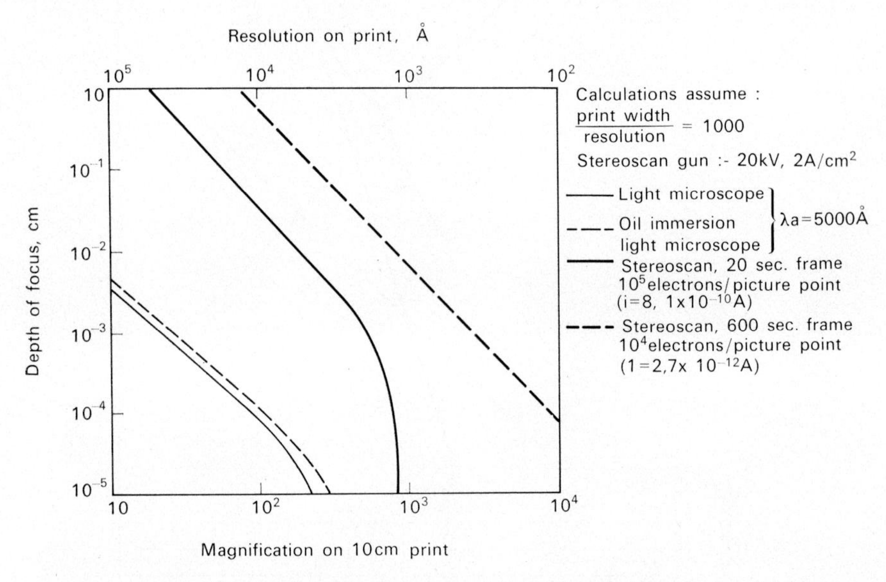

FIG. 3.6. Resolution as a function of depth of focus for the Stereoscan. (Courtesy of Cambridge Inst. Co. Ltd.)

desirable to sacrifice resolution in order to use the remarkable depth of focus which is attainable with SEM. In this case the magnification required should be obtained using a value of α close to α_{opt}. Then the aperture can be decreased in stages until the noise level becomes unacceptable. Guidance to the choice of α_{opt} is usually given in the operator's handbook as a graphical solution to equation (3.18) for the particular instrument. The radii of the final aperture are usually in the range 50 to 400 μm.

3.3.3. The specimen

As was emphasized in the previous chapter it is not, in fact, permissible to regard the specimen as a passive reflector of an incident beam as we have just done. The interactions of the electron beam with the solid which produce the signals are not confined to the surface. Rather, the electron beam enters the specimen and we may define a "sampled volume" from which the signal will be obtained. The size of the sampled volume varies, as is shown in Table 2.2, with the energy of the incident beam of electrons and inversely with the atomic number of the solid. From the point of view of resolution, however, we are concerned not with the size of the sampled

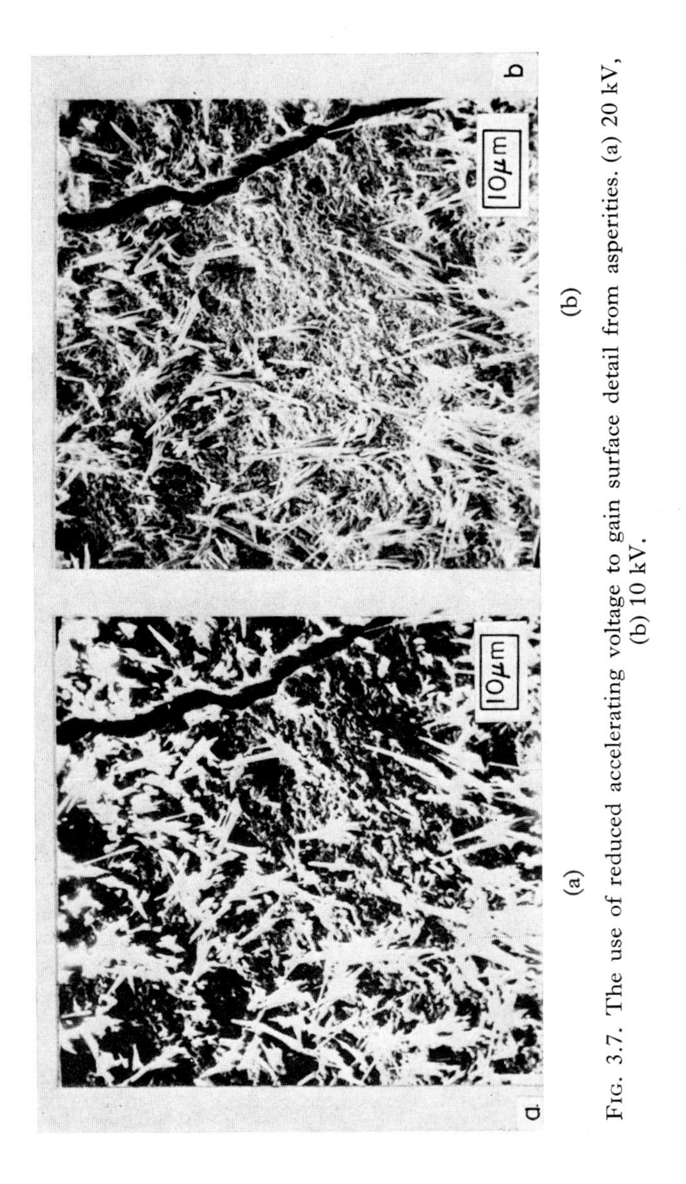

FIG. 3.7. The use of reduced accelerating voltage to gain surface detail from asperities. (a) 20 kV, (b) 10 kV.

volume but with the area from which a signal is emitted, that is, the intersection of the sampled volume with the surface being examined. It is no accident that the scanning electron microscope is used most frequently in the secondary emission mode, since these electrons have such lower energy (*ca.* 100 eV) that their penetration of most materials is less than 100 Å. This may be taken as the dimension of the area from which the secondary emission signal is received. At the other end of the signal range are the X-rays, which have such great penetrative power that they are received from the whole cross-section of the sampled volume: i.e. the dimension of the area from

FIG. 3.8. Examination of thin surface deposits using reduced accelerating voltage. (a) 20 kV, (b) 5 kV.

which X-rays are received will be of the order of 1 micron. The re-emergent, or back-scattered, high-energy electrons are derived from an area intermediate between these two extremes. Thus the operator will be influenced in his choice of electron-probe size by the operating mode in which he has chosen to work, having regard also to the accelerating voltage used in forming the probe and to the atomic number of the material being examined. As a rough guide, the probe size should be similar to that of the area from which the signal is obtained; the use of very much smaller probes than this provides no increase in resolution but will reduce the clarity of micrographs by lack of signal strength.

The minimum size of the sampled area, using secondary emission as the signal, is about 100 Å when they are back-scattered from a bulk solid. When the microscope is operated so as to collect secondaries by transmission through thin specimens the signal may be obtained from much smaller areas. There is then an incentive to obtain a smaller probe size than that possible with the tungsten filament. The manner in which this can be achieved has been pioneered by Crewe (1966), who has worked to increase the value of J_c remarkably by the use of field emission sources instead of thermionic sources. Recent reports suggest that resolutions of 5 Å are attainable by the method.

Whilst the resolution of a surface being examined in the secondary emission mode by reflection is not markedly dependent on specimen thickness and its penetration by the electron beam, the local noise level is greatly dependent on this parameter. It is possible for the signal-to-noise ratio at surface asperities to be so degraded that no detail can be discerned. The effect arises because secondary electrons are produced wherever back-scattered high-energy electrons emerge from the surface. On a plane surface the number of electrons produced by such re-emergence remains fairly constant and the signal merely contributes randomly to the background noise. Where

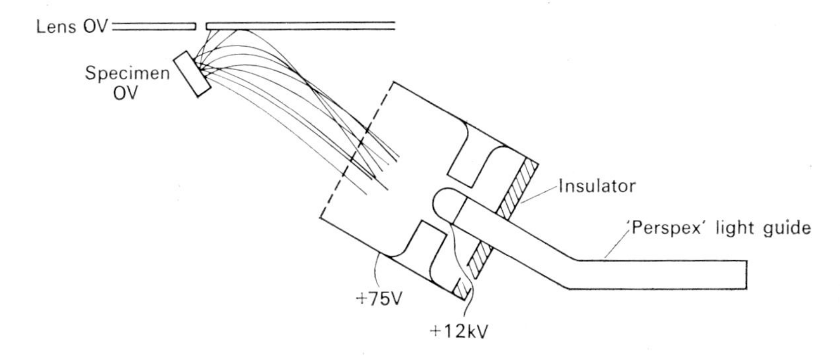

FIG. 3.9. A scintillation-type detector set up to receive secondary electrons. The trajectories shown are those plotted for this arrangement by Everhart (1968). The electron transit time is 5×10^{-7} sec.

the sampled volume intersects an edge, or in the case of a thin specimen the reverse surface, a much larger number of secondaries are produced, the collector is overloaded and the micrograph appears to be "burned out" in that region Fig. 3.7 (a). The same effect is obtained at edges and in granular samples the individual particles may appear to be surrounded by a bright halo. To some extent these effects can be minimized by reduction of the accelerating voltage (Fig. 3.7 (b)) as foreshadowed by Table 2.2. A related effect occurs when the specimen is so thin that the primary beam is able to elicit a signal from underlying surfaces. The surface on being examined may then appear partly transparent (Fig. 3.8 and Fig. 2.19).

3.3.4. The collector

Rather less than one secondary electron is produced on average for each electron in the probing beam (Thornton, 1968). It is thus essential to have an efficient means of collection if a usable signal is to be obtained. The development of collectors of the type illustrated in Fig. 3.9 by the Electrical Engineering Laboratories at Cambridge is regarded by many as the key stage in the development of the Scanning Electron Microscope. A grid in front of the collector is maintained at a small positive potential in relation to that of the specimen. This causes the emitted electrons to be accelerated along a curved trajectory towards the collector surface. Once through the grid the electrons are accelerated through a high potential (e.g. 12 kV as used in the Stereoscan collector) by means of a thin electrically conducting layer on the surface of the scintillator. When the electrons strike the scintillator, light is produced which is transmitted by means of a light guide into a photomultiplier mounted outside the specimen chamber. By this means the collector can be mounted close to the specimen and thus collect electrons from a large solid angle, whilst at the same time the highly efficient process of amplication in a photomultiplier can be used. This collection system is sufficiently efficient for it to be regarded as having no influence on the minimum detectable contrast observable in the scanning electron microscope or on the signal-to-noise ratio.

Measurements of the time required for electrons to travel from the specimen to the collector have been published by Everhart (1968) and show that with a collector potential of 75 V the delay is about 5×10^{-7} s. The results of Everhart and Thornley (1960) also suggest that processes within the collector system itself also occupy a similar length of time since they found the maximum rate of working to be 10 MHz. It is only when the scanning rate reaches the values necessary for television-type presentation that information from successive picture points is sent out at this sort of rate: in a 625-line monitor operating on a 50-Hz supply the scan rate is 10^{-7} s/point. The measurements quoted above suggest that this rate represents just about the limit of the present technology.

A scan-rate limitation is already found when the microscope is used in the absorbed current mode. In this mode, although excellent micrographs can be obtained at the scan speeds generally used for photography, the maximum permissible scan speed is rather too low for display purposes.

Whilst small delays in electron transit will result in confusion on the left-hand edge of the screen the resolution of the display as a whole will be unimpaired. Resolution will only be influenced by the detector if the number of lost counts due to "simultaneous" events rises to levels of several per cent or if the number of extraneous electrons collected is sufficient to seriously "fog" the image. The former possibility has been disposed of by Everhart and Thornley who showed that the collector could cope with currents of up to 1 A/cm^2, i.e. enormously in excess of the signal strength. Everhart *et al.* (1959) showed that of the total signal some 65% was due to secondary electrons derived from the probe area on the specimen, 30% was "background" secondary electrons derived from elsewhere in the system and 5% was reflected primary electrons.

It might be thought that the examination of radioactive specimens would be precluded by the high background levels of β-emitted electrons; this is not the case. Daniel (1969) has shown that even intensely β-active sources placed close to the scintillator do not interfere with the normal brightness and contrast exhibited by specimens.

The success of the scintillator/photomultiplier combination has given little incentive for the adoption of solid-state devices or channel multipliers for electron detection. These collectors are useful in special situations, for example they can be mounted in matched pairs to enhance or to eliminate topographic effects (JEOL Co., 1970). Circuits for the utilization of channel multipliers have been given by Thornton (1968) who considers that some 60% of the electrons entering these devices are counted. The channel multiplier has a gain of around 10^6 which is a major contribution to the total amplification.

3.4. THE BASIC FACILITIES REQUIRED IN AN OPERATING MICROSCOPE

Up to this point we have dealt with principles only, in considering how an "image" can be produced from an electron-beam apparatus. In this section we consider the facilities required in a working microscope in order to realize in practice the resolution which is theoretically obtainable.

3.4.1. The vacuum system

An electron beam can only be produced and manipulated in a vacuum. An ordinary vacuum of about 10^{-5}–10^{-6} torr residual pressure is sufficient, but higher pressures than this can be expected to produce electron scattering in the main beam, to yield spurious secondaries, to encourage contamination of the lens apertures, and to lead to premature failure of the tungsten filament. The vacuum is produced by a combination of rotary and diffusion pumps, usually with suitable baffling to prevent oil streaming into the electron optical column. Both for speed of specimen change and to keep contamination to a minimum it is good practice to have the specimen chamber separately pumped. This necessitates a closure valve between the final lens and the specimen stage, but even during operation, when this valve is open, the gas flow from specimen chamber to column is very small because of the small size of the final aperture (50–400 μm). The pumping capacity provided by all the manufacturers is sufficient to remove products of specimen out-gassing and on the best instruments is good enough to permit the use of gas probes for the maintenance of gaseous atmospheres in the vicinity of the specimen (see Chapter 6). Because of the poor interconnection of the chamber and column the pressures should be monitored by pirani and ionization gauges in each section.

3.4.2. The specimen stage

Advantage can be taken of the long working distance to design specimen stages in which a great deal of specimen movement is possible. "Stereoscan" stages, for example, permit rotation, tilt and a three-translational movement. Other special stages are available which permit tensile or compressive stressing of specimens and other mechanical manipulations. These movements are normally controlled by mechanical feed-throughs, but it is the difficulty of making such rotary vacuum seals which precludes the ready conversion of microscopes to ultra-high vacuum operation.

The other main features of the specimen chamber are the ports for mounting the electron detector and other signal receivers. These will be set up to collect signals from the largest possible solid angle of the specimen but are not normally adjusted by the casual user and do not need external mechanical adjustment. It is useful, however, to be able to centre the final aperture in the electron beam and controls are usually provided to enable this operation to be carried out. An aid which is frequently useful is a tag board for bringing external electric circuits into the chamber.

3.4.3. The electron-optical column

The column in reality is a stack of electron lenses which are demountable for cleaning (Fig. 3.10). The apertures in all but the final lens are usually fixed in size although there are exceptions when provision is made for selected area electron diffraction or for use in the channelling pattern mode (see below). The column terminates at one end in the specimen chamber and at the other end in the gun chamber. It is again beneficial to be able to isolate the gun chamber from the column so that routine

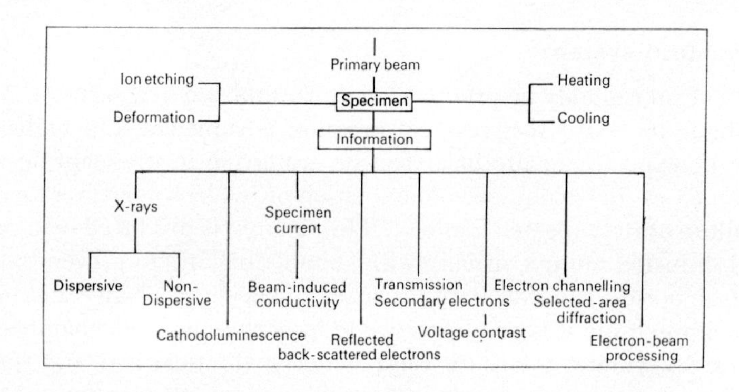

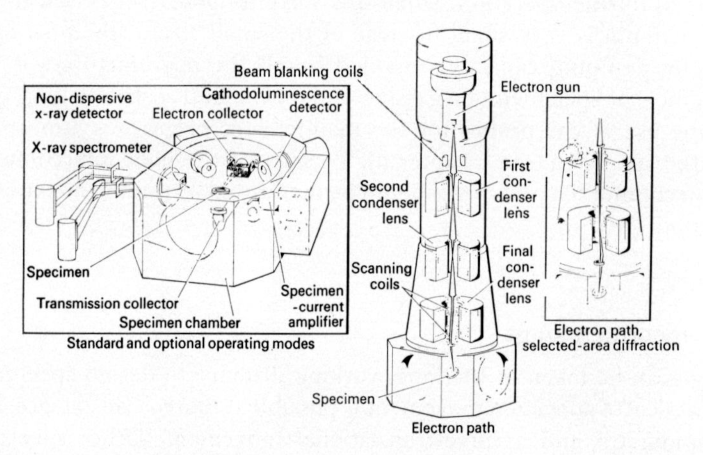

FIG. 3.10. Diagrams of the "Stereoscan S4", showing lens positions and the arrangement of collectors for the modes listed above. (Courtesy of Cambridge Instruments Co. Ltd.)

filament changes can be made without breaking the general vacuum. To get the maximum beam current from a given gun it is essential to be able to adjust the filament position in relation to the grid and anode positions. The filament is at the full accelerating potential used in the microscope, however, and properly insulated drives are provided by the manufacturer for this adjustment. On the simplified microscopes an equivalent alignment of the beam in the lenses is performed by subsidiary electromagnetic deflection coils.

Controls and the appropriate meters are necessary for the operator to be able to control the lens currents, filament current, and grid bias potentials. The astigmatism of the lens system tends to vary with the state of the lens apertures and with the presence of any electrostatic charge on the specimen itself. It is therefore necessary for the current to the astigmator to be under the control of the operator also. The currents to the scanning coils are normally provided automatically according to the settings made by the operator on the scan generator. For many purposes either during setting-up of the microscope or when making X-ray or electron diffraction analyses of the surface, it is necessary to have the electron beam describing a line scan across the surface of the specimen or remaining stationary at a single point. In order to control the position of the beam across the specimen surface under these modes the currents to each of the deflection coils must be individually under the operator's control.

3.4.4. Data treatment

As we have implied throughout this chapter, the normal method of displaying the stream of signals obtained from the electron detector is to use its output voltage to modulate the brightness of a cathode-ray oscilloscope scanning in unison with the probing beam. The requirements of the display oscilloscope—persistence and moderate resolution—differ from those of the recording oscilloscope—short persistence and high resolution—and therefore it is usual to provide one of each oscilloscope. The brilliance of the display and the contrast and its contrast range can be altered at will by adjusting the amplification factors of the photomultiplier. The contrast range found on the

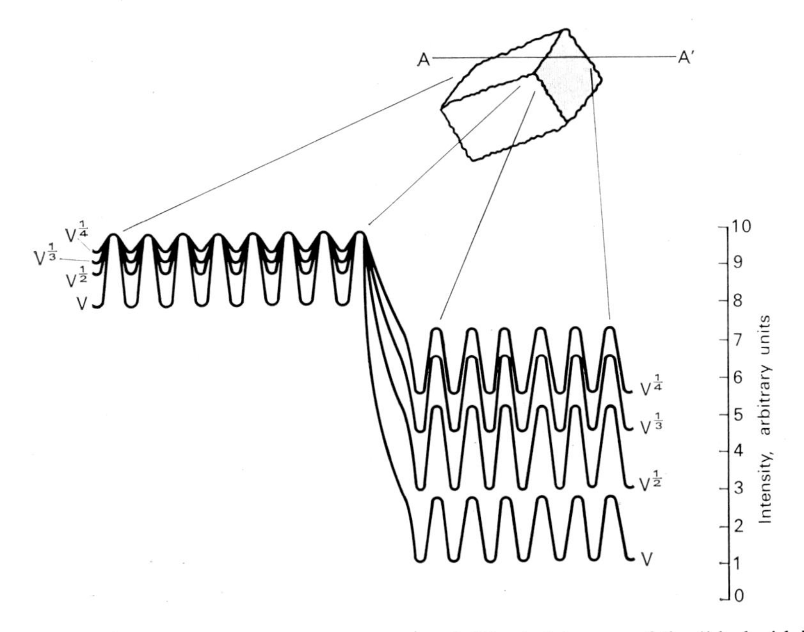

FIG. 3.11. The function of the gamma control. The brightness of the "dark side" of the cubic crystal is increased at the expense of the contrast range on the bright side.

specimens is often sharply distorted, for example, by shadows of surface asperities or by deep surface convolutions. Neither increasing the general brightness nor contracting the contrast range can satisfactorily deal with this situation. The problem can be overcome by making the output voltage a non-linear function of the input voltage. The manner in which this manipulation benefits the final image can be seen in Fig. 3.11. Here the imaginary, sinusoidal, brightness signal from a cubic crystal oriented as shown in the inset drawing is plotted as a linear function and for comparison as the input voltage raised to the powers $\frac{1}{2}$, $\frac{1}{3}$ and $\frac{1}{4}$ respectively. This is the function of the "gamma" control and, whilst brightness levels close to zero and to the maximum (saturation) level are distorted, the brightness and contrast in dark areas of a micrograph are considerably improved.

In Fig. 3.11 the surface brightness is plotted on the y-axis as a function of the position of the probing beam on the x-axis of the specimen. Clearly this represents an alternative mode of data presentation to that in which the output voltage is used to modulate the brightness of the CRT beam. Presentation of the data in this form is useful in providing a quantitative assessment of the noise level, of the resolution, and of the contrast range. It is extensively used in setting up the specimen for photography and provides one of the most reliable indications of correct focus. If the base line of such an xy display is shifted in unison with the raster line a form of display known as "y-modulation" is obtained. This combines some of the quantitative advantages recorded above with an image display of the surface.

3.5. MAIN FEATURES OF COMMERCIALLY AVAILABLE MICROSCOPES

The brief descriptions here are not intended as critical comparisons of the available instruments, nor is it practicable to list the up-to-date specifications in a book of this character. In many cases the instruments are not directly competitive but are optimized in slightly different ways to meet differing requirements. Equally, however, it would be wrong not to describe the instruments offered by present-day manufacturers since this throws an interesting light on the approach to optimization. The wide range of extras available are discussed below.

3.5.1. High-resolution microscopes

There are three instruments currently available which set out to offer resolution at the limits of that which is theoretically attainable: the Stereoscan S4 manufactured by the Cambridge Instrument Co. Ltd., U.K.; the J.S.M.-U3 of the Japan Electron Optical Laboratories Co. Ltd. and Model 900 of the Advanced Metals Research Corporation, U.S.A. The S4 is the latest of the Stereoscan series first manufactured in 1965 and continues the tradition of the three lens column. It has all the facilities which could be regarded as basic requirements as described in the previous section. Thus the accelerating voltage is fully variable from 1 to 30 kV, which enables ceramics and other insulators to be examined below the 2nd cross-over point, i.e. as described in Section 2.3.5. This, it will be recalled, is one way of eliminating surface charging of the specimen. At the highest accelerating voltage there is adequate X-ray emission from the surfaces to permit analysis of all elements above boron ($z = 5$). In common with the other high-resolution microscopes the vacuum sequences associated with specimen entry or removal are fully automatic; two diffusion pumps provide separate pumping for the specimen chamber and the lens column. This appears to be the only instrument which offers a fully variable working distance over the range 0–10 mm. Throughout this range the magnification in use can be read directly from an illuminated dial. Magnification is normally adjusted in discrete steps but zoom attachments are available.

The J.S.M.-U3 is designed as a combined scanning electron microscope and electron microprobe analyser. It uses the two electron lens column which is typical of microprobe analysers. Nevertheless, those micrographs which have been published show convincingly that resolutions of the order of 100 Å are attainable. The J.S.M.-U3 offers alternative fixed values of the working distance at 13 and 32 mm respectively. The accelerating voltage is variable between 5 and 50 kV which precludes low voltage examination of insulators but is optimized a little more closely to the requirements of X-ray analysis. In other respects the specifications are similar to those of the Cambridge S4 and allow all the operations suggested in Section 3.4.

The special features of the AMR 900 instrument are the goniometer stage and the interchangeable apertures on all lenses. Stereoscopy should be achievable on this instrument to a very high precision. But it is doubtful whether the interchangeable apertures can give an improvement in the ultimate resolution which should be similar to that of the other two instruments.

3.5.2. Simplified instruments

Two other instruments* on the market demonstrate how many of the standard features in high resolution microscopes can be eliminated in deference to simplicity of operation and maintenance. Even so the J.S.M.-S1 of JEOL Co. Ltd. and the "Scanscope" of Hitashi–Perkin Elmer both offer resolutions of around 250 Å with a worth-while reduction in primary and running costs. Each instrument uses a single diffusion pumped system, has fixed or dual value accelerating voltage, has fixed working distances, limited gun bias controls and only a limited number of scanning speeds on the display and recording oscilloscopes. In keeping with the limited evacuation system pirani gauges only are used but in both cases pumpdown is automatically controlled. The novel feature in the Hitashi machine is the use of an inverted column which gives the operator excellent access to the specimen chamber. The resolution offered by these microscopes is still far superior to that of the optical microscope and has all the advantages of depth of focus associated with the scanning electron microscope. They are, however, microscopes and not the versatile experimental systems provided in addition by the high-resolution instruments.

3.5.3. Scanning attachments for transmission electron microscopes

Several manufacturers of transmission electron microscopes offer scanning attachments. For example, Philips market a scanning attachment for their EM 200 and EM

* *Note added in proof.* The "Miniscan", manufactured by Vacuum Generators Ltd., also falls within this category.

300 transmission electron microscopes which is fixed to the normal goniometer stage of the microscope. The condenser lenses which normally provide uniform illumination of the transmission specimens are modified to provide a probe-forming function and the goniometer lens is used as the objective. The instrument is normally used in the primary electron mode which probably accounts for the high noise levels and sharp shadows in the published micrographs. These are the expected features of any micrographs taken in this mode.

3.5.4. Optional extras

All the instruments mentioned above provide absorbed, emitted and primary electron images permitting atomic number, surface voltage, and topographic contrast to be obtained. Photoluminescent images are readily obtained with suitable samples by use of a simple light guide without the scintillating material. The following modes of use are more properly regarded as specialized techniques and are not generally available.

(a) *X-ray analysis*

X-ray production, particularly from elements of high atomic number, requires higher accelerating voltages (Elion, 1966) than is necessary for secondary electron emission (Section 2.6). It is for this reason, rather than for the secondary electron emission characteristics, that Cambridge used a 30-kV accelerating voltage whilst JEOL Co. Ltd. offer a 50-kV gun as standard. The Cambridge gun is able to excite the characteristic X-ray emission of the *K*-shell to the point where it overlaps with useful X-ray emission from the *L*-shell. It is not quite able to excite the characteristic emission from the *K*-shell of barium ($z = 56$) which is usually taken as the optimum element for transfer to the *L*-shell excitation which has improved peak to background ratio for higher atomic numbers.

There are two techniques available for the analysis of the X-ray emission spectrum from a specimen bombarded with electrons. These are dispersion spectrometry and energy analysis spectrometry respectively. In the former method the X-rays are allowed to fall on a crystal spectrometer. The beam is diffracted by the crystal and the resultant spectrum can be scanned by movement of a counting device through its plane of focus. In order to cover the full wavelength range of the *K* and *L* emissions for all elements it is necessary to use crystals of different atomic spacings. Hence with this form of spectrometer it is usual to have interchangeable crystals. The range of crystals offered by JEOL Co. Ltd., together with the elemental range for which they are most useful, is given in Table 3.1. A similar range of eight crystals is offered by Cambridge Ltd. In energy dispersive analysis the complete X-ray emission from the test piece is allowed to fall upon a solid-state device, such as lithium-drifted silicon, which will measure the energy in each photon of X-ray radiation. The individual quanta of energy can be electronically sorted and fed to the appropriate channels of a multichannel analyser. In this way a histogram of the number of quanta versus the energy of the quanta can be built up. Thus a simultaneous display emerges of the characteristic X-ray emission of all elements within the electron beam. The resolution attainable

TABLE 3.1. THE X-RAY ANALYSER CRYSTALS OFFERED BY JEOL CO. LTD. FOR USE WITH THE J.S.M.-U3 MICROSCOPE

Analysing crystal	Chemical formula	Spacing (2d, Å)	Range of wavelength Å	Range of elements		
				K	*L*	*M*
Lithium fluoride	LiF	(200) 4·03	1·0~3·5	$_{20}Ca \sim _{34}Se$	$_{51}Sb \sim _{86}Rn$	
Quartz	SiO_2	(1011) 6·7	1·7~5·8	$_{16}S \sim _{27}Co$	$_{41}Nb \sim _{68}Er$	$_{80}Hg \sim _{92}U$
PET (Pentaerythritol)	$C_5H_{12}O_4$	(002) 8·8	2·3~7·6	$_{14}Si \sim _{24}Cr$	$_{37}Rb \sim _{61}Pm$	$_{72}Hf \sim _{92}U$
KAP (potassium acid phthalate)	$C_8H_5O_4K$	(1010) 26·6	6·9~23	$_{9}F \sim _{14}Si$	$_{24}Cr \sim _{37}Rb$	$_{47}Ag \sim _{74}W$
Myristate	$M(C_{14}H_{27}O_2)_2$	79	20~68	$_{5}B \sim _{8}O$	$_{17}Ci \sim _{24}Cr$	
Stearate	$M(C_{18}H_{35}O_2)_2$	98	25~85	$_{5}B \sim _{7}N$	$_{16}S \sim _{22}Ti$	
Lignocerate	$M(C_{24}H_{47}O_2)_2$	125	32~108	$_{5}B \sim _{6}C$	$_{15}P \sim _{20}Ca$	
Cerotate	$M(C_{26}H_{51}O_2)_2$	137	25~119	$_{4}Be \sim _{6}C$	$_{15}P \sim _{20}Ca$	
Mellissate	$M(C_{30}H_{59}O_2)_2$	156	40~135	$_{4}Be \sim _{6}C$	$_{14}Si \sim _{19}K$	

by this method of analysis depends on the resolving power of the solid-state detector. At present commercial non-dispersive X-ray detectors have a resolving power of 140–200 eV and are obtainable from several nucleonics firms. This is not quite as good as the resolving power of the best dispersive X-ray spectrometers.

(b) *Transmitted electron detectors*

With thin specimens a strong signal is transmitted and can be detected by a collector placed below the specimen and aligned along the beam axis. The image is similar to a conventional transmitted image and resolution can be very high since it is not limited by beam spreading within the specimen. The equivalent of dark-field illumination can be used by using a diffracted beam as a signal. Moreover, by using a stationary beam selected area diffraction patterns are produced. These can be presented in visual display by employing subsidiary scanning coils and traversing the transmitted beam past a small aperture in front of the detector. Thus the intensity of the pattern is used to modulate the brightness of a cathode-ray oscilloscope scanning in unison with the subsidiary coils. This is the technique of scanning electron diffraction (Tompsett and Grigson, 1966). Transmitted electron devices are available for the S4 and the J.S.M.-U3 microscopes.

(c) *Electron channelling patterns*

Electron channelling is probably the weakest of the effects which can give rise to contrast variations in the SEM. The contrast range is of the order of 5% which is the figure we took earlier for the noise variation likely to be observed in secondary electron emission images (Booker, 1970). Thus even if the rather special conditions required to observe electron channelling contrast were complied with it is unlikely that channelling patterns would ever interfere with the usual topographic contrast. In fact, electron channelling contrast depends on the interaction between the incident beam of electrons and planes of atoms lying in the Bragg angles of the crystalline solid. When the angle of incidence of the beam is less than the Bragg angle more electrons are scattered through large angles and hence greater numbers of primary electrons emerge from the specimen and may be collected by a suitably placed detector. Conversely, if the angle of incidence is greater than the Bragg angle fewer electrons are back scattered and the collector sees a weaker signal (Booker, 1970). When an ordinary raster at fairly low magnification is described on the surface of a single crystal the angle of incidence varies through several degrees across the surface (Fig. 3.12(a)). This is sufficient to give rise to bands of high and low intensity parallel to the y-axis of the display. Such bands were first observed by Coates (1967). Coates (1969) has since developed this technique by rocking the specimen in both the x- and the y-plane so that a complete angular raster is displayed which indicates the position of the characteristic Bragg angles of the crystal lattice in three dimensions. Booker (1970) has given an analysis of the ideal electron optical conditions for obtaining channelling patterns and shows that the beam should preferably be parallel, the spot size should conform to the rules of resolution which we have already discussed and the electron beam current must be some two orders of magnitude greater than that used for

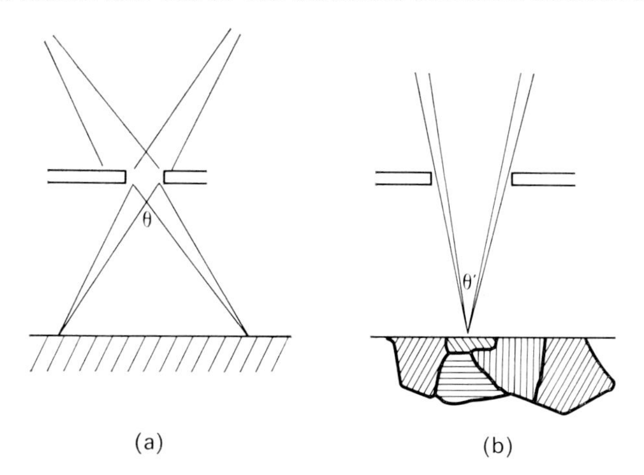

FIG. 3.12. Electron channelling contrast mechanism. (a) In normal scanning mode the angle between the Bragg directions of large single crystals is scanned through an angle θ. (b) The rocking-beam method gives the same result on a polycrystalline surface by direct variation of θ, using the scanning coils with a stationary spot.

secondary electron patterns in order to obtain contrast at such low levels. He also shows, however, that the range of Bragg angles which are obtained in the display depends on the total angle of incidence through which the electron beam scans. In the early work this necessitated the use of low magnifications and large single crystals so that a large angle of incidence range was obtained by the normal scanning process. It is now practicable, however, to vary the angle of incidence of the beam about a stationary spot on the specimen surface by the use of subsidiary double scanning coils (Fig. 3.12(b)). This is the method known as rocking beam selected area channelling patterns and has been used on both Stereoscan scanning electron microscopes and the J.S.M.-U3. Selected area channelling patterns by this method have been obtained from areas as small as 2 microns across and there seems no reason why the technique should not be developed to give selected area patterns from areas as small as 1000 Å. A supplementary deflection coil to enable this method to be used on the J.S.M.-U3 microscope is available from the makers.

(d) *Solid-state detectors*

Solid-state detectors and devices such as the channel multiplier are small enough to be mounted diametrically opposite the specimen in matched pairs. This has the particular advantage that the signals can be added or subtracted and thus the topographic contrast either heightened or reduced according to the requirements of the examination. This is particularly useful in revealing atomic number contrast.

(e) *Specimen surface treatment in the SEM*

There are two main techniques used for surface treatment in the SEM: in the first method the electron beam may be programmed to described a particular configuration on the surface and thus carry out microscale electron beam machining, or treatment of photoresist devices. For this technique to be carried out effectively it is necessary to switch the beam on and off according to the requirements of the programme control. The Cambridge instruments offer a beam blanking device for fitment to the Stereoscan which will perform this function. The other major form of surface treatment which can be carried out within the specimen chamber, but not while the instrument is running, is argon ion etching of the surface. Both JEOL Co. Ltd. and Cambridge offer argon ion guns which may be fitted within the specimen chamber. The power of argon ion sputtering for revelation of surface detail was shown by Pease and Ploc (1965) in a study of iron oxide surfaces. The method has since been used for a wide range of inorganic and organic substances.

3.6. CONCLUSION

Whilst it has been the primary aim of this chapter to introduce the reader to the problems of resolution and contrast in the SEM in order that he can use it at its optimum in his particular application, the inherent value of the microscope for *in situ* studies should also have been apparent. It is in this area that progress is being made at the present moment and no doubt many *in situ* experiments will be referred to in the succeeding chapters on practical applications.

REFERENCES

BOOKER, G. R. (1970) *Proc. 3rd Annual SEM Symposium*, IIT Research Institute, Chicago, Illinois, p. 490.

COATES, D. G. (1967) *Phil. Mag.* **16,** 1179.

COATES, D. G. (1969) *Proc. 2nd Annual SEM Symposium*, IIT Research Institute, Chicago, Illinois, p. 29.

COSSLETT, V. E. (1950) *Introduction to Electron Optics*, O.U.P., Oxford, pp. 68, 155.

CREWE, A. V. (1966) *Science* **154,** 729.

DANIEL, J. L. (1969) *Proc. 2nd Annual SEM Symposium*, IIT Research Institute, Chicago, Illinois, p. 295.

ELION, H. A. (1966) Instrument and chemical analysis, *Aspects of Electron Microanalysis and Macroanalysis*, Pergamon Press, Oxford, p. 48.

EVERHART, T. E. (1968) *Proc. 1st Annual SEM Symposium*, IIT Research Institute, Chicago, Illinois, p. 1.

EVERHART, T. E., WELLS, O. C. and OATLEY, C. W. (1959) *J. Electron Control* **7,** 97.

EVERHART, T. E. and THORNLEY, R. F. M. (1960) *J. Sci. Instrum.* **37,** 246.

GRIVET, P. (1965) *Electron Optics*, Pergamon, Oxford, p. 419.

JEOL Co. (1970) Descriptive Material on JSM-U3.

NIXON, W. C. (1969) *Proc. 2nd Annual SEM Symposium*, IIT Research Institute, Chicago, Illinois, p. 1.

MOSS, H. (1968) *Advances in Electronics and Electron Physics*, Sup. 3. *Narrow Angle Electron Guns and CRT's*, Academic Press, London & New York.

OATLEY, C. W., NIXON, W. C. and PEASE, R. F. W. (1965) *Advances in Electronics and Electron Physics* **21,** 181.

PEASE, R. F. W. and NIXON, W. C. (1965) *J. Sci. Instrum.* **42,** 81.

PEASE, R. F. W. and PLOC, R. A. (1965) *Trans. AIME* **233,** 1949.

ROSE, A. (1948) *Advances in Electronics*, Academic Press, New York, **1,** 131.

SMITH, K. C. A. and OATLEY, C. W. (1955) *Brit. J. Appl. Phys.* **6,** 391.

THORNTON, P. R. (1968) *Scanning Electron Microscopy*, Chapman & Hall, London, p. 96.

TOMPSETT, M. F. and GRIGSON, C. W. B. (1966) *J. Sci. Instrum.* **43,** 430.

CHAPTER 4

Specimen Preparation

MRS. P. M. CROSS

4.1. INTRODUCTION

One of the greatest strengths of scanning electron microscopy is the ease of preparation of the samples. Since specimen preparation is so easy some investigators tend to be careless when preparing specimens for examination. It should be pointed out that poor preparation and mounting techniques will result in micrographs of low quality due to poor electrical conduction from the specimen to the holder. Highest resolution is obtained with proper mounting of the specimen in conjunction with proper use of the microscope. What follows here is mainly a general discussion of preparation and associated techniques for samples which are to be viewed in the emissive mode using secondary electrons, when topography is the only interesting feature. Other techniques may be necessary to view a specimen in other modes and they will be mentioned in Chapter 5.

It should be stressed that the simple preparation procedures described in the next section, or the slightly more complicated ones in Section 4.2, cover the vast majority of specimens. The procedures discussed at greater lengths apply only to a small minority of difficult specimens.

4.1.1. Simple preparation

Conducting samples usually need nothing more than cutting to an appropriate size, sticking to a convenient metal mount making sure there is good electrical conduction between the specimen and the mount. The size of sample taken varies widely with the model of SEM used but even in the small Cambridge stage the order of magnitude of the upper limit of size is 1 cm^3 and larger specimens can be accommodated if the operator is prepared to lose one or more degrees of freedom in the movement when examining the specimen. The type of glue should be one that does not seriously outgas in the vacuum used in the instrument (about 10^{-5} torr), is prepared easily and dries quickly. For delicate samples (e.g. powders) a suspension of carbon in water, sold commercially as Aquadag, or any silver glue, including Silverdag, may be used; but heavier metal samples need a stronger glue and most of these are non-conductive. If a non-conductive glue is used it should be painted all over with a conductive paint like Aquadag, or on one good conduction line from specimen to mount. It is better to use a non-conductive glue, out of a tube, which dries in 2 minutes and can then be painted and ready in 1 minute, than to use a conductive glue which needs preparation or takes 5 minutes or more to dry. Another useful technique is to mount specimens with a double-sided adhesive tape (Twinstik) making sure that there is a good conduction from the specimen to the holder by using Aquadag. Muir and Rampley (1969) have made a study of the effect of the electron beam on various mounting and coating media.

4.1.2. More complex preparation

Unfortunately, some samples, mainly the non-conductive ones, do not give a clear picture when prepared in the above manner and viewed under normal conditions; they may also be damaged by conditions in the microscope. The four obstacles to clear pictures which may necessitate more complicated preparation procedures are poor electron emission, charging, beam damage and vacuum damage. These effects are discussed in Chapter 10 with illustrations so that they may be recognized when they occur.

Specimens which contain volatile liquids, including water, usually lose these substances under vacuum and the effect becomes more pronounced as the beam is rastered over the sample. Tissues and other non-rigid structures may collapse so that the topography becomes changed. The effects may be overcome by conventional methods of fixing, by air drying, freeze drying, rapid drying or critical point drying or by using a controlled stage to reduce evaporation. None of these methods will be discussed here as they are given a detailed presentation in Chapter 9, on Biological Applications.

4.2. SPECIMEN COATING

4.2.1. Coating with a thin layer of metal or carbon

The application of a conducting layer of carbon or metal is the most popular method of suppressing charging and increasing electron emission but, while simple to perform, has a number of potential drawbacks. The purpose of coating is to put on a uniform covering of conductive material, so that the surface of the coating is, as nearly as possible, an exact positive replica of the surface of the underlying material. This is in contrast to the techniques used in transmission electron microscopy where the topography is enhanced by "shadowing" techniques; and the coating material is put on from one side.

For scanning electron microscopy, the material may be deposited on the sample surface by vacuum deposition; commercial apparatus for this process is easily available. The production of a continuous uniform layer requires that each face of the sample is presented to the source for the same amount of time. A fairly flat surface may well be uniformly coated by spinning it about its own axis in one plane at an angle to the source during deposition, but the more irregular specimens will certainly need a more complex movement and even then one may well doubt if the coating is truly uniform. The problem is not simple because in varying the angles of deposition compensation must be made for (a) variations in specimen/source distance, since the thickness of an evaporated layer depends on the Inverse Square Law, and (b) changes in the nominal angle of deposition (see Cosslett, 1951, and Kay, 1967).

A key consideration to the uniformity of the coating is the type of coating material used. Amongst the metals the noble metals have good electron emission and contouring ability; Tolansky (1954) has shown that silver is easily the best for contouring. Gold is often used for coating, as it has such good electron emission, and gold–palladium is used even more often, but the ability to contour accurately is such a prime consideration that many people use silver exclusively, and it is the opinion of the author that this is the best course. The disadvantage of silver is that it tarnishes and in many climates cannot be viewed more than a few days after preparation unless kept in a closed environment. The "grainyness" of silver that prevents it from being used in transmission electron microscopy is not apparent at the magnifications used in the SEM.

With very irregular specimens with a number of re-entrant surfaces (such as a tangle of fibres or complex biological structures) charging may not be suppressed because the coating is not continuous. In these cases it is often useful to use carbon as

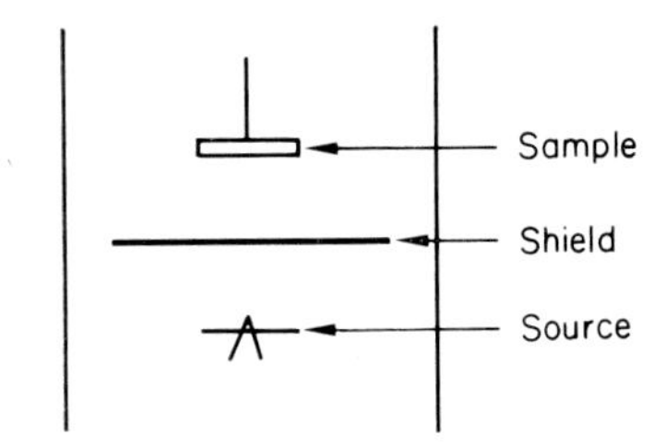

FIG. 4.1. A method of shielding the sample from primary paths of the evaporating metal.

a first coat, since carbon tends to bounce off the sides of the chamber and reapproach the sample from all sides, more completely coating the under and recessive surfaces. An attempt can be made to make metals behave in the same way by shielding the primary paths between the source and the sample as indicated in Fig. 4.1, and by coating the sides of the chamber with Teflon (polytetrafluoroethylene). A lot of work has been done by E. H. Boult (unpublished) on residual gas scattering using argon. If carbon is used as a first coat, it is usual to put a light metal coating on top of it, so that the electron emission is improved. Only a small amount of each coating material is used so that the thickness of the combined layers is kept to a minimum. An alternative procedure is to deposit carbon and platinum together, either by using commercially prepared rods with platinum running through the centre of carbon rods or by winding platinum wire around standard carbon rods. Chatterji *et al.* (1972) have recently shown that an 80 nm (800 Å) thick carbon film can be used as an alternative to a metal film.

For work at very high magnifications the uniformity of the coating layer becomes even more important, and elaborate precautions must be taken to ensure that it is so. This not only includes spinning the specimen about as many axes as possible during coating but also keeping the thickness of the layer down to a minimum to avoid blotting out the smaller features. The aim should be the thinnest possible layer which will suppress charging; this will vary from specimen to specimen and will depend both on the electrical properties of the specimen and on its real surface area, which may be very great with some specimens of complex shape. The tendency to make such statements as "a layer of 30 nm (300 Å) of silver was applied" should be avoided, partly because the layer is unlikely to be so uniform that an average thickness has any meaning, and partly because the surface thickness on any but the flattest specimens cannot be accurately estimated. In practice, the best method is to put on a small amount of metal, check and see if it suppresses charging; if it does not, add more,

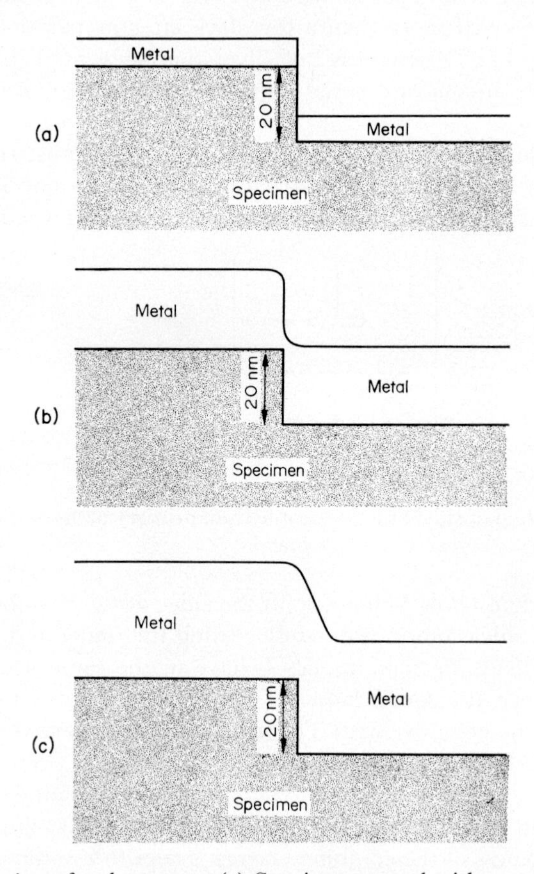

FIG. 4.2. Coating of a sharp step. (a) Specimen coated with a very thin layer of metal. (b) A coating of metal which has the same thickness as the step. (c) A much thicker coating.

remembering that the more complex the surface the more metal will have to be used. Even when all these precautions are taken the layer is inevitably, with any non-flat specimen, non-uniform and thus a source of misinformation.

Consider a simple 20-nm (200 Å) step in a flat specimen as illustrated in Fig. 4.2. A thin layer, Fig. 4.2(a), would not be continuous. A layer 20 nm, deposited normally to the surface, might appear as in Fig. 4.2(b). With a thicker layer as in Fig. 4.2(c), the sharpness of the step would be lost. Even if the sample is spun around its own axis the metal will tend to build up on the lip of the step. If one considers a slightly more

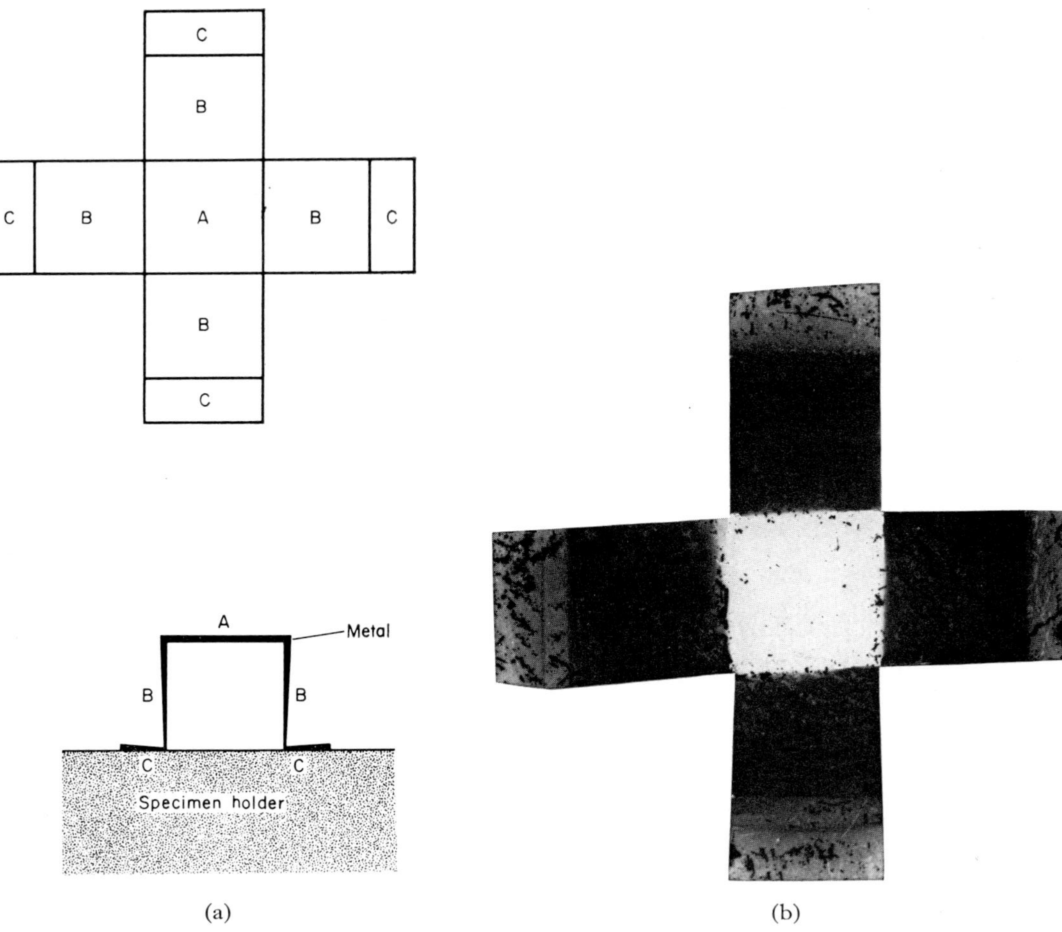

(a) (b)

FIG. 4.3. (a) Geometry of cross before folding and during evaporation. (b) Unfolded cross of cellulose acetate sheeting after coating with gold on Boult/Brabazon (1968) coating device using a single filament placed 2 in. vertically above the specimen. This is a negative of the actual cross; therefore, the lighter the area the heavier the metal coating. (Compliments of E. H. Boult, School of Chemistry, University of Newcastle, Newcastle-upon-Tyne.)

complex geometry, say a cube, the situation becomes more grave. Boult (unpublished) took a cruciform of thin transparent film, and folded it into a cube as in Fig. 4.3 (a). This was mounted on the specimen stub, and coated with gold while the stub was rotated about its own and two other axes of rotation. After coating the cruciform was opened again and Fig. 4.3 (b) shows the result. If even a simple cube cannot be coated uniformly, we can hold out very little hope with irregular specimens of achieving a layer uniform enough to give a true picture of the actual specimen surface at the highest magnifications. At magnifications up to 2000 × it is safe to use antistatic agents or to reduce the beam energy to suppress charging. Metallic and carbon coatings are quick, convenient and suit all but the most delicate of samples. Not all metals are suitable coating materials and to achieve good contouring some care must be taken in the deposition of even those which are capable of contouring well. It is interesting and instructive to try out various metals and at various thicknesses on similar samples so that the loss of detail can be appreciated. If an attempt is made to vaporize the metal too quickly it tends to move off in aggregates and settle on the sample in blobs, obscuring detail. The vaporization should be done slowly and in a controlled manner, watching the vaporizing metal all the time.

4.2.2. Practical coating procedure

The usual apparatus for vaporizing metals consists of a glass bell-jar which is connected to a vacuum system capable of bringing the pressure in the bell-jar to about 10^{-5} torr. Inside the bell-jar there should be several electric terminals capable of supplying sufficient energy to vaporize metals and carbon and a position to mount the sample. The complication needed in a rotating device depends on the type of samples normally viewed, the flatter the samples the less the necessity for sample movement. Ideally the sample should rotate about three mutually perpendicular axes, one of which is the axis of the sample mount, in the centre of a sphere of sources. In practice, however, the set-up is likely to be less complex. At the very simplest level the sample is inclined at an angle to one source and spun about its own axis (see Fig. 4.4 (a)). It is considerably better to have two or more sources with the sample rotating about two axes (see Fig. 4.4 (b) and (c)) or have the sample moving laterally between the sources (Fig. 4.4 (d)). The power for the movement of the sample mount can be electric or manual with the rotating shaft brought outside the vacuum system through a suitable vacuum seal, or the electric power can be fed to a motor inside the bell-jar.

The sources have been shown in Fig. 4.4 as a loop of wire coating material hung over a tungsten filament which is heated electrically so that the coating material is vaporized. The current should be slowly increased until the loop of wire suddenly melts and hangs in a droplet on the filament and then turned up just a fraction more so that the metal vaporizes off slowly. If the increase in current is too sudden, the metal will fall off the wire when it melts, or vaporize in aggregates. It is best to buy the coating material as wire with the smallest possible diameter, since it is easier to cut off more precise amounts, and the thin wire melts and vaporizes more easily than large-diameter wire.

The sample should be far enough from the source so that it is not affected by radiant heat and near enough so that the expensive wire is not being wasted. Boult (unpublished) has used his rotating device (Boult and Brabazon, 1968) to coat low melting point (70°C) wax specimens placed only 1 cm from the tungsten filament without the specimen being damaged. The evaporant source was operated at higher temperatures and for longer times than normally used. Some materials that are very sensitive to radiant heat could not be coated in this manner. The point being made is that one should adjust the specimen/source distance so that the specimen is not damaged and the expensive metal is not wasted.

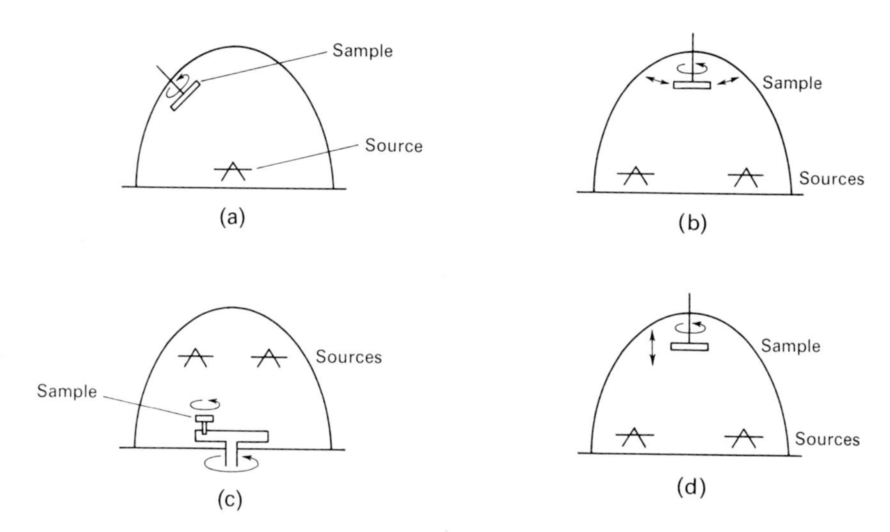

FIG. 4.4. Orientation of sample in relation to evaporation sources. (a) Specimen rotated about its axis at an angle to the source. (b and c) Specimen rotation about two axes with two sources. (d) Specimen rotating about its axis while moving vertically between two sources.

The filament and loop method is not the only way to hold metals for vaporization although it is probably the best way for gold which has a high melting point. Alternatively the coating material can be placed in a tungsten wire spiral, a boat made out of moybdenum sheet or a ceramic boat; in the latter case considerable current must be supplied to warm up the boat. For silver, the most convenient method is a molybdenum boat which lasts longer than the tungsten filament and from which there is less likelihood of losing the coating material. Again the current must be turned up slowly until the silver is seen to melt, and then turned up a little more for slow vaporization.

It is not safe to view the vaporizing metals through a clear bell-jar and anti-glare glasses should be used, or the bell-jar left with some metal from the last evaporation on it. One useful trick to make it easier to keep the bell-jar clean enough to see through is to wipe the inside of the clean bell-jar with diffusion pump oil; this makes later cleaning easier and also tends to make the metal vapour paths a little more random, so that the coating is more complete. For this reason, too, the vacuum should never be too high during coating—10^{-4} torr is quite high enough—so that the metal atoms bounce off the remaining air molecules and their paths become less direct. For some samples it is best to strengthen this effect by placing shields in the direct line of sight between the sources and the samples.

Much more power is needed to vaporize carbon than metals and special terminals are usually provided often with a push-button switch so that the carbon may be applied in short bursts, to prevent the vaporization of aggregates.

When the coating is completed the samples should be left to cool down for a few minutes before air is admitted into the bell-jar. Almost certainly the bottom edges of the specimen will not have been coated even with the most careful coating system and these and the glue should be completely covered with conducting paint (Aquadag or Silverdag), and it is wise to continue this up the sides of the samples, if these are not to be viewed, especially for difficult samples like polymers.

4.2.3. The use of antistatic agents

J. Sikorski *et al.* (1968, 1969) have used an antistatic agent as a coating when examining specimens in the SEM. Antistatic agents are solutions of long-chain organic liquids which have been found to suppress the formation of static electricity at the surface of some insulators when applied very thinly. The process of static electrification is still a matter of scientific controversy and the action of antistatic agents is similarly ill-understood. However, after a century and a half of industrial problems from static electricity (mainly in the textile industry) antistatic agents are well developed empirically. Their interest to scanning electron microscope users lies in the fact that the layer needed to suppress charging is very much thinner (maybe in some cases a monomolecular layer) than that needed if the layer is metal or carbon so there is less change in the topography of the treated sample; also it is sometimes possible to introduce the antistatic into the body of the sample so that *in situ* experiments which change the surface will not have charging on the new surface areas.

Many antistatic agents are produced in aerosol containers and for low magnification work (below 2000× instrument magnification) spraying the sample in air from about 3 feet suppresses the charging quite well. Sometimes artefacts, such as droplets and films, are formed but they are usually quite obvious. For specimens with a deep pile or with a delicate fabric that might be disturbed by metal coating or soaking, for instance dental floss, spray-on antistatics and lowering of the beam voltage are often the only way to suppress charging. Spraying does not, however, produce a film continuous enough to suppress charging at magnifications higher than 1000 times. For higher magnifications, the antistatic is best dissolved in a solvent to a very dilute solution; the specimen is soaked in this solution for a few hours and later the excess antistatic from the surface is washed off with the solvent and the sample left to dry. This system works particularly well with porous materials, in some cases successful work is claimed with solutions of a few parts per million. For samples which require fixing the antistatic can be introduced into the fixative. One of the best all-round spray antistatic agents is "Duron".

In general antistatic agents can be divided into three groups—cationic, anionic and nonionic—and one of each of these groups at least should be tried on each new material viewed, since there can be a decided difference in the ability to suppress charging. The manufacturer of the material being examined often has specialist knowledge of which antistatic is most effective. One form of antistatic agent is osmium tetroxide vapour; while in general this is used as a fixative for soft tissues it can also control charging when used with carbon and metal, sometimes in a spectacular way (Pfefferkorn, 1970).

When the specimen has been treated with antistatic agent and mounted, the glue and edges of the specimen should be thoroughly coated with conductive paint so that a good electrical conduction path is provided to the stub.

The use of antistatic agents is at present rather limited, mainly because the effects of different types, concentrations and methods of application have not yet been studied systematically; but as this information becomes available the use of antistatic agents may well become the most important and usual method of suppressing charging.

4.3. LOW BEAM ENERGY WORK

It is not possible to decide on the best method of preparation, without thinking about the conditions of examination. There are two ways of lowering the beam energy;

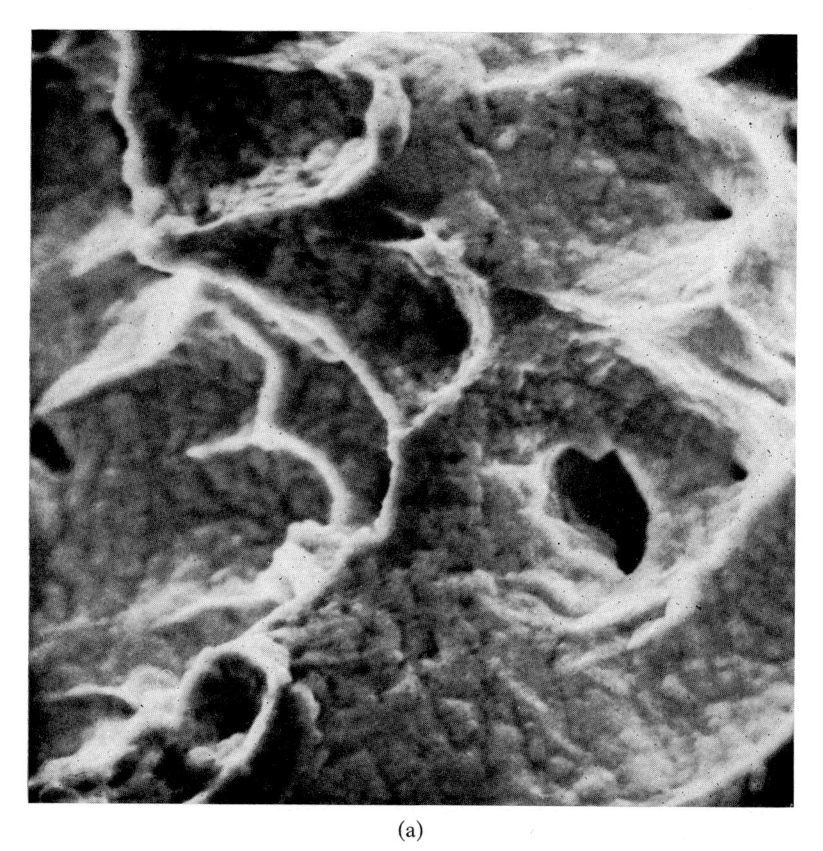

(a)

FIG. 4.5. Fractured silver wire (a) taken at 20 kV.

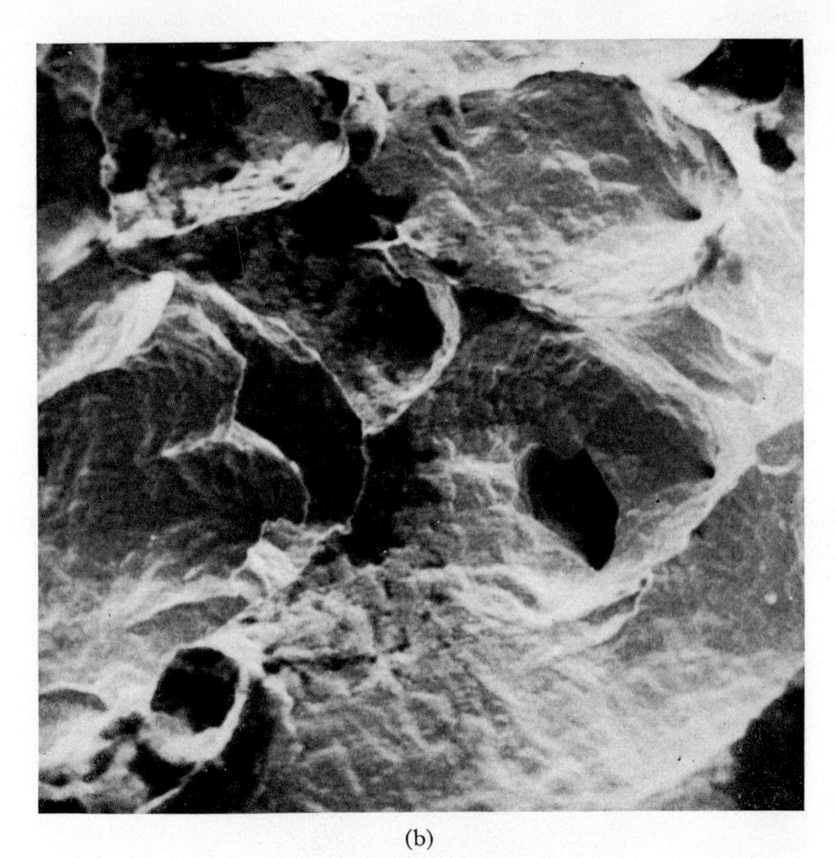

(b)

Fig. 4.5. (b) at 5 kV.

one is by dropping the beam voltage, the other by chopping the sides of the beam by strengthening the first and second lens and decreasing the final lens aperture. In the latter case the signal strength drops drastically and so the signal-to-noise ratio is lower and the image is very noisy. The effect may be lessened on the photomicrograph by photographing at a very slow frame speed, 100 to 200 seconds per frame (see Fig. 10.11). Reduction of the beam energy by dropping the beam voltage theoretically reduces the resolution but also decreases the penetration and this has an important practical effect, as discussed by Boyde and Wood (1969) and shown by Hearle *et al.* (1970) and De Mets and Lagasse (1969). One of the complications of interpreting SEM pictures lies in the question not only of what caused the contrast but also of where the information shown came from. Necessarily the beam penetrates the sample slightly so that the information comes from the top layer of the sample, the lower the beam voltage the thinner the layer of excitation. If therefore the topography of the surface layer is different from the bulk of the material (as in the case of an oxide film on a metal, see Fig. 2.19) and it is the surface layer which is to be viewed, the beam voltage must be kept very low. This effect is also very marked with light organic materials with a lot of surface decoration. It is more disturbing to find there is a clear difference in information obtained at different voltages even with metals. Figures 3.8 and 4.5 (a and b) show the effect. In general then it is better to use a low voltage in order to reduce the charging and beam damage, and increase surface detail unless resolutions greater than 50 nm (500 Å) are required.

4.4. REPLICATION

It sometimes happens that the actual specimen to be studied is not suitable for direct observation in the microscope, perhaps because the specimen is delicate and the topography might change due to vacuum or beam damage or perhaps because the area to be studied cannot be moved to the specimen chamber, as, for instance, in the study of continuing wear on working machinery. In these cases it is convenient to replicate the specimen. Replication for scanning electron microscopy is much easier than for transmission electron microscopy as there is no necessity to make the replica thin, nor is there any need to reproduce details less than 20 nm (200 Å) in size. Various plastic films and silicone rubbers can be used, most of which are easy to apply and also to strip from the surface. The first replica gives, of course, an inverse view of the surface, and this is very often extremely useful in examining the details down long holes. To get a positive replica it is necessary to replicate the first replica.

Eckert and Caveney (1970) have successfully used a water-soluble sodium carboxymethyl-cellulose (Cellofas B50, ICI) to replicate surfaces of diamond grinding wheels for examination in the SEM or TEM. Carteaud (1970) has made negative replicas of human skin by using a synthetic elastomer. The most suitable for SEM work was found to be Silflo (Plexico Developments Ltd., England). Positive replicas were obtained from the negative by using Araldite (Ciba) as a moulding agent. There was some loss of detail in preparing the positive replica. Two other replicating materials which have been used with the greatest success are Formvar and Tensol. These and other replicating materials used in transmission electron microscopy will probably be found to be suitable although some may be damaged by the beam and low accelerating voltages should be used. In most cases the replica will require coating with metal for observation in the SEM but the use of low voltages may eliminate the need for coating.

4.5. CONCLUSION

As one of the advantages of the SEM is the speed and ease of viewing samples, every attempt should be made to keep the complication of preparation down. However, especially in the case of non-conductive and delicate samples, it becomes necessary to prepare them, usually by fixing, coating with metal or carbon or with an antistatic, and using conductive paint to get at least one good conduction line to the stub. It may also be necessary to reduce the beam energy. With the more difficult samples all these methods may need to be used simultaneously and special care taken with each step so that the optimum conditions for each treatment are obtained. Even so one could estimate that the preparation time, not including fixing or drying times, would be about 15 to 20 minutes, which compares favourably with preparation times for difficult specimens with most other types of microscope.

REFERENCES

Boult, E. H. and Brabazon, E. J. (1968) *J. Sci. Inst.*, series 2, **1,** 565.
Boyde, A. and Wood, C. (1969) *J. Micros.* **90,** pt. 3, 221.
Carteaud, A. J. P. (1970) *Proc. 3rd Annual SEM Symposium*, IIT Research Institute, Chicago, Ill., p. 217.
Chatterji, S., Moore, N. and Jeffery, J. W. (1972) *J. Phys. E: Sci. Inst.* **5,** 118.
Cosslett, V. E. (1951) *Practical Electron Microscopy*, Butterworths, London.
De Mets, M. and Lagasse, A. (1969) *J. Phys. E.*, series 2, **2,** 813.
Eckert, J. D. and Caveney, R. J. (1970) *J. Phys. Earth* **3,** 413.
Hearle, J. W. S., Lomas, B. and Sparrow, J. T. (1970) *J. Micros.* **92,** pt. 3, 205.
Kay, Desmond (Ed.) (1967) *Techniques for Electron Microscopy*, Blackwell Scientific Publications, Oxford.
Muir, M. D. and Rampley, D. N. (1969) *J. Micros.* **90,** pt. 2, 145.
Pfefferkorn, G. E. (1970) *Proc. 3rd Annual SEM Symposium*, IIT Research Institute, Chicago, Ill., p. 89.
Sikorski, J., Moss, J. S., Hepworth, A. and Buckley, T. (1968) *J. Sci. Inst.*, series 2, **1,** 29.
Sikorski, J., Moss, J. S., Hepworth, A. and Buckley, T. (1969) *Proc. 2nd Stereoscan Collog.*, Engis Equipment Co., Morton Grove, Ill., p. 25.
Tolansky, S. (1954) *Vacuum* **4,** 456.

CHAPTER 5

Procedures for Using a Scanning Electron Microscope

MRS. P. M. CROSS

5.1. INTRODUCTION

The operating modes available in a scanning electron microscope are numerous. Consequently it is worth taking the time to consider carefully what information is needed and which mode can reveal this information before any practical work is done. Most users merely want to study the topography of the surface, perhaps coupled with a knowledge of the chemical composition of the surface layer. In more specialist applications the electrostatic and magnetic fields associated with the surface, the current carrying patterns of the bulk material or the crystallographic planes in the bulk material may be studied. It is also possible to use the transmission mode, or to use the instrument as an electronic device fabricator. In order to aid in interpreting the information, SEMs are increasingly being coupled to computers both by computer control of the instrument operation and by digital storing of the output: this approach is facilitated by instruments with a digital scan so that each raster point has an address which may be accurately "dialed".

5.1.1. Techniques to gain topographical information

The most useful mode for viewing the topography of a sample is the emissive mode, collecting secondary electrons with a scintillator in a positively biased shield. The scintillator is mounted on a light guide which is connected to a photomultiplier and amplification system. Under these circumstances the contrast comes mainly from changes over the surface of the sample in the angle of tilt to the collector and also to the beam. This contrast is often heightened by negatively biasing the Faraday cage so that mainly primary (reflected) electrons are collected, although details inside holes and recessive surfaces are often lost. The resolution quoted for these modes is commonly 15 nm (150 Å) but it should be remembered that this is under ideal conditions with a perfect conducting specimen; with insulators one might hope to get to 50 nm (500 Å) under perfect conditions. The standard method of using the information from secondary or primary electrons is to intensity modulate a raster on a CRT which has a point-to-point correspondence with the raster on the sample and this produces an image which has many features in common with that produced by reflected light and can generally be interpreted in the same way. There is, however, another method of using this signal and that is deflection modulation (Y-modulation) of the display screen raster. The map so produced has to be rather more carefully interpreted but often gives additional topographic information which cannot easily be obtained from the intensity modulated image. With very delicate samples which are easily damaged by a normal beam with voltages from 5 to 30 kV, it is sometimes useful to use the microscope in the mirror mode: the specimen itself is maintained at a controlled potential such that the electrons are decelerated as they approach it and then either reaccelerated in the opposite direction or permitted to hit the sample with about 1 keV energy difference between the beam and the sample. In both these modes some topographical information is given of a very thin layer of the surface (although the techniques are more commonly used to give electric and magnetic field mapping or chemical composition) and the surface is either not touched at all by the beam or suffers a minimum amount of beam damage.

5.1.2. Techniques to gain chemical composition information

A certain amount of information about the chemical composition of the surface layer can be found by increasing the ratio of primary to secondary electrons detected. The reflection of the beam is greatly affected by the atomic number of the element viewed and some element contrast is introduced into the image. If the sample contains cathodoluminescent materials, interesting and informative pictures may be taken using some form of light detector and, although the resolution varies from specimen to specimen, it is typically of the order of a few microns. Cathodoluminescence may for certain samples yield more qualitative information than X-ray analysis. However, the information given by both these modes may be obscured by topographical detail and is not precise, especially as to the amount of each element present. One more accurate method is to examine the X-rays given off in the interaction between the beam and the sample. The electrons given off when a very low-energy beam (about a kilovolt) interacts with a solid have recently been studied for the chemical information they carry, these electrons are known as Auger electrons. The main differences between X-ray spectroscopy and Auger spectroscopy are that X-ray spectroscopy samples a volume of the order 1 $(\mu m)^3$ while Auger spectroscopy samples a surface layer of only a few angstroms in depth. The resolution of the X-ray analysis is of the order of the penetration of the beam (1–20,000 Å, depending upon the sample) while with Auger analysis, resolution (for electrons with energies of less than 1 keV and beam diameters of the order of thousands of angstroms) is determined by the diameter of the beam; with an SEM this may be as low as 200 Å. Auger spectroscopy is very much more sensitive to low atomic number elements ($Z < 11$) (MacDonald *et al.*, 1970).

5.1.3. Techniques to gain other information

Electric and magnetic field mapping is generally produced by detecting the emitted and reflected electrons from the surface through some type of energy analyser, either very simply by restricting the entrance to the collector or by more complex energy filters. Some work has also been done (MacDonald, 1970) on the shift of Auger electron peaks due to high electric or magnetic fields. Until recently most work has been done qualitatively but sophisticated methods of handling the output (modified detectors, Y-modulation, computer processing of input and output) have made quantitative measurements a viable possibility. The resolutions obtained, both spatial and potential, vary widely depending on the complexity of the dectection system and the baseline potential around which potential differences are to be measured. For instance, with a restricted collector entrance one might hope to get resolutions of 1 V for voltages in the order of 10 V but with very sophisticated equipment it is possible to resolve 250 μV in the same circumstances.

If the electrons used for modulating the visual display raster are not those emitted or reflected but those induced by the beam inside the specimen, the map produced will contain information as to the changes in conductivity and potential fields inside the specimen. This method involves detecting the electrons directly by an electrical lead from the specimen to a sensitive amplification system and thence to modulating the raster. The results have some difficulties in interpretation. This approach is not necessary if the internal crystallographic orientation of a single crystal specimen is required; reflected electrons are used and a normal detector. The only variation on the normal reflective mode involves the conditions of the electron optical column and, sometimes, movement of the specimen or beam. The patterns produced are known variously as electron channelling patterns (ECP), Coates patterns or pseudo-Kikuchi patterns and these can often be interpreted using Kikuchi patterns prepared by transmission electron microscopy workers. Pattern resolution of up to 10^{-4} radian have been obtained. One of the real limits of this type of work is the area which can be scanned. With a stationary specimen and the beam simply scanning, the sweep must be 1 to 2 mm across so a large single crystal must be used. With rocking devices on the sample or beam the sweep drops to 10 μm and the limitation here is the brightness of the electron optical system. With a good electron source it has been estimated (Booker, 1970) that the sweep may be as small as 10 to 100 nm.

At first sight transmission scanning electron microscopy might not seem to be worth the effort since it could combine the disadvantage of the very difficult specimen preparation of the direct electron microscope (TEM) with the relatively low resolution of the SEM. Nevertheless there are significant advantages even when the simplest possible methods of obtaining TSEM pictures are used. The relationship of low magnification (20$\times$) to medium magnification (20,000$\times$) for thin slices can be established, which is impossible in direct electron microscopes; the area of the samples can be much greater than in TEM (up to 5 mm^2); thick, specimens, up to 1 μm thick, have also been examined (Swift and Brown, 1970) and in some cases stereopair observation has led to an understanding of the three-dimensional structure of the section. The resolutions obtained by siting a scintillator below a thin specimen and using a normal gun are of the order of 7 nm (70 Å), dropping somewhat as the specimen gets thicker.

One aspect of work in all modes which is receiving a great deal of attention is data processing, either by storing it in digital form on magnetic tape or using computer operations on the received signal so that the resulting micrograph is less noisy or has fine detail enhanced.

5.1.4. Learning how to use the instrument

In the next sections the practical details of using the SEM are discussed. The main emphasis is on the emissive mode used to view topographical details which is the main use of the microscope. Some practical details of using the other modes are described, more to indicate the complexity (or otherwise) of the mode to an interested operator than to provide a detailed handbook of operation. For further details, users should consult the manufacturer or read research papers on the topic. It is well worth while joining the mailing list of Oliver Wells and Miss V. Johnston who compile a bibliography of SEM papers, with abstracts and cross and forward references in subject and author groups. The address is:

Oliver C. Wells,
I.B.M. Watson Research Center,
Yorktown Heights,
New York, 10598, U.S.A.

What will be described here is general information applicable to all scanning electron microscopes. The user should, of course, study the operating manual of the particular instrument, or receive personal instruction, in order to get to know its controls and adjustments.

5.2. USE PROCEDURE OF THE EMISSIVE MODE

5.2.1. Conditions for optimum topographical detail

In the emissive mode a scanning electron beam is focused on the sample with the Faraday cage of the normal detector system biased positively so that secondary electrons are attracted by the collector and form the main instruments of contrast. Some primary electrons which are back-scattered from the specimen may also enter the collection system and affect the contrast.

Before instrument or sample are touched the results expected should be carefully considered, the most interesting magnification range decided upon and other points of interest, such as the depth of focus required and the type of sample being used, noted. The sample should be prepared, with this information in mind, by the methods described in the preceding chapter, and placed on the stage. In most instances, the microscope will be used in a standard set-up suitable for the majority of uses of a particular instrument. However, apart from the various dials and switches, there are several other adjustments which can be made, if necessary.

Choice of aperture and working distance

If low magnifications are required and the specimen has a very rough surface the smallest final aperture should be used along with a large working distance since with a rough surface we are interested in having as much depth of focus as possible. At low magnifications the spot size does not have to be at its smallest diameter and, therefore, the lens currents can be reduced so that there is an increase in signal and the noise (resulting from using a small aperture) is removed from the image. On the other hand, if high magnifications are required the largest aperture should be used along with a short working distance since the spot size is directly related to the working distance. At high magnifications the spot size must be at a minimum, therefore high lens currents are required. This has the effect of reducing the signal and longer exposure times are required to reduce the amount of noise on the final micrograph. A smaller aperture should also be used for better resolution if noise is not a limiting factor.

Choice of beam voltage

The beam voltage chosen will depend mainly on the type of specimen used and, to a lesser extent, on the magnification required. High beam voltage may be chosen where very high magnification is required and the specimen is not easily penetrated by the beam; but in general beam voltages of about 10 kV are a safer alternative since at high beam penetration one may receive a picture of a sub-surface layer which may be substantially different in topographical detail from the surface layer (see Figs. 2.19 4.5, 7.1 and 10.1). If the specimen shows evidence of beam damage or charging it may be necessary to drop the beam voltage still lower.

Choice of tilt

The tilt of the specimen to the beam at which the best electron emission is given varies with the make of instrument but for most of them it is 45°, this gives a foreshortened image unless a tilt correction unit is used, and 30° has been found a useful compromise. For viewing certain features of the specimen the angle of tilt is crucial. For instance, surface hairs or extrusions normal to the surface are best viewed with the specimen at almost 90° to the beam. Very smooth surfaces probably require angles of tilt greater than 45°.

Choice of scan parameters

The choice of line and frame speed on the visual display raster depends entirely on the individual operator. The faster the frame speed the easier it is to judge the focusing and movement of the specimen, since there is a shorter delay time between moving the controls and seeing the results. On the other hand, the picture becomes less clear as the frame speed rises. The line speed is determined by the frame speed chosen. The scan is most commonly set horizontally, but if the detail being studied is horizontal, it is convenient to set the scan to vertical so that the scan line and detail are not confused.

5.2.2. Operational sequence

With the specimen in place and imaged on the screen, the magnification should be set low and the image coarsely focused. The magnification should then always be turned up in stages, to a final high value, so that the focus point is quite obvious and not a matter of debate. Intermediate focusing takes place at about 1000×, and fine focusing somewhere above that. It is often useful to use a reduced raster size while focusing since not only is the operator's attention concentrated at that place but also the raster speed is increased so that the exact point of focus can be determined easily. The position and size of the reduced raster can usually be chosen by the operator and should normally be at the centre of the normal raster square, especially if the specimen is non-planar since that is the area shown in the next highest magnification. Occasionally it happens that the contrast is bad at the centre and then an off-centre position is chosen where a sharp contrast is available.

At each magnification range the signal amplification should be set below the flattening level as seen on the Y-modulated oscilloscope. If a very low amplification is chosen detail is lost due to smoothing; if the amplification is taken too high the detail of higher intensity than the flattening level is chopped off.

The visual screen has contrast and brightness controls, which do not affect the Y-modulated oscilloscope or the record screen, and are set for the operator's own convenience. It is easier on the whole to focus on a screen which is not very bright. If the contrast is not suitable, one should consider using the black level or gamma controls. Increasing the black level will increase the contrast and make a flat surface look less flat; increasing the gamma control is often used for looking at the bottom of holes or getting a good picture when the specimen is charging. Both the black level and the gamma control affect the image on the visual and record screens and, in a sense, give an artificial picture, so the fact that they have been used should always be noted and remembered.

For magnifications above 2000 ×, precautions should be taken against astigmatism. The magnification should be increased to 10,000 × or 20,000 ×, focused, and the fine focusing control passed through the focusing point. If the picture appears to have "smear" marks which lie at an angle to the scanning axes on one side of the focus point and at the opposite angle on the other side of the focus point, the beam is affected by astigmatic fields (see Fig. 10.12). This is corrected by using the stigmator, adjusting amplitude and angle, while swinging the focus control through the true focusing point until the "smear" marks have disappeared or are at a minimum. The procedure is not easy and needs practice. Astigmatism is normally caused by dirt in the column or final apertures and when it cannot be controlled by the stigmator the column should be cleaned or the final apertures changed.

Focusing and astigmatic adjustments should be checked at the highest possible magnifications, or at least one step higher magnification than is to be recorded. If at any stage the beam voltage, aperture or lens currents are altered the image must be refocused and the astigmatic corrections reapplied; this may well occur when checking to see which voltage gives maximum surface detail or when altering the lens currents to give maximum resolution.

5.2.3. Photography

When the image on the visual display screen is satisfactory it may be decided to take a photomicrograph. The line and frame speed of the record raster will depend upon the type of image displayed. The signal is integrated over time for each picture element so that less noise appears on a micrograph taken at a slow frame speed than one taken at a fast frame speed. As a general rule, 40-second exposures are usual. However, a very noisy image may require 100- or 200-second exposures to give clear micrographs; information may then be apparent which could not be seen in the visual display screen. For transient effects, it may sometimes be necessary to change to 20-second exposure times but a noisy micrograph may well result. Beam damage can also be reduced during exposure if fast scanning times are used. The line speed should be adjusted so that it is 1000th of the frame speed giving 1000 lines per frame. More lines per frame may overlap and blur the detail; less lines per frame may cause the lines to show on the final photomicrograph. If lines are showing while the ratio of frame to line speed is set to 1000, it is best to adjust the fine controls for the frame's line speed.

Many instruments nowadays have automatic cameras which relieve the operator of the necessity of making difficult decisions as to the brightness and contrast levels of the record screen. However, these cameras adjust to the overall image, and it may be that the operator is only interested in a small part having optimum brightness and contrast, in which case the automatic controls can be overridden by a manual control. Some instruments, too, may not have the automatic cameras. In these cases a subjective judgement must be made as to the brightness and contrast needed and this is usually performed by studying a line across the interesting part of the image while the room is darkened. A knowledge of the effect on a specific type of film of changing the brightness and contrast controls is only built up by trial and error, although using Polaroid film matched to the ASA rating of the negative film used speeds the process up. The scan line should never be left very bright on the record screen as the screen very easily burns and is expensive to replace.

It is well to note here that when a particular area of interest is to be recorded it would, in some cases, be better to make a series of overlapping micrographs at a high magnification than to make one micrograph of the area at a low magnification. After printing the micrographs can be fitted together into a montage which will cover the complete area and give greater detail at the high magnification. It was found that this method was particularly useful when recording the fractures of cotton fibres. Scan rotation units make it easier to arrange the image in a more pleasing position.

Some microscopes can be equipped with these units which enable the operator to rotate the scan from the normal operating position through 360° to correct for the effects of scan rotation apparent when changing the working distance. It is also useful when preparing true stereomicrographic pairs to use the scan rotation to align the beam scan correctly relative to that on the CRT and to the movements of the specimen stage. The scan rotation can also be used to change the position of the image in relation to the screen so that more pleasing micrographs can be made or so that (as in the case of fibres) more effective use can be made of the scanned area. An example would be to rotate the scan so that a linear sample would be diagonal across the screen and not perpendicular to one of the sides.

When several different sets of information are required—for example, the distribution of various elements in the X-ray mode described later—it may be useful to use colour film with different filters inserted for successive exposures.

5.2.4. Summary of procedure

1. Note: Type of sample and preparation.
   Magnification range expected to be most interesting.
   Depth of focus needed.

2. Choose: Working distance.
   Aperture size.
   Lens currents.
   Beam voltage.
   Tilt of specimen.

3. Shut the isolation valve between column and chamber and let air into the specimen chamber.
   Position specimen.
   Shut stage and pump down. On instruments that do not have a safety circuit make quite sure that the isolation valve is not opened when air is in the specimen chamber. Stick a notice on the instrument to remind casual operators to be careful of this slip. Check magnification is at lowest, collector voltage is at maximum positive value.
   When Vac. Ready light is on, open isolation valve.

4. On visual display screen put line speed to fastest, frame speed to suit yourself.
   Focus starting with coarse focusing at the lowest magnification. Use reduced raster if wished.
   Check signal amplification is just below flattening level on Y-modulated oscilloscope.
   Set brightness and contrast levels on visual display to suit yourself.
   Use black level or gamma control if necessary.
   Check for and reduce astigmatism.

5. Alter beam voltage and lens currents to give maximum detail with tolerable noise, refocus and reset astigmatic controls.

6. When a photomicrograph is wanted: Choose line and frame speed for record raster to suit image displayed. Set brightness and contrast controls on record screen, take photograph.

It is instructive to note here that each area studied on the visual display screen may not give enough new information to warrant a photomicrograph but that photomicrographs at high magnifications should always be accompanied by others of the same area at lower magnifications so that the position of the area with relation to the rest of the sample is established. Several micrographs at various rotation and tilts relative to the collector often yield information unsuspected at first viewing.

5.2.5. Y-modulation of data

If the electrons collected are used to deflection-modulate the display raster instead of intensity-modulating it, a map results with many of the intensity-modulated image's features, but often with increased detail (see Fig. 5.1). The difference in signal strengths from point to point can also be estimated quantitatively and this is particularly useful when studying electric or magnetic contrast maps or images which carry chemical composition information. The capacity to display data by Y-modulation is now a standard feature or optional extra on all instruments and includes variation of the degree of modulation up to 100%. As the modulation is increased the illusion of relief

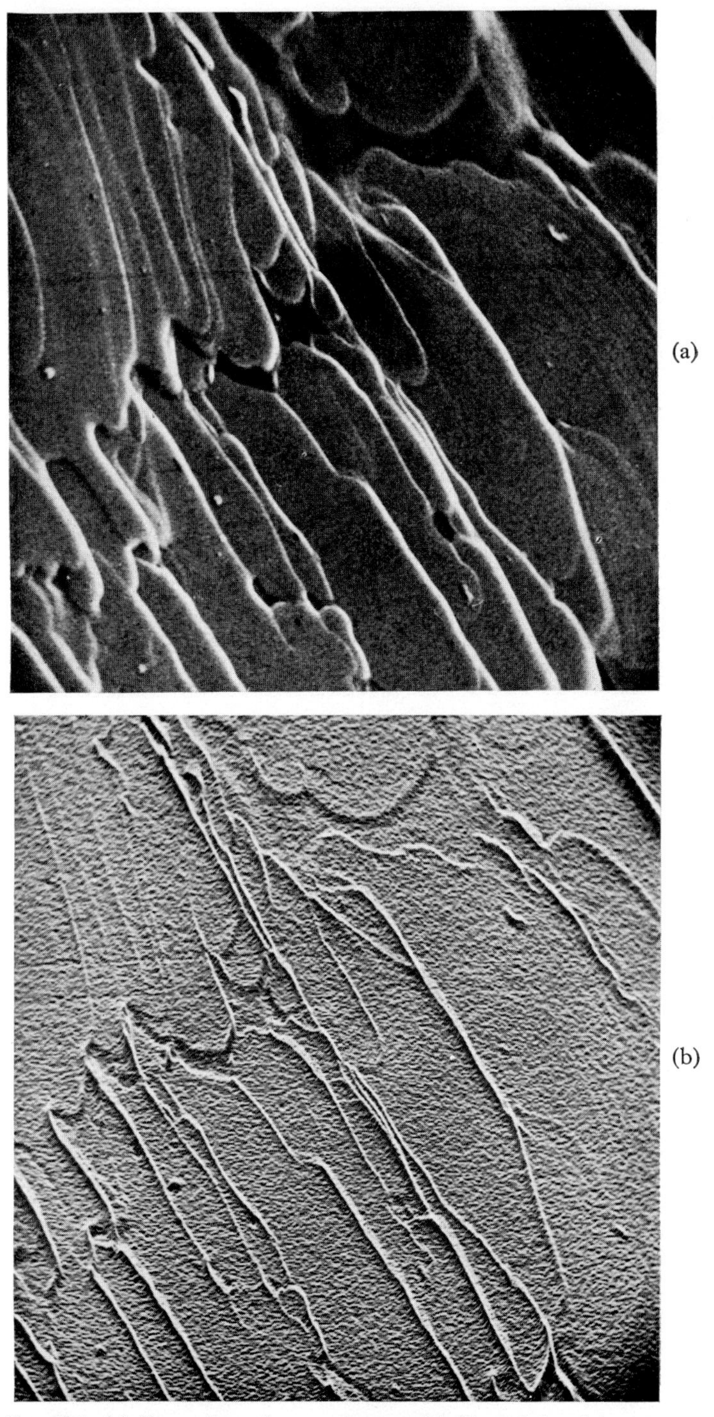

FIG. 5.1. (a) Secondary electron image. (b) Y-modulated image, 500 lines per frame.

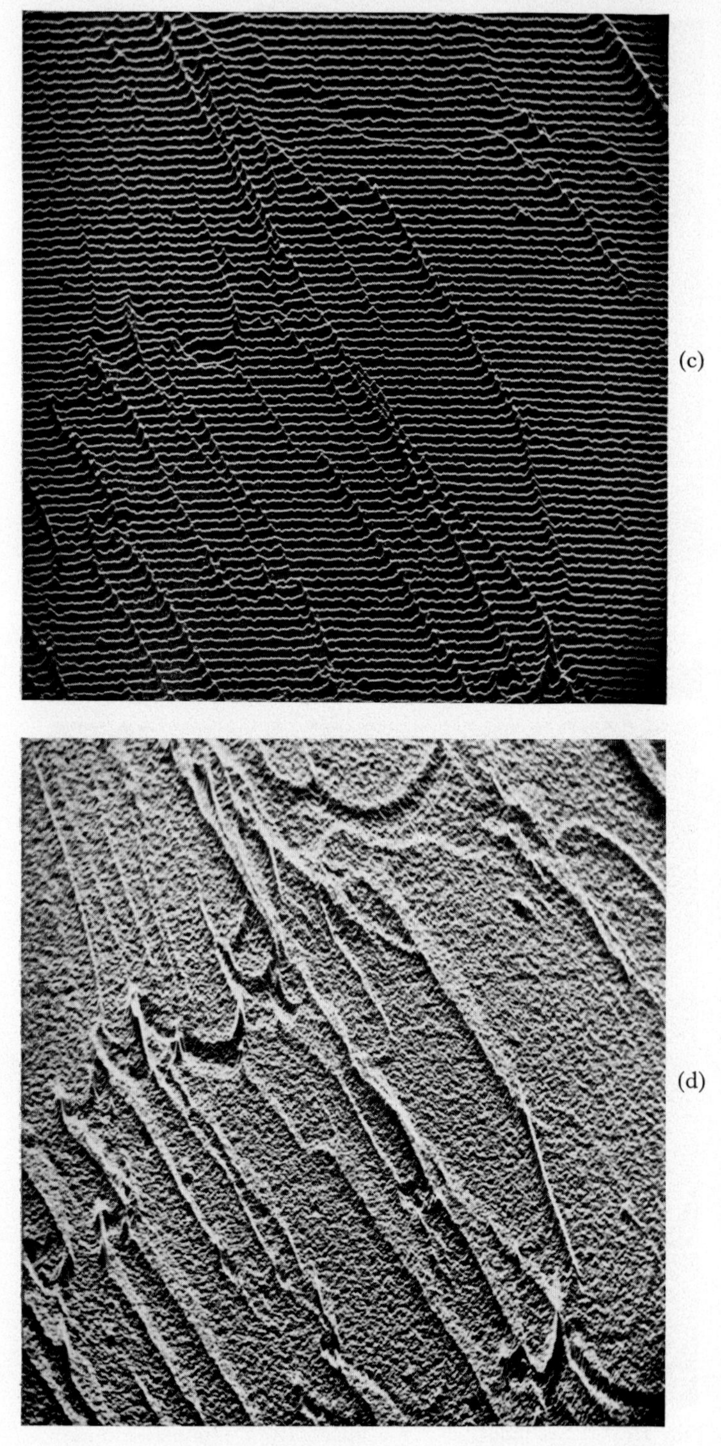

FIG. 5.1. (c d) Y-modulated images. (c) 100 lines per frame. (d) Increased amplitude.

becomes more pronounced, unfortunately this is accompanied by an apparent Y-shift in features on the image so that interpretation of the micrograph must be approached with care. Similarly decreasing the number of lines on the micrograph from 1000 to 500 or even lower produces a greater illusion of depth but this may involve some loss of detail. Y-modulation should be considered whenever details of an intensity modulated image are hazy or indistinct, for whatever reason. It has been shown, for instance (Kelly and Lindqvist, 1969), that topography of magnetic materials is more easily deciphered in this manner.

5.2.6. Electric and magnetic field mapping

To see the effect of surface electric or magnetic fields it is necessary to introduce some type of energy analyser in front of the electron detector. The very simplest method is to restrict the entrance to the collector by replacing the mesh on the Faraday cage with a thin metal foil having a small aperture, Banbury and Nixon (1967). For electric-field contrast a central horizontal slit is cut and for magnetic-field contrast a triangle slightly above centre. Useful qualitative work can be performed with both these devices, but the signal differences may be small and difficult to distinguish using intensity modulation. Y-modulation gives a clearer picture of where signal differences are occurring, and if the facility is not available the line may be swept slowly over the frame while observing the Y-modulated oscilloscope.

5.2.7. Use of Auger electrons

Auger electrons are secondary electrons of under 2 keV energy emitted when the primary beam is at about 3 kV. They depend upon the atomic number of the elements present at the point of impact to a depth of a few angstrom units, and are sensitive to electric and magnetic fields which may be at or near the surface of the sample. To a certain extent they are affected by topographical details of not more than a few angstroms in depth (thin coatings, absorbed layers, etc.). The latter may prove a disadvantage if the specimen being examined has been contaminated in some way and forces the operator to use a very clean, low-pressure vacuum system at about 10^{-11} torr. If the gun is held at 1 kV so that the primary beam is at a low energy throughout its passage through the electron optical column, the beam is severely affected by bad alignment of components of the column and by any astigmatic effects. A system specifically designed for working with Auger electrons would have an ultra-high vacuum system with all column and gun components capable of accurate alignment. One of the commercial instruments available (AMR) does have an alignable column, but the vacuum system is average. One way of getting around the necessity for careful alignment is to use the microscope in the mirror mode (see Section 5.7).

The Auger electrons are usually detected by a solid-state detector after being passed through some form of energy analyser. The signal obtained is very weak so that record times are very long and the data is often processed to reduce noise and produce the results in a digital form. MacDonald (1971) has illustrated the usefulness of this technique.

5.3. USE PROCEDURE OF THE REFLECTIVE MODE

5.3.1. General observation

The reflection mode has a focused scanning beam on the sample, but the Faraday cage is slightly negatively biased so that the secondary electrons are repulsed and only the high-energy primary electrons are collected. A solid-state detector with a larger acceptance angle can also be used to increase the signal, for with the conventional collector system the signal is weak and the picture is noisy.

The use procedure is the same as the emissive mode apart from the biasing of the Faraday cage and the increased precautions to keep the noise on the image low. The contrast will be sharper than in the emissive mode but details not in the "line of sight" of the collector will be lost (see Fig. 1.9).

The topographic contrast normally swamps the element contrast so that the specimen must be flat and polished if element contrast is to be seen. If may also be necessary to use Y-modulation, so as to accentuate the contrast.

5.3.2. Electron channelling patterns

The scattering of electrons in a homogeneous specimen is controlled by the topographical detail and the atomic number of the elements present. If however, the specimen is a single crystal, the scattering is also controlled by the angle between the incident beam and the Bragg planes of the specimen. If this angle changes over the sample (for instance, if the sample is bent) bands of contrast result. The same effect is shown in the direct electron microscope, but the greatest intensity there is given where the electrons have been scattered least, while the greatest intensity with the SEM is given where the electrons have been scattered the most (and so are able to escape back out of the sample). The two patterns are complementary. The patterns obtained with the direct electron microscope are also known as electron channelling patterns or Kikuchi patterns.

Another method by which the angle between the beam and the Bragg planes change is if the beam is scanned over the sample at a very low magnification so that the angle through which the beam moves is about $\pm 5°$. The spot size and electron current in the beam must also be adjusted and this may necessitate operating the electron optical column under non-standard conditions. This method involves using a single crystal at least 1 mm^2 in area; for smaller areas the sample itself may be rocked so that the stationary beam performs a raster over it or the beam may be electrically rocked over a small area. The area covered and the resolution obtained are then governed by the brightness of the electron source and with high-voltage pointed-tip filaments the sweep might be as low as 10 nm (100 Å) across. For these "rocking" procedures the beam is focused on the specimen at the centre of the electron optical system with one of the lower lenses switched off and the final aperture removed. For more detailed information see Booker (1970).

5.4. CATHODOLUMINESCENT MODE

Some materials emit light on being struck by an electron beam, the wavelength and intensity depending upon the material. To detect this radiation one can use the standard light pipe with the scintillator removed (use Perspex polish) or detectors developed for this purpose which are sensitive to specific wavelengths. On instruments fitted with an optical microscope one can observe the cathodoluminescence directly. Coloured micrographs are the best way of displaying the information; they are prepared by taking successive exposures using filters in front of the camera matched to specific detectors or, if one detector which covers a range of wavelengths is used, matched with filters placed in front of the detector. The last method makes the signal still weaker than it was in the first place.

The resolution depends on the intensity of light emitted and the efficiency of light collection. The first is always quite good, often as high as 1000 times better than secondary electron emission, but the efficiency of the light collection is usually low, so for detailed work specific detectors with large acceptance angles should be placed as near as possible to the sample. The effect of a low collection efficiency is to give a noisy image, so that record scan times must be very long. Another reason for long scan times is that many cathodoluminescent materials have a marked decay time.

The types of material which commonly show cathodoluminescence (phosphors, rocks, polymers, biological materials) often have a tendency to charge; this may be overcome by coating with a thin layer of metal or carbon, but changes in the thickness of the layer causes marked changes in the absorbtion of the emitted light, so special care should be taken to see that the layer is uniform and, if possible, the sample should be flat, if not polished, before metallization. Muir *et al.* (1971) discuss cathodoluminescence spectra and modifications which they made to the SEM where they use a single photon counting technique.

5.5. X-RAY SPECTROSCOPY

The X-rays emitted when the electron beam strikes a sample can be analysed to yield information as to the chemical composition of the top layer of the sample. The data may be displayed as intensity modulated or Y-modulated maps, but is considerably more complicated to interpret than topographical maps, or it may be printed out digitally.

There are two distinct methods of analysing the X-rays and two types of detectors available. In the dispersive (also known as wavelength dispersive or WD) system the X-rays fall on a crystal sited to allow photons with a certain range of wavelengths to reach the detector. At the detector some amplification and some energy discrimination may take place and a plot for that particular wavelength over the sample can be produced. To look at another range involves changing the crystal, and it can be a slow and tiresome process to look at a whole spectrum. In the non-dispersive (or energy dispersive, ED) system the detector is sited close to the sample and the energy discrimination takes place within it. The geometry of the two systems is shown in Fig. 5.2. It will be seen that the exact position of the X-ray source is extremely critical in the dispersive system and less critical in the non-dispersive system. With the dispersive

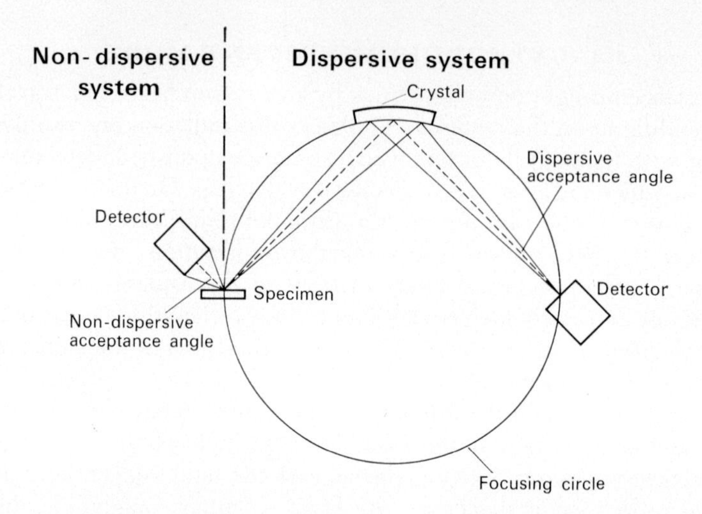

FIG. 5.2. Diagram of dispersive and non-dispersive X-ray spectrometers.

system the position of the sample must be finely adjusted using a light microscope and a very rough SEM sample may produce anomalous results. Another effect of the necessary geometry is that while the detector in the non-dispersive system can be put close to the sample so that the acceptance angle is quite large, in the dispersive system the acceptance angle must be smaller and therefore the images are noisier.

The older type of detector is the gas-flow proportional counter which has very little energy discrimination but considerable amplification and is generally used in the dispersive system. It has very thin windows and will detect even the light elements. The newer type of detector is the solid-state counter which, although providing no amplification, has good energy discrimination over most of the X-ray spectrum and has led to the rise of non-dispersive X-ray analysis. The lower end of its range is 1 keV so that elements lighter than sodium cannot be detected. If the non-dispersive system is used to detect these elements the gas-flow proportional counter must be used as the detector. At very high count-rates the efficiency of both types of detectors will drop but this will occur (for a rising X-ray signal strength) earlier with the solid-state detector than with the proportional counter, since it must accept the whole spectrum of counts while the proportional counter only accepts a narrow range of wavelengths.

For SEM samples with rough surfaces and low signal strengths the non-dispersive system is significantly better than the dispersive system and it has the added advantages of quickness and convenience. For light elements, however, it may be necessary to move to the dispersive system or to Auger spectroscopy. Much of the information and conclusions given here are taken from Sutfin and Ogilvie (1970).

With the detector set to pick up only X-rays of a particular energy, the picture displayed shows the location of a particular element in the specimen. However, another, and very useful, application of the X-ray mode involves switching off the scan and adjusting the electron beam to a particular point of interest, such as a foreign body, which remains visible in an emissive mode picture on the long delay tube. The whole range of X-rays is collected, and the energy discrimination causes a spectrum to be displayed on an auxiliary graphic display screen. The elements present at the particular point are thus identified. An example is shown in Figs. 7.20 and 7.21.

5.6. SPECIMEN CURRENT MODE

If the specimen is connected by an electrically conducting lead to a sensitive noise-free amplifier and if the signal so produced, as the beam scans the surface, is used to modulate a CRT raster synchronous with the beam raster, a map is produced which contains information as to the internal electrical phenomona of the sample. On some instruments the specimen touching alarm lead makes a good lead to the outside amplifier. The maps are rather difficult to interpret without a lot of prior knowledge as to the probable conduction paths within the sample. The depth in the sample from which the information came can only be estimated, and images often show disturbing ghosting effects probably due to the long delay times in internal circuits. The currents so produced are very small and need an extremely sensitive low noise-amplification system; even so the image may be noisy, so that focusing is difficult. More detailed information on this mode is given in Chapter 8.

5.7. THE MIRROR MODE

In this mode a normal high-voltage primary beam is emitted from the gun and traverses the electron optical column. The specimen is, however, biased at or around the gun potential so that the electrons are either reversed or allowed to hit the sample with a very small energy difference between it and the beam. In the latter case Auger electrons are produced, as discussed in Section 5.2.4, and in the former primary electrons of any chosen energy can be detected using a scintillator and light guide or a solid-state detector. These reflected electrons are very sensitive to small topographical changes and to electric or magnetic field changes. They have the advantage that as they never touch the surface of the specimen, they never cause charging or beam damage. At the time of writing this facility is not available commercially, but should be shortly. Cox (1971) gives a discussion of the technique along with some applications.

5.8. THE TRANSMISSION MODE

In the very simplest method the scintillator and light pipe are sited below a mesh capable of carrying a thin section of the sample. This is the usual apparatus offered commercially; but it can be improved by using a block of scintillator material mounted directly on a photomultiplier and by using more complex detectors involving a degree of energy discrimination, with high scattered and low scattered electrons being collected separately, so that fine details given by the highly scattered electrons at low signal strength need not be swamped out by the high strength signal from the less scattered electrons.

One way to improve resolution and picture clarity is to use high-voltage pointed-tip filaments in the gun which have a much higher electron yield. This has the side advantage that much thicker specimens (up to 1 μm) can be viewed, but the disadvantage that the possibility of beam damage is increased. An example of TSEM with a thick section is given in Fig. 5.3.

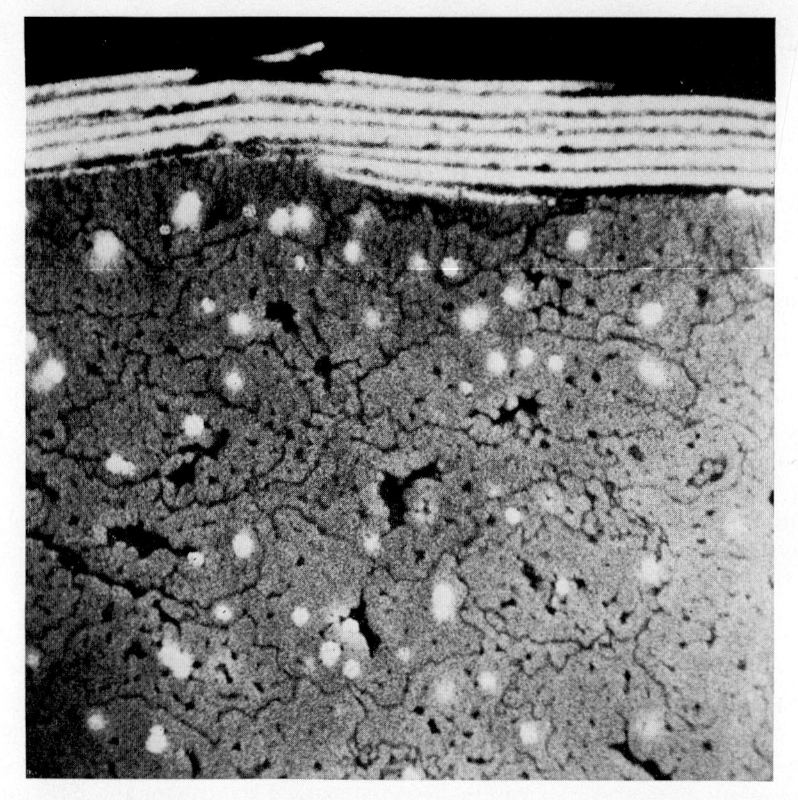

FIG. 5.3. A transmission scanning electron micrograph of a $\frac{1}{2}$-μm-thick section of human negro hair stained with silver. (A. C. Brown and J. A. Swift, *Beitr. elektronemikroskop Direkstabb. Oberfl.*, 1970, **3,** 299–306.) (Compliments of J. A. Swift, Unilever Research Laboratories, Isleworth, Middlesex.)

5.9. CONCLUSIONS

The array of information and techniques available to an SEM user is bewildering, and the attempt to keep the art simple is often abandoned as the prospect of new, especially quantitative, information is opened. Nevertheless, for the routine worker interested mainly in topographical detail, with some chemical composition information, the main advantage of the SEM lies in its simplicity; therefore, though all these techniques may be tried for fun, it is to be hoped that the simplest one that fits the operator's purpose will be adopted as standard, and that more complex procedures left to workers with specialist problems or those with a particular interest in techniques for their own sake.

REFERENCES

BANBURY, J. R. and NIXON, W. C. (1967) *J. Sci. Instrum.* **44,** 889.

BOOKER, G. R. (1970) *Proc. 3rd Annual SEM Symposium,* IIT Research Institute, Chicago, Ill., p. 489.

Cox, S. B. (1971) *Proc. 4th Annual SEM Symposium*, IIT Research Institute, Chicago, Ill., p. 433.

Kelly, T. K., Lindqvist, W. K. and Muir, M. D. (1969) *Science* **165,** 283.

MacDonald, N. C. (1970) *Proc. 3rd Annual SEM Symposium*, IIT Research Institute, Chicago, Ill., p. 481.

MacDonald, N. C., Marcus, H. L. and Palmberg, P. W. (1970) *Proc. 3rd Annual SEM Symposium*, IIT Research Institute, Chicago, Ill., p. 27.

MacDonald, N. C. (1971) *Proc. 4th Annual SEM Symposium*, IIT Research Institute, Chicago, Ill., p. 89.

Muir, M. D., Grant, P. R., Hubbard, G. and Mundell, J. (1971) *Proc. 4th Annual SEM Symposium*, IIT Research Institute, Chicago, Ill., p. 401.

Sutfin, L. V. and Ogilvie, R. E. (1970) *Proc. 3rd Annual SEM Symposium*, IIT Research Institute, Chicago, Ill., p. 17.

Swift, J. A. and Brown, A. C. (1970) *Proc.* 3rd *Annual SEM Symposium*, IIT Research Institute, Chicago, Ill., p. 113.

CHAPTER 6

Applications to Metallurgy

J. E. CASTLE

6.1. INTRODUCTION

Scanning electron microscopy joins a battery of tried and perfected metallographic and microscopic techniques in the field of metallurgy. Most metallurgical departments or laboratories are well equipped with transmission electron microscopes as well as microprobe analysers, diffraction cameras, and optical metallographic microscopes. Notwithstanding this expertise in the traditional methods, the scanning electron microscope has been universally welcomed and micrographs are now regularly seen in all the metallurgical journals. In this chapter we shall explore, under a number of headings, the manner in which this new technique is being used.

Microscopy as an adjunct to metallurgy has several discreet uses. Firstly, it is the basic non-destructive technique which we use to examine a surface and thus form the preliminary conclusions. We may be interested in the structure or form of deposits or of corrosion pits; we may wish to relate corrosion, plating efficiency, or weld penetration to the overall shape of an article; we may wish to examine fracture surfaces, surface grain structure or rolled in defects. All these examinations and a host of similar studies require the best possible "magnifying glass" and this, with light, reaches its magnification limit in the stereoscopic optical microscope at about 150 times. Secondly, the microscope is used in the examination, in great detail, of plane sections taken through important directions in the workpiece.

Metallography describes the art and science by which these sections are polished or etched to reveal their features which may then be observed in the optical microscope at magnifications up to 1500 times or may be replicated for transmission electron microscopy. By these means we arrive at the equivalent of an engineer's or an architect's knowledge of the workpiece. That is, we have the plan and elevation views from low-power optical microscopy and a detailed knowledge of a limited number of sections from the metallography. It is important to notice that the resolution of these cross-sections is some 10 times greater than that of the meaningful general views possible with light photography. Thus there were, prior to the development of the scanning electron microscope, many features the structure of which could only be pieced together from the cross-sections and which had never been examined in three dimensions. Contrast, in Fig. 6.1, an optical metallographic cross-section with the scanning electron micrographs at similar magnification. The material is Bi 5% Ag eutectic alloy with a precipitated phase of almost pure silver: it is soft and difficult to prepare for metallography and even after etching little relationship can be seen between precipitate and matrix. The electron micrographs show the surface after deep etching in nitric acid: from these we get an immediate feel for the aspect ratio of the precipitate and its relationship with the grain structure of the matrix. An excellent comparison of this type is to be found in the paper of Thompson and Lemkey (1970) on directionally precipitated carbide phases. One simple reason therefore for the popularity of the scanning electron microscope is that general views of the surface, or of internal features, can now be obtained which relate in magnification to cross-sections obtained by optical metallography.

The third major way in which microscopy helps the metallurgist is in the field of analysis. The use of selected area diffraction has become a standard technique on test pieces prepared for transmission electron microscopy; the use of a hot stage on the optical microscope enables the composition of certain microscopic phases to be arrived at by measurement of the melting point; and microscopy is invaluable in the study of the epitaxial relationships between deposits and the substrate. Examples of the use of the scanning electron microscope to achieve each of these ends are now to be found in the metallurgical literature. Perhaps the only example missing from the SEM literature is that of the observation of crystallographic dislocations. Such observations are still made using the transmission electron microscope, although no doubt once the transmission stage for the SEM is widely available, dislocation studies will be made on this instrument.

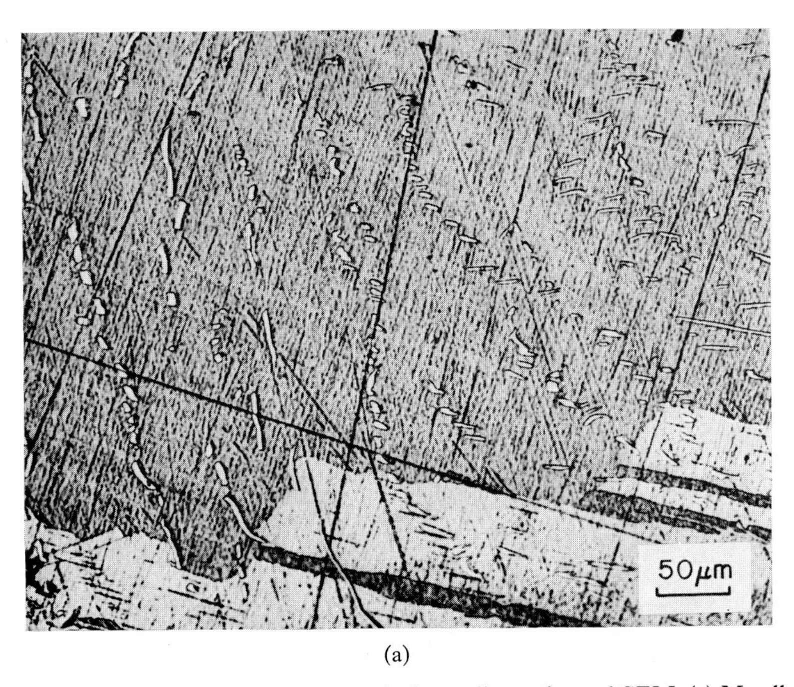

(a)

FIG. 6.1. The relationship between optical metallography and SEM. (a) Metallographic section of Bi 5% Ag alloy showing eutectic precipitate of silver rich phase.

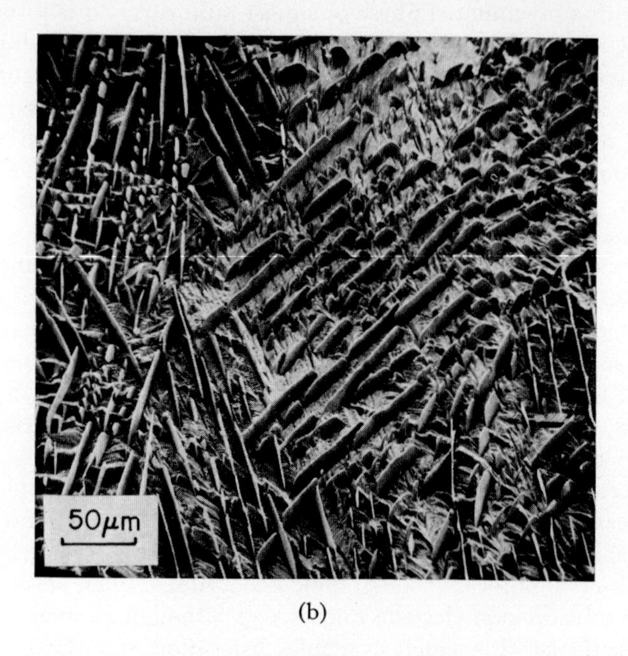

(b)

(c)

FIG. 6.1 (*cont.*). (b–e) SEM of eutectic phase illustrating use of deep-etch technique.

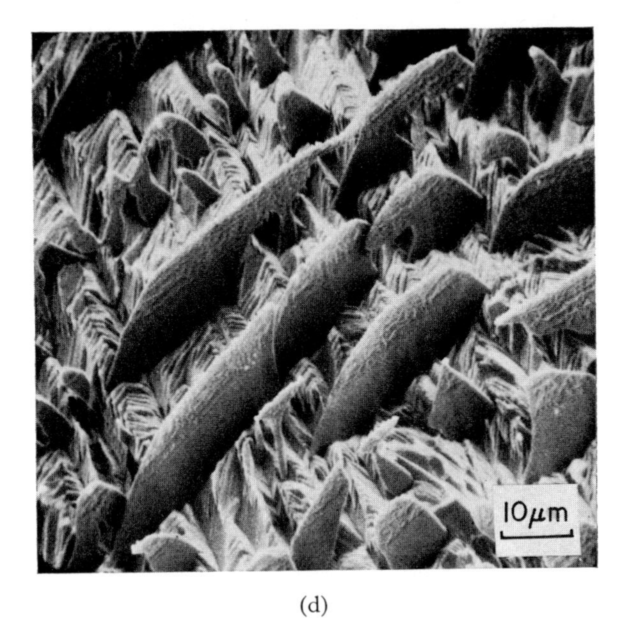

(d)

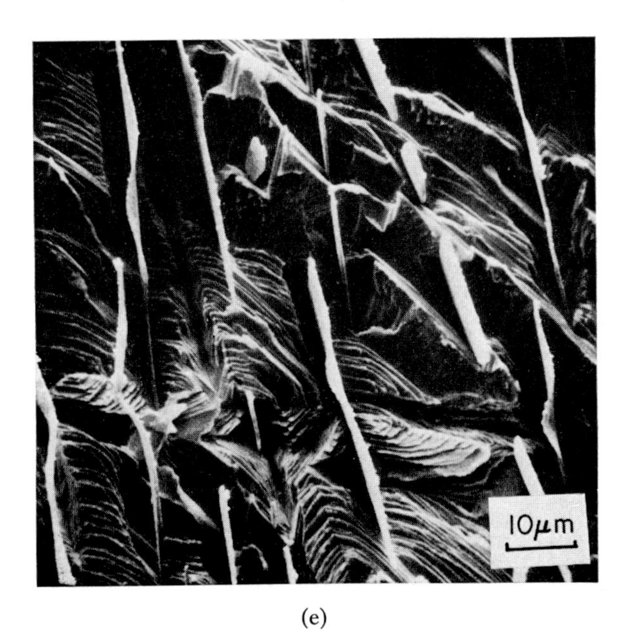

(e)

FIG. 6.1 (*cont.*)

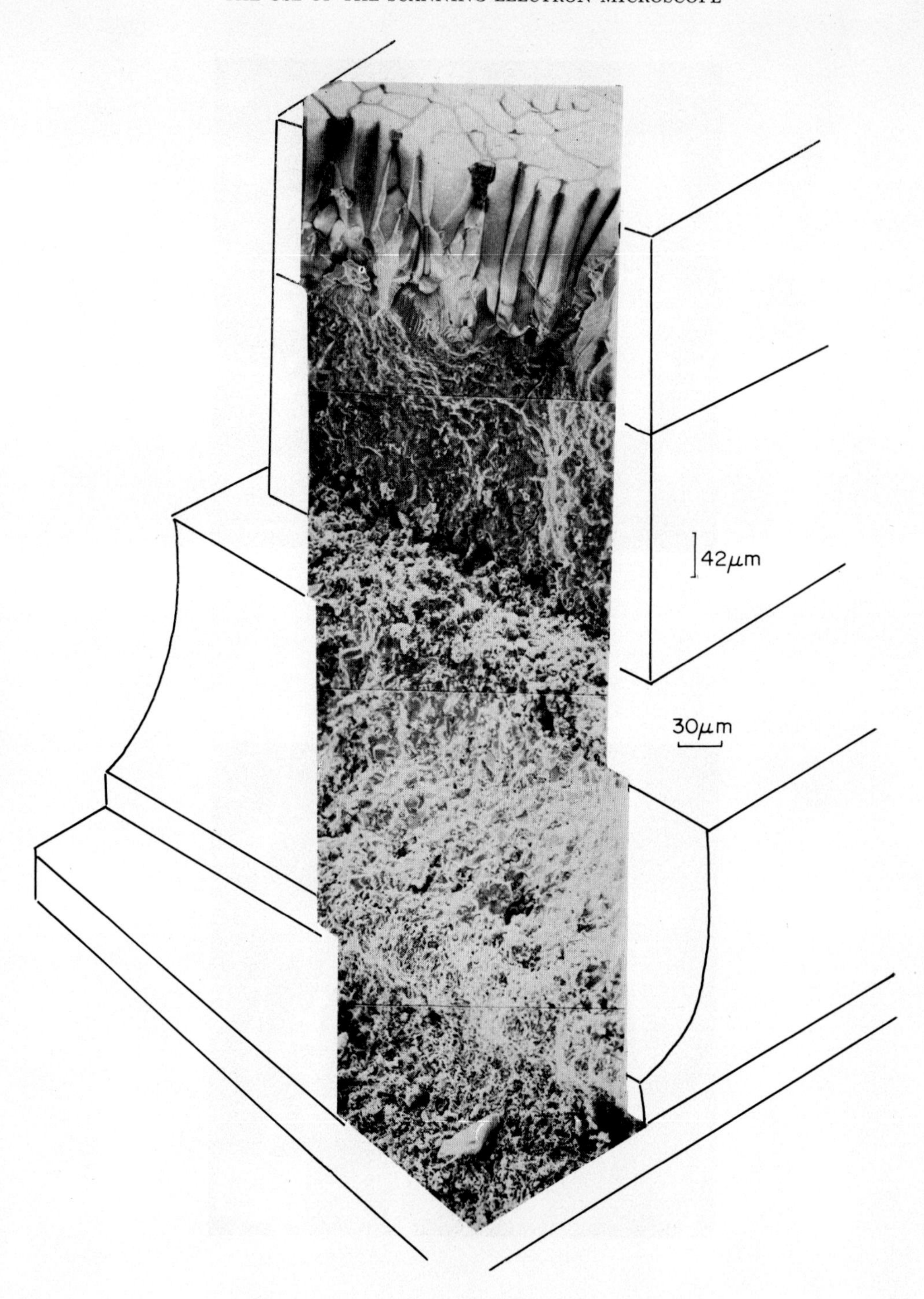

FIG. 6.2. The oxide layers on a chromium steel: micrographs mounted in montage form to assist in "reader orientation".

Many authors favour the use of a single scanning electron micrograph, often at low magnification, in order to rapidly orient the reader into the context of the succeeding material. Sometimes it is useful to include the micrograph in a montage presentation. The montage in Fig. 6.2, for example, acquaints the reader with the layer structure of the oxide formed on nickel/10% chromium alloy as well as allowing the author to indicate the composition of the layers and their stereographic relationship. A further example of the use of the strong pictorial powers of the SEM may be found in the paper of Vennett *et al.* (1970) on multiple necking of the fibre in a tungsten/brass composite. Here a view of the broken end of a composite rod at 165 × magnification gives the reader an immediate grasp of the orientation and size of the fibres, of the nature of the fracture and of the incidence of multiple necks.

Each microscopic development encourages the use of new preparative techniques: optical microscopy of metallurgical sectioning; transmission electron microscopy of replication and preparation of very thin metal sections. The technique of SEM, likewise, has encouraged the development of two metallographic techniques which were previously of little value, i.e. fractography and deep etching. These two techniques are almost invariably used for the examination, in the SEM, of metals in the bulk, and shallow metallographic etching is scarcely, if ever, used. Many examples of these new techniques will be found in the succeeding sections.

It would not be right in an introduction in which the attributes of the SEM are being explored to ignore the feature which is mentioned by most authors sooner or later: the ease with which specimens can be prepared for examination. In this respect the SEM is closer to the simple magnifying glass than any other optical or electron optical microscopic technique.

6.2. FRACTOGRAPHY IN THE STUDY OF FAILURE MECHANISMS

6.2.1. Introduction

Whereas fractography, or the study of fractured surfaces, at high resolution had been one of the technically more difficult areas of metallography it is now as easy as any other observation in the SEM. If we were to generalize the difficulties we should associate them with the small cross-section, heavy convolution and re-entrant porosity of many fractured surfaces. These features all combine to make replication difficult and tedious for transmission electron microscopy whereas observation in the optical microscope is strictly limited by the small depth of field of this technique. Moreover, as Beacham (1968) has pointed out, the "reading" of micrographs is by no means simple. One of the points which we must consider in this section is therefore whether scanning electron microscopy has merely made the preparation of fractographs a less tedious business or whether indeed it is able to produce more easily interpreted results than the older technique of transmission electron microscopy. Because of its limitations fractography was rarely used as a preferred technique for studying the internal structure of bulk metals. Rather, it appears to have been exclusively used for the study of failure mechanisms such as the initiation of cleavage steps, the development of ductile failure dimples, or of the mode of crack propagation in fatigue or stress corrosion failures. The situation has so changed with scanning electron microscopy that the preparation of fracture surfaces as a method of investigation has become a useful and valuable tool. Whilst we shall review in this section the use of the SEM to study the mechanism of fracture we must devote the next section to the study of metallurgical structure by the use of prepared fracture surfaces.

Each of the main mechanisms by which metals fracture has its characteristic hallmark. We shall consider the following: transgranular cleavage; ductile failure under tension; fatigue failure under cyclic stresses and then the various forms of chemically assisted failures, i.e. inter- or transcrystalline stress corrosion; corrosion fatigue; and hydrogen embrittlement. The main features of many of these fracture surfaces are visible in the low-power light microscope. The necking and surface dimples of ductile failure, the flat reflecting surfaces of cleavage and the faceted interfaces of the various intergranular fracture mechanisms are all easily seen. Transmission electron fractography (see, for example, ASTM No. 436, 1967) can yield much more information, however, and for this reason the skill of replica-making from broken surfaces has been developed, since the 1940s, to a most sophisticated level. By taking a replica directly from the fracture surface, surface details of as little as 20 Å to 50 Å can be revealed (Branmer and Dewey, 1966) and clearly SEM cannot yet compete with this. Unfortunately such direct replication is not usually possible and it is often necessary to use one or more intermediate replicas en route to the production of the final one for viewing in the transmission microscope. Such work is obviously skilled and tedious, and inevitably detail will be lost. Moreover, the replica covers only a small area of the fractured surface so that the rest of it must either be replicated in turn or omitted from the examination. Again, if the surface is particularly jagged or contains re-entrant pores or cavities, it may prove impossible to remove the plastic cast, or intermediate replica, from the surface. Interpretation of replicas also requires considerable experience. Density differences can arise from spurious differences in the thickness of the replica, from folds, bubbles or tears, as well as from the surface topography which it is desired to observe. A sense of depth in the micrograph must be imparted artificially by the use of heavy metal shadowing: there is no help from the contrast between grains or differing phases in the original metal surfaces (Beacham, 1968).

The preceding considerations add up to a very good case for the adoption of the scanning electron microscope for fractography. Probably because of unfamiliarity with the technique and the still relatively few instruments available, the 1970 journals in the metallurgical field carried far more transmission electron than scanning electron fractographs. Nevertheless, these have been sufficient for the value of the technique to be assessed. We shall do this in turn for the various types of failure mentioned above.

6.2.2. Cleavage fracture

Most metals and alloys have a ductile-to-brittle transition below a given temperature. Cleavage fracture occurs in the brittle temperature range and is characterized by the fixed orientation which the fracture plane makes with the crystal structure. In iron and steel the cleavage plane is in the 100 direction of the iron lattice. In a polycrystalline material the plane of a crack initiated at the edge in the 100 direction of a surface grain will be misorientated with respect to most of the other grains. As the crack passes into an adjacent grain, which is twisted with respect to the first, characteristic cleavage steps are formed as cleavage attempts to occur in the 100 direction (Fig. 6.3). As the crack propagates through the twisted grain individual cleavage steps tend to merge with each other producing larger, more pronounced steps. The direction of crack propagation can therefore be determined since cleavage steps are fine at the point of entry and coarsest at the exit boundary. The cleavage steps have the appearance of a river and its tributaries and are often known as river marks. They are readily replicated and there is an extensive literature of published transmission electron fractographs (Beacham, 1968).

The earliest published scanning micrographs of river marks appeared to be those of Tipper *et al.* (1959) on α-iron fractured by bending at -196°C. Wells reported in this paper some measurements of the angle between the 100 fracture plane and the plane of a projecting tongue produced by fracture through mechanical twins. The value found using a special rotational stage in the SEM was given as 20–29° $\pm$ 5° compared with the accepted value of 35° 16′ but this measurement must still represent one of the first metallurgical measurements in the scanning electron microscope. The example shown in Fig. 6.3 is of an area of cleavage in mild steel fractured at -196°C. The twist grain boundary runs from top to bottom of the micrograph and the crack propagation is parallel to the steps, i.e. diagonally down to the bottom left-hand corner.

One of the big advantages of scanning electron microscopy is the possibility of low-power surveillance of the whole of a fractured surface. By this means the area of cleavage shown in Fig. 6.3 is seen to be an isolated area surrounded by ductile, dimple-type, failure (Fig. 6.4). This facility is particularly important when it is desired to select an area in which cleavage traverses a special feature such as a grain boundary, inclusion, or oxide/metal interface (see, for example, Section 6.5.1). In accord with their importance in metallurgy, ferrous materials have provided the subjects for most of the detailed, comparative studies of SEM fractography with other methods. In three papers alone, those of Meny *et al.* (1970), Kuroda and Takada (1970) and Kanai and Uchibori (1969), we are able to examine some 170 micrographs obtained on carbon steels (0·01, 0·22, 0·45, 0·89, 0·9, 1·2% carbon) and cast irons. Kanai and Uchibori use a range of specified heat treatments and give known Vickers hardness values for all specimens. Each was broken by notched bar impact at both -196°C and 100°C. The authors concluded that river markings were always observed in normalized carbon steels below their brittle/ductile transition, but became far more complex at high carbon levels because of interference by pearlite colonies. Cleavage facets are not found in the quenched or tempered steels, instead the steps look like feathers or petals. The similar range of steels catalogued by Meny *et al.* (1970) were fractured by impact at -196°C. They concluded that resolution in the transmission microscope was better than in the scanning electron microscope although a comparison of their figs. 29 and 30 does show the extent to which bridging and in-filling of deep cracks and fissures can occur in replication. More important, however, they draw attention to the advantage of the transmission microscope that the influence of structural defects or of precipitates on crack propagation is more readily established because of the possibility of selected area diffraction. Nevertheless, their micrographs show that cleavage patterns

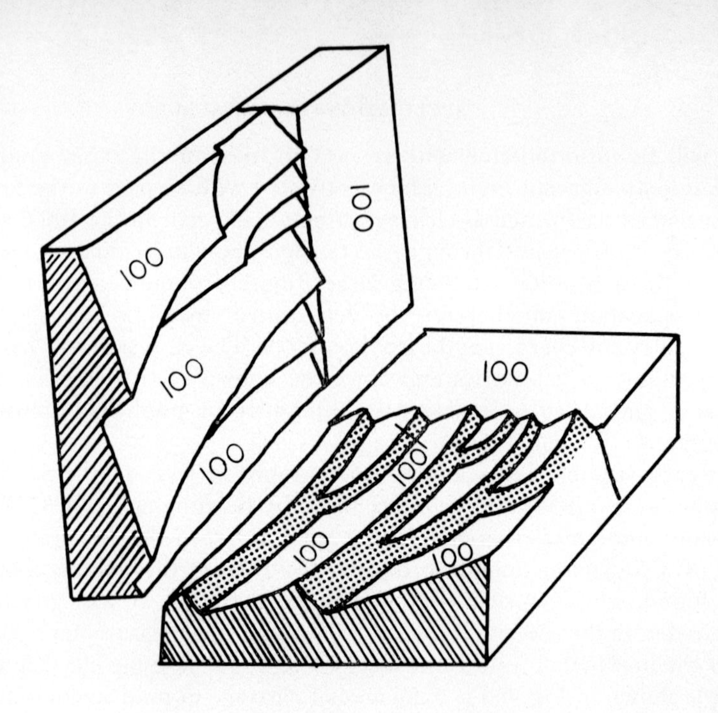

FIG. 6.3. Cleavage fracture: sketch showing cleavage through a bicrystal to illustrate the origin of "river marks" on mating surfaces. The micrograph shows cleavage through a grain boundary in mild steel: the height of the finest steps is 200 Å.

FIG. 6.4. Low-power surveillance of the fracture surface shows the area in Fig. 6.3 to be an isolated example of cleavage: the characteristic dimples of ductile failure can be seen in surrounding areas, e.g. at *A*.

are readily examined in fine detail, both for niobium and for the range of carbon steels, in the scanning electron microscope.

The pearlitic matrix iron of cast iron shows a typical cleavage failure even at room temperature (Kuroda and Takada, 1970) but in general pearlite appears to be easily recognized because of the presence of individual lamella of cementite (Kanai and Uchibori, 1969; Meny *et al.*, 1970).

One of the few measurements made using the SEM in this area was made by Heyes and Langdon (1969). They measured the velocity of cleavage fracture in tungsten by determination of the spacing of ripples formed by exposure of the test piece to ultrasonic vibration during fracture.

6.2.3. Ductile failure

There are two important modes of non-brittle failure of metals under stress: microvoid coalescence with the formation of characteristic dimples at the surfaces

(Fig. 6.4) and simple stretching of the metal to failure with slip occurring along dislocations. Dimple formation results from the growth of microvoids formed by diffusion of vacancies to the area of maximum tensile stress. Second-phase particles often act as local stress raisers and the size and distribution of dimples can be related to the microstructure. Kanai and Uchibori give examples of the relationship between dimple structure and the distribution of carbide particles in high-carbon steel, and between the rolling directions in tempered mid-carbon steel and the distribution of dimples. Meny *et al.* have illustrated the initiation of dimples on the pearlitic lamella in low carbon steel whilst Dalal and Grant (1970) have related dimple formation to the distribution of second-phase particles in stainless steel. Mixed dimple and cleavage type fracture in Ti/Mo alloys has been illustrated by Murr (1970) and in Ti/V alloys by Koul and Breedis (1970).

Since the orientation of the dimple wall can be used to indicate the direction of an applied shear stress (Beacham, ASTM, STP 436, 1967) this is another area in which the scanning electron microscope can excel in the investigation of a plant failure (see, for example, Parker and Morrison, 1969; Klein *et al.*, 1969). The mode of dimple growth itself has been studied using a scanning electron microscope by Taplin (1970) who observed the dendritic growth of dimples in stressed α-iron and by Gittins (1970) who related the density of dimples to the density of helium bubbles in irradiated stainless steel. Taplin's micrographs in particular are revealing in that although the transmission electron micrographs are superficially of higher magnification a good deal of crack filling has occurred.

In contrast to dimple formation which is frequently and easily seen the surface markings arising from slip have occasioned little comment. Kotval (1969) has discussed the surface shear steps arising from the martensitic transformation of a copper 12·5% germanium alloy in quantitative terms. He used the unusual (for scanning electron microscopy) technique of directionally shadowing gold onto the surface to improve the contrast and visibility of the surface steps. Micrographs of slip bands in a cobalt single crystal whisker and in a superplastic alloy may be seen in Lee's paper (1970) and in the review by Lifshin *et al.* (1969).

6.2.4. Fatigue failure

The failure of materials by fatigue is associated with the repeated stressing of the metal at a crack tip so that crack propagation is stepwise. In many materials failure by this mechanism is revealed by the presence of striations on the fracture surface. It was shown in 1962 by McGrath, Buchanan and Thurston that the striations could be seen in excellent detail in the scanning electron microscope. Working with copper fatigued at an alternating stress of 10,000 lb/in^2 they were able to show that the ripples became more widely spaced as the crack propagated down the slip plane with increasing velocity. Micrographs of opposing faces showed that groove striations mated with ridge striations. The frequency and orientation of striations reveals the direction and velocity of the local crack front and the ease with which striations are identified, counted and matched on opposing surfaces in the SEM makes this the ideal technique for the study of fatigue. Recent studies have been by Whitehead and Noble (1970) in Mg_3Cd and by Wright and Argon (1970) in Fe/Si alloys. Measurement of striation spacings in the scanning electron microscope were used as primary data in this latter study.

6.2.5. Stress corrosion cracking

The mechanism of stress corrosion cracking differs from the other mechanisms of fracture mentioned above in that it is intimately interwoven with the environment (usually aqueous) in which failure occurs. Even with a single material two distinct types of failure, inter- or transgranular cracking, may be observed and hitherto failures have been classified into one or other of these. Scully (1970) has remarked, however, that the fracture morphology revealed by the SEM is in all systems more complex than is commonly described in the stress corrosion literature. In general, both trans- and intergranular features can be seen on any fracture surface. Examinations of some of the published micrographs (Scully, 1970; Stavros and Paxton, 1970; Scully and Powell, 1970; Marek and Hochman, 1970; Harston and Scully, 1970) shows that the power of SEM in this field is not the ease with which repetitive features can be counted and sized as with dimples, fatigue ripples, or cleavage river-marks, but rather the very possibility of examining large areas of very complex fracture surfaces for the tell-tale signs of the way in which the cracks were initiated. One can observe crack initiation by electrochemical tunnelling or pitting from the surface, by anodic dissolution of grain boundaries, or by repeated failure of a surface tarnish film as in season cracking of brass or of 304 stainless steel (Baker *et al.*, 1970). Other features which might occur only in isolated regions of the fracture can be intergranular micro-cracks or characteristic angles which give a clue to the mechanism of failure. Most of these features would be difficult, if not impossible, to replicate for transmission microscopy. The determination of characteristic angles in regions of transgranular stress corrosion cracking (s.c.c.) is important because of the possibility that it might lead to identification of the fracture plane. In making such measurements the calibrated "tilt" and "rotate" movements on a stage are usually satisfactory providing the orientation of the specimen to the incident electron beam is established in some way. Scully gives examples of two methods.

(a) If the two intercepting planes are oriented so that the axis of intersection is parallel to the tilt axis of the stage and perpendicular to the incident beam, the angle between them is equal to the angle between the tilt positions at which each face just reaches a maximum. In each case correct alignment of the axis of intersection can be obtained by rotating it to the position of maximum length. To determine this position, and those of the other maxima, micrographs should be taken at 1° intervals. (b) The angle can be measured directly by viewing the specimen with the axis of intersection parallel to the incident beam. In this case both faces just disappear simultaneously (Fig. 6.5 (a, b)). In the end, however, it is only angles that are determined and these must be compared with the angles between lattice planes in the crystal structure of the substrate in order to guess the correct crack orientation. The presence of emergent slip lines is a further valuable indicator and may be used to confirm the orientation. Scully's papers give a detailed description of surface features which lead to the overall impression that, by use of the SEM, s.c.c. fractures can be read in a way which hitherto has been possible only with the more uniform fractures such as cleavage and dimple failure.

Although there have been many suggestions that the use of a tensile stage could enable stress-corrosion cracking to be studied in the microscope, nobody yet seems to have surmounted the difficulty of providing the environment within the specimen chamber. The enormous possibilities of *in situ* tensile testing are illustrated by a series of micrographs in the descriptive literature of JEOL's JSM U-3 showing the ductile failure of a chain link. One elegant study has been carried out (Ugianski *et al.*, 1968) to show that cracks may be present in metals during the so-called induction period of SCC. The point was settled by applying stress to a specimen in the specimen chamber of the SEM. This enabled grain boundaries and cracks to be differentiated from each other.

(a)

(b)

FIG. 6.5. The determination of characteristic angles in a stress corrosion surface. (a) Austenite fracture: H_2SO_4/NaCl Environment. (b) Part of (a), angle: 106°. (Courtesy of J. Scully, University of Leeds (Scully, 1970).)

6.2.6. Embrittlement

The possibility of observing fracture surfaces in detail using SEM has led several authors to compare s.c.c. fractures with those from other forms of embrittlement. Hughes *et al.* (1970) showed that there were many similarities in the fracture surfaces, obtained when uranium was exposed to water (s.c.c.), when it was embrittled with hydrogen, or when it was merely failed in air at a temperature below the brittle/ductile transition. Orman (1969) used SEM alone for his discussion of a rapid test for stress corrosion cracking. SEM has also been used in studies of hydrogen embrittlement of stainless steels (Okada *et al.*, 1970) and of oxygen embrittlement of iron (Rellick and McMahon, 1970). Joshi and Stern (1970) have published a study of intergranular brittleness in tungsten using both SEM and Auger spectroscopy, although not in this case combined in the same instrument. The scanning electron microscope was used for grain sizing and confirmation of fracture mechanism and the Auger spectra used to measure grain boundary concentrations of carbon and oxygen in the fracture surface. Neither of these elements were found although concentrations of phosphorus in the fracture surface were confirmed. The influence of atomic ordering on the transition from trans- to intergranular failure has been studied in Ni_4Mo alloys (Chakravali, 1970) and for a range of materials by Marankowski (1970). The mode of failure of polycrystalline magnesium at low temperatures has been established by Wilson (1970) using scanning electron microscopy.

6.2.7. Conclusions

With the exceptions of dislocation density and distribution all the important features of fractured surfaces can be studied in the scanning electron microscope. The surfaces are examined directly since there is no need even to apply conducting coatings and, in contrast to the preparation of direct replicas, non-destructively. The major part of a fractured surface can be included in the microscope and scanned at any convenient magnification. Moreover, in many cases differences in the secondary emission coefficient give grain/grain and phase/phase contrasts. There is excellent topographic contrast and there is no difficulty in the examination of the re-entrant cavaties. All these features must be rated as positive advantages over the transmission electron microscope. The magnification range of the scanning electron microscope is quite adequate for fractography. For example, in the chapter on transmission electron fractography by Beacham (1968) the magnification used for the various micrographs fall into the following distribution: 4 less than 1000 ×; 38 between 1000 × and 10,000 ×; 28 between 10,000 × and 25,000 ×; and only 3 at greater than 25,000 ×. Thus on this sample it may be that some 4% of desirable micrographs are beyond the reach of scanning electron microscopy.

6.3. PHYSICAL STRUCTURE OF METALS, ALLOYS AND COMPOSITE MATERIALS

6.3.1. Physical structure revealed by fracture

In this section we deal with the deliberate fracture of metals in order to reveal the structure and distribution of internal phases. As we have said it is a young branch of the science and, for example, it finds no mention in the ASTM special technical publication *50 Years of Progress in Metallographic Techniques*, published in 1966, and only one paragraph in a subsequent ASTM publication *Electron Fractography*, published in 1968. However, as we have seen in the preceding section, there are several mechanisms of metal rupture each of which is potentially useful for this form of metallography. We must immediately discard chemically induced fractures from the list since this could itself interfere with the observed phases. Similarly fracture mechanisms involving mainly surface slips are unlikely to reveal much of the internal phase. The three most useful are brittle failure by cleavage or intergranular cracking and ductile failure by microvoid coalescence (dimple failure).

Brittle failure can be induced by flexing a metal of small cross-section at a temperature below its ductile/brittle transition. In practice, this usually means while immersed in liquid nitrogen, but where higher temperatures are suitable the bend testing stage of

Clarke *et al.* (1970) could be useful for the *in situ* production of fractured surfaces. In some cases even immersion in liquid nitrogen is not sufficient to induce a brittle fracture. For example, McGrath *et al.* (1962) obtained dimple failures at this temperature with Al_2CuMg alloys and Castle (1971) obtained dimple failure with superposed fatigue ripples (due to multiple flexing) with Cu/30% Ni alloys. In each case, however, second-phase particles were revealed with good clarity. In the latter case they were of NiO formed by internal oxidation and could be usefully compared with the layered NiO in the oxide scale.

Brittle materials such as grey cast iron can be fractured by impact at room temperature. Kanai has used a notched bar impact method to achieve this and Kuroda and Takada (1970) report and exemplify the results of using several simple stress systems

TABLE 6.1

Etching time (sec)	No. of pearlite colonies	Area as pearlite (%)
0	0	0
10	0	0
20	2	0·5
30	5	0·5
50	18	1·5
70	34	3·0
100	48	4·0
300	166	12·0
result from conventional metallography		12·0

on cast irons. From their SEM micrographs they were able to give very good descriptions of the graphite particles in cast irons. These had flake crystalline interiors with an apparently amorphous surface layer within which failure always occurred during tension or bending fracture. In some cases the graphite was similar in appearance to that obtained from a matrix of iron/nickel alloys by complete dissolution of the metal phase (Minkoff and Nixon, 1966). Compressive or torsional stresses resulted in failure of the internal structure of the graphite. They also showed that inclusions were located in the eutectic cell boundary and played a role in the development of the crack.

Whilst pearlite colonies can often be seen in fracture sections of steels (Meny, Kuroda, Kanai) Inckle (1970) has shown the benefit to be gained from careful etching of the fractured surfaces. The number of colonies observed in sections produced by fatigue failure and etched in 2% nital solution was determined using large composite photographs. The results are indicated in Table 6.1. It is seen that a 300-second exposure of the surfaces to nital was necessary in order to reveal the total coverage of pearlite as determined from conventional metallography. Exposure to the etchant for this length of time did give severe deterioration of the metal surface, however. Lower exposures, although not revealing all the pearlite, did give good details of its structure. Johari (1969) has also examined pearlite in the SEM as part of a wider examination, using SEM, of the action of metallographic etchants on surfaces.

In some cases it is advantageous to select conditions known to produce dimple failure for fractography. Thus Dalal and Grant (1970) fractured oxide dispersed stainless steel at $-13°C$ in order to study bonding between Cr_2O_3 and ThO_2 particles and the matrix. The oxide showed up well at the base of the dimples and the workers concluded (a) that there was no sign of oxide/matrix bonding and (b) that there were no particle fractures.

6.3.2. Deep etching techniques: eutectic alloys

Deep etching of surfaces to expose internal structure has been nearly as popular as fractography. It is particularly valuable for the study of stringer-like phases or dendritic precipitations (Fig. 6.1). Asbury and Baker (1967) show several examples of etched stainless-steel surfaces with the $M_{23}C_6$ carbides standing in high relief. Beautiful micrographs of side-branched filamentary growth of SiO_2 resulting from internal oxidation of Ni/Si have been published by Barlow and Grundy (1970). No details of the etchant were given in either of these papers. Day and Hellawell (1967), however, used dilute HCl to reveal for the first time the silicon structure of a sodium-modified Al/Si alloy. The silicon was present as an interwoven network of branched filamentary crystals. The work has since been extended to a study of the iron/carbon system (Day, 1969). The work of Thompson and Lemkey on CoCrC eutectic was referred to in the introduction. The carbides were revealed by deep etching in aqua regia. Similar micrographs of an etched Cu/Pb eutectic alloy are included in Lifshin's review. Rege *et al.* (1970) used a 6-minute etch in a 5% solution of bromine in methanol to reveal in superb detail clusters of Al_2O_3 particles in Al-killed steel. In some cases it has been possible to relate the eutectic structure revealed by deep etching to the grain structure of the matrix material. For example, in the work of Walker and Clive (1970) on the effect of solidification rate on the high temperature strength of Ni/Al/Cr it is possible to see that the eutectic rods terminate at grain boundaries in the matrix. The etch used by these authors was 100 cm^3 water, 40 cm^3 HCl plus 5 g of CrO_3.

There appears to be only one reported study of filamentary metal in a non-metal matrix. This is by Gerdes *et al.* (1970) who studied the structure and distribution of tungsten fibres precipitated from a UO_2 matrix. The tungsten fibres had a radius of about 1 micron and are seen in good detail.

6.3.3. Study of composite materials

The problem here has a different emphasis to the study of stringer and dendritic growth described above. In this case the shape of the fibrous material is not unknown since it is readily examined in the raw before being merged with the matrix material of the final composite. Factors which must be sought, and in which the SEM has proved useful, are packing density and alignment of the fibre reinforcing, the nature and compatibility of the fibre/matrix bond, the degree of bonding. This can be estimated by measurement of the length of fibre pulled out, or standing proud of the matrix surface after a tensile failure, or from the degree of ductile necking of the fibre prior to failure. A useful paper on the scanning electron microscopy of composites has been published by Fairing (1967) but does not deal expressly with metallic materials.

Papers on the properties of the original fibre material have been produced by Perry *et al.* (1970) and by workers at the Fulmer Research Institute (1969). These studies of carbon coated with nickel or tungsten, respectively, have used SEM for reader orientation only. A similar study in which the SEM has been used to full advantage has been published by Jackson *et al.* (1970). Here *in situ* ion etching was used to look at the degradation of nickel or cobalt coatings on carbon fibres during the course of heat treatment. By this method it was possible to detect the penetration of coating metal into the carbon filament. The properties of tungsten filament composite material has been examined, using SEM, in α-brass (Vemett, 1970) and silver matrices (Johari, 1968 (a, b); Bhattacharyya *et al.*, 1970). Features such as fibre pull-out, necking, and fracture mode could all be readily established. The failure mechanism of boron fibres in an aluminium matrix has been discussed on the basis of SEM observations, in terms of edge initiation of a crack in a fibre (Breinan, 1970).

6.3.4. Structure of sinters and compacts

Scanning electron microscopy could find an important use in the production control of metal components by the powder metallurgy processes of compaction and sintering. In this technology the size range of the starting metal powder as well as the grain size and porosity of the compact are the starting points for discussion of the optimum sintering time, temperature or compaction pressure. Lardner (1970) has pointed out that in the past powders have been characterized by compaction and impregnation (in his case of tungsten carbide with molten copper) followed by metallographic sectioning. In his opinion direct observation of the powders in the SEM is now the most effective method of carrying out observations on the starting material. During the course of sintering the scanning electron microscope can be used to indicate the range of grain sizes: the comparison micrographs (Fig. 6.6) of tungsten carbide seen in metallographic cross-sections and in the scanning electron microscope are taken from Lardner's work. Chance (1970) has published scanning electron microscope observations of metallized ceramics which were assessed for grain size, openness of structure and interface bonding. The impressions received from the micrograph compared well with those from adhesion tests. Castle and Wood (1968) have discussed porosity in oxide layers (Section 6.5) and concluded that microscopic observations of porosity should always be corroborated by other techniques. The point is thus discussed further below.

Murrell and Enoch have shown that the SEM is of particular use where it is desired to produce alloys by powder metallurgy. Each component in powdered form of a 77Ni–14Fe–5Cu–4Mo (wt. %) alloy was characterized by scanning electron microscopy. The characteristic forms of particles in each of these powders enabled the progress of sintering and alloying to be monitored by scanning electron microscopy alone without recourse to microanalysis. The series of micrographs (Fig. 6.7) taken from their paper shows (a) how little change occurs from the original starting materials prior to melting of the copper component, when large pores are also produced. These also show the elimination of microporosity by prolonged sintering at temperatures above the melting point of copper.

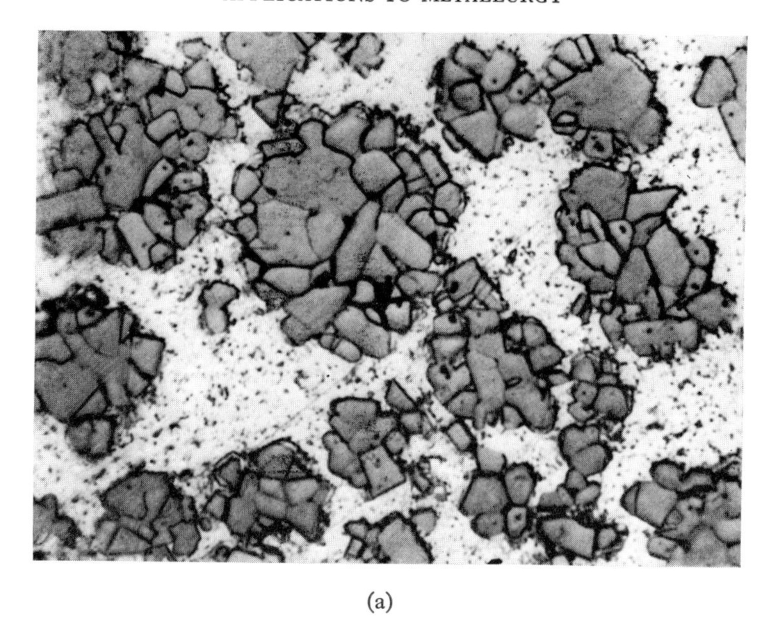

(a)

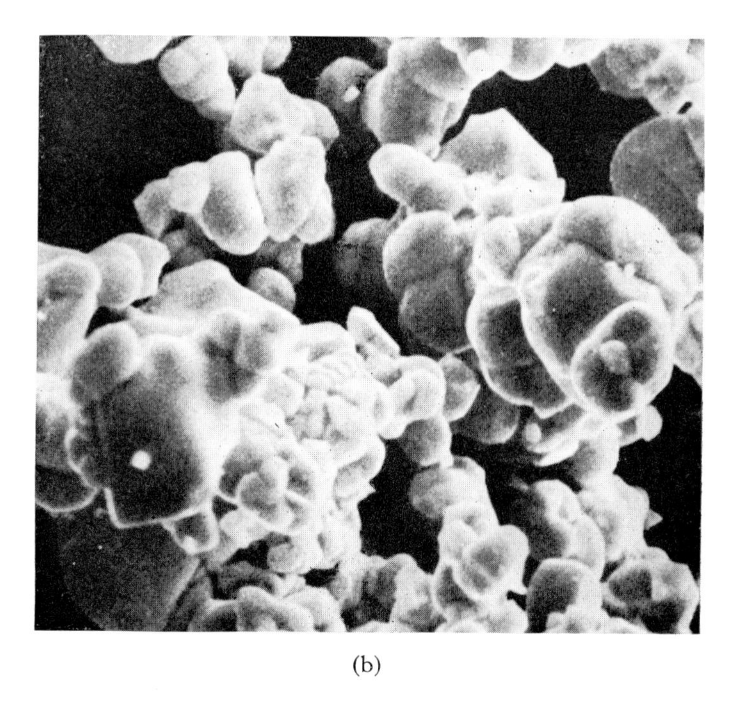

(b)

FIG. 6.6. The use of SEM in the appraisal of starting material in powder metallurgy: (a) optical micrograph of tungsten carbide mounted in copper, (b) SEM of unmounted powder. (Courtesy of Lardner (1970), Wickman, Wimet Ltd.)

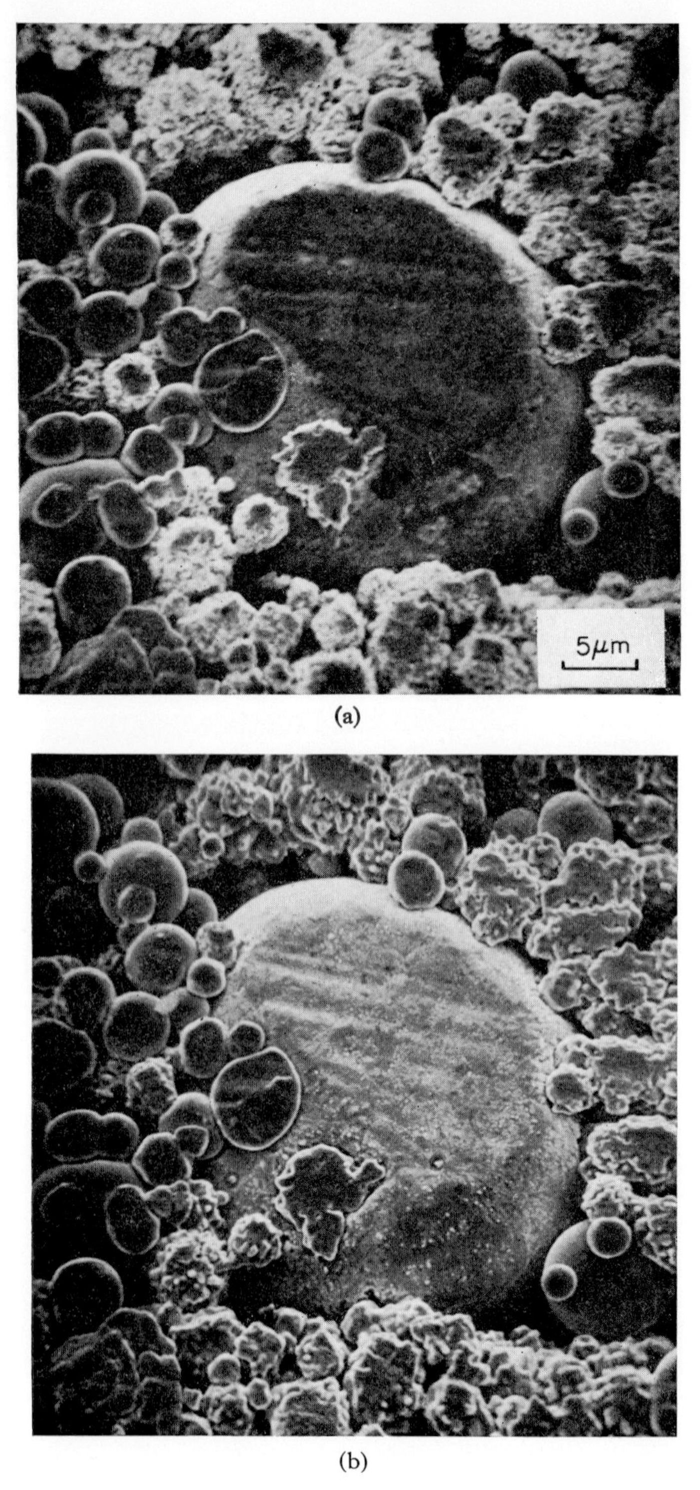

FIG. 6.7. Sequence showing the behaviour of a pressed compact of the components of a 77Ni–14Fe–5Cu–4Mo w/o alloy: (a) as pressed, (b) after heating to 700°C.

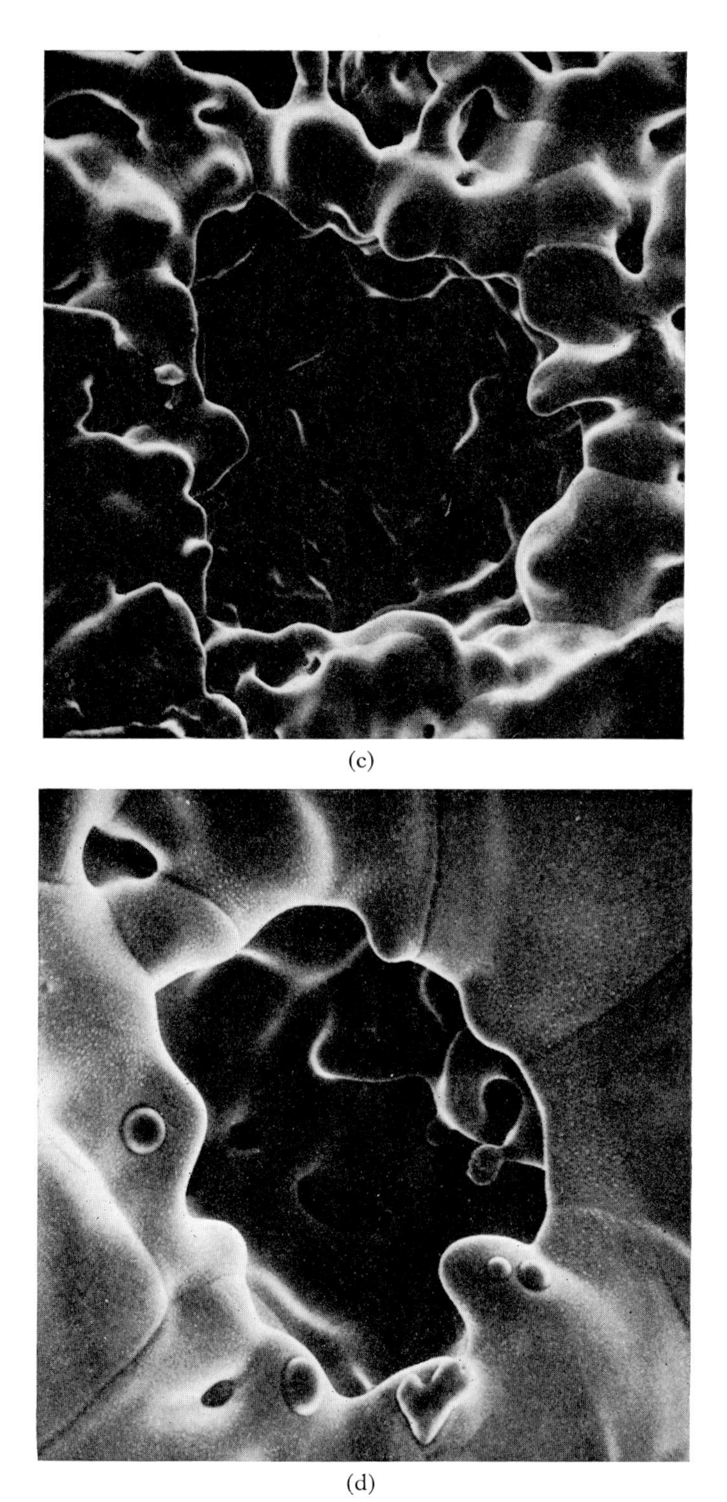

(c)

(d)

FIG. 6.7 (*cont.*). (c) After heating to 1100°C, (d) to 1300°C. The central pore has been formed by melting of the large particle of copper. (Courtesy of Murrell and Enoch (1970), The Post Office Corporation.)

Other non-metallurgical papers which may nevertheless be of use to workers in powder metallurgy are those of Evans *et al.* (1970) on the fracture mechanism of magnesium oxide compacts, of Johari and Bhattacharyya (1968) on the characterization of powders and of Bjerle on bonding of foundry sands (1970).

6.4. STUDIES OF SURFACE MORPHOLOGY

6.4.1. Pitting, etching and surface damage

SEM was first used 10 years ago (Cole *et al.*, 1961) to study surface damage in general: on this occasion to assess a spark machining technique. It is probably used in this capacity in many organizations on routine troubleshooting which is rarely published. A typical example of the use of the microscope in this manner is the examination of surfaces produced by penetration of aluminium alloys by projectiles (Stock and Thompson, 1970). Phelps and Kemp (1970), in evaluating techniques for diagnosis of corrosion failures, conclude that all-important features are observable and that

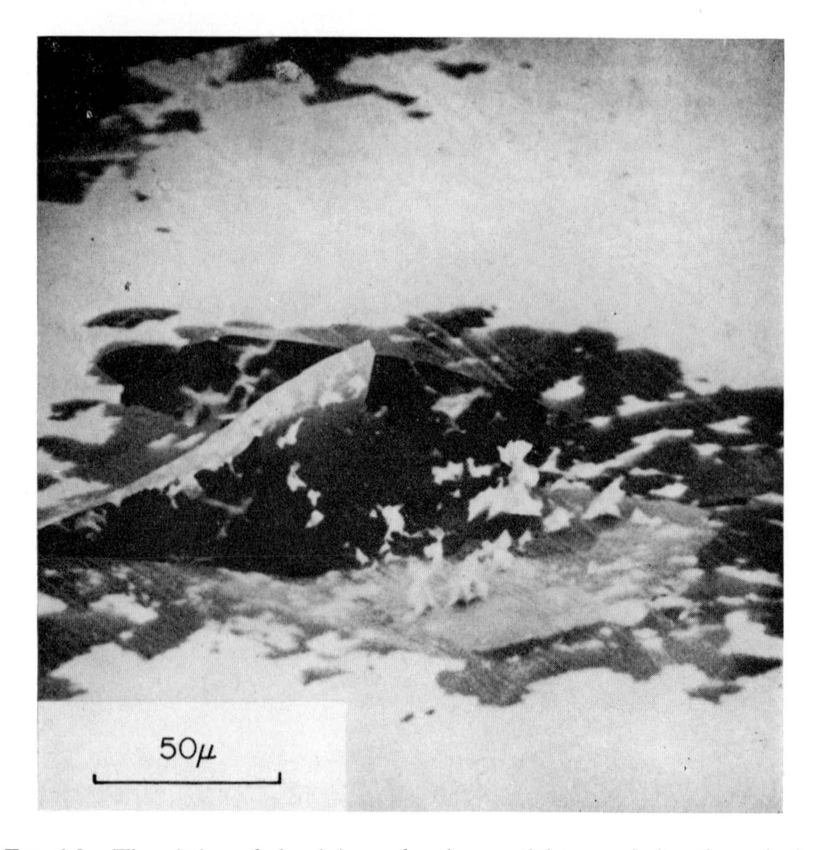

FIG. 6.8a. The pitting of aluminium, showing partial transmission through the surface oxide layer. (Courtesy of Richardson and Wood (1970), University of Manchester, Institute of Science and Technology.)

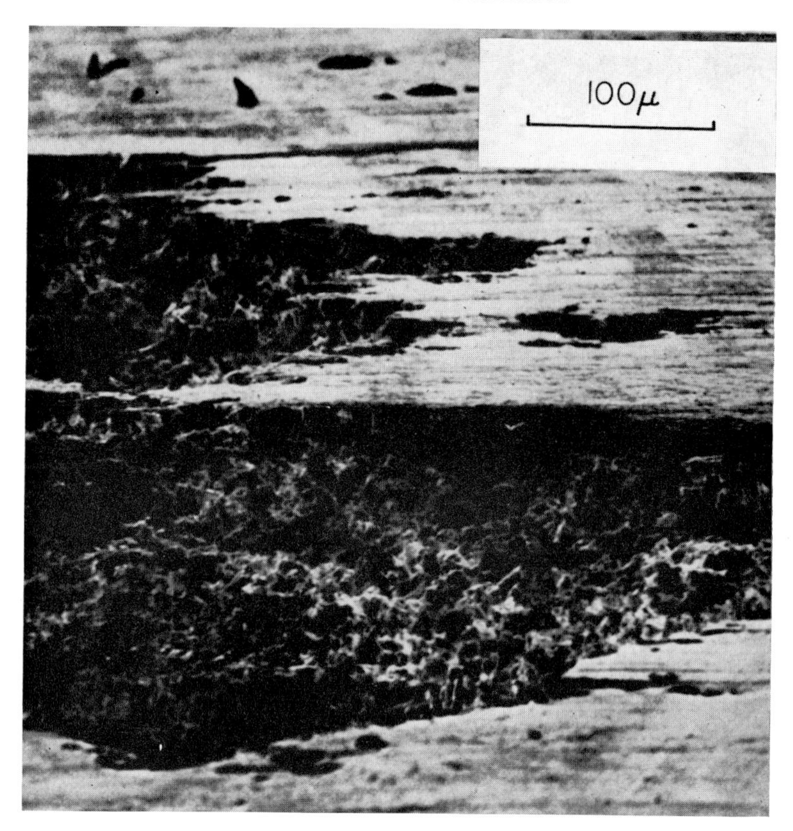

FIG. 6.8b. The pitting of aluminium, showing the internal surface of the pit. (Courtesy of Richardson and Wood (1970), UMIST).

SEM is a powerful tool because of its speed vis-à-vis transmission electron microscopy. Certainly the microscope is often useful to discover whether surface melting has occurred or the direction of material transport in situations of abnormal wear and has been used in friction research by Bowden and Childs (1968). Tucker and Meyerhoff (1969) have made a useful comparison of surface roughness, as measured by a "Talysurf", with the topography as seen in the SEM. A general study of wear surfaces has been published by Bhattacharyya *et al.* (1970); example of water-droplet erosion may be found in the review by Asbury and Baker (1967) and the effects of laser irradiation are shown in a note by Baston and Bowden (1968).

Several examples of the value of scanning electron microscopy in the study of oxide morphologies were given by Castle and Masterson in 1966 but, surprisingly, few studies have been published in which SEM has played a major role. One notable exception is the study of aluminium pitting corrosion by Richardson and Wood (1970). Their paper amounts to an evaluation of SEM as applied to testing the salient points of their theory, viz. that flaws in the oxide film represent the initial site of pit development; that the surface oxide film plays an inert role during pitting; that pits propagate immediately on immersion of aluminium in an aggressive medium and that the

FIG. 6.8c. As Fig. 6.8b.

localization and topography of a pit are dependent upon the local, true, corrosion current. They found that the SEM was unable to detect the starting flaws, *ca.* 1000 Å in diameter, which could be observed in the transmission electron microscope. When the pits started a further advantage of the SEM became apparent: that the surface oxide layer of Al_2O_3 was sufficiently transparent to electrons to enable the outline of the pit to be discerned beneath the surface oxide (Fig. 6.8). Also there was good grey contrast between the corrosion product derived from the pit and the original oxide layer. By observations of the extent of damage to the oxide layer overlaying a pit the authors were able to conclude that its rupture after pit initiation was only by H_2 evolution from the pit. The pits themselves varied in shape according to the applied pitting potential from faceted crystallographic outline to almost hemispherical ones. Richardson and Wood concluded that the SEM could provide crucial evidence in the formulation of a mechanism of attack.

Although the transmission of primary electrons through the oxide layer was of particular use to Richardson and Wood this is not always the case: Castle (1971) has given examples of a case in which a considerable amount of surface detail was lost because thin surface-corrosion products became partly transparent. The simple solution is to use a much lower accelerating potential for the examination of thin surface deposits: a useful setting for general observations is 7 kV. The effects of variation in viewing conditions are shown in Fig. 2.19.

SEM has also been used in a general survey of the features associated with hydrogen evolution (Toy and Phillips, 1970) whilst Ehman and Faust (1970) have used SEM to study the morphology of pits in aluminium. This time the metal was etched in acid with no corrosion products present and the variable was the applied level of ultrasonic vibration. This tended to give rectangular pits with the 100 faces exposed. The pitting behaviour of mild steel in chloride-rich solutions is closely similar to that described for aluminium. Again the pits assume a faceted shape (Ashworth *et al.*, 1970) although in this case there was no evidence of the presence of an overlaying oxide layer. Corrosion products derived from the pit formed a dome.

A very successful use of SEM in association with a separately mounted non-dispersive X-ray spectrometer may be found in the series of quarterly reports (to NACE) on clarification of de-alloying phenomena by Heidersbach and Verink (1970 a, b and c; Verink and Parrish, 1970).

6.4.2. Electrodeposits

To describe the impact of SEM in electrodeposition studies is again to emphasize the importance of depth of field combined with a large magnification in many technological processes. Fukubayashi *et al.* (1969) have put the case for SEM in studies of the electrowinning of zinc and since several of the papers on electrodeposition are concerned with this topic (Naybour, 1969; Mansfield and Gilman 1970 a and b) it is worth discussing in some detail. In the final stage of refining, zinc is deposited from acid solution and this is possible only because of the high over-voltage of hydrogen on the metal. Impurities can decrease the hydrogen over-voltage or modify the structure and morphology of the electrodeposit. Any one of these effects will influence the current efficiency of the electroprocess and thus of industrial production. Detailed studies of the morphology have, prior to SEM, dealt only with the initial stage of the process since the characteristically rough or dendritic form produced in the later stages are impossible to replicate. Fukubayashi *et al.* were able to correlate the morphologies produced with commercial additives to plating baths, such as animal glue, gum arabic or sodium silicate, singly or in combination, with the current efficiency. They concluded that surface examination by scanning electron microscopy was good enough to assess the benefit of additives and use the method to compare results of laboratory trials with those of commercial baths. Naybour (1969), using alkaline solutions of zinc, was able to classify the type of deposit as dendritic, bulbous or flat, and correlated these with the current density and Reynolds number corresponding to flow past the zinc surface.

Similar studies of plating morphology, particularly identification of dendrite growth, have been made for copper (Robinson and Gabe, 1970) and nickel (Dennis and Fuggle, 1970). Typical examples of dendritic growth taken from the work of Dennis and Fuggle are shown in Fig. 6.9. As a series they tell all that is necessary concerning the power of the technique for surface morphology.

SEM studies of precious metal plating have appeared for platinum (Reid, 1970) and gold (Craig, Jr. *et al.*, 1970; Cooksey and Campbell, 1970). Cooksey and Campbell

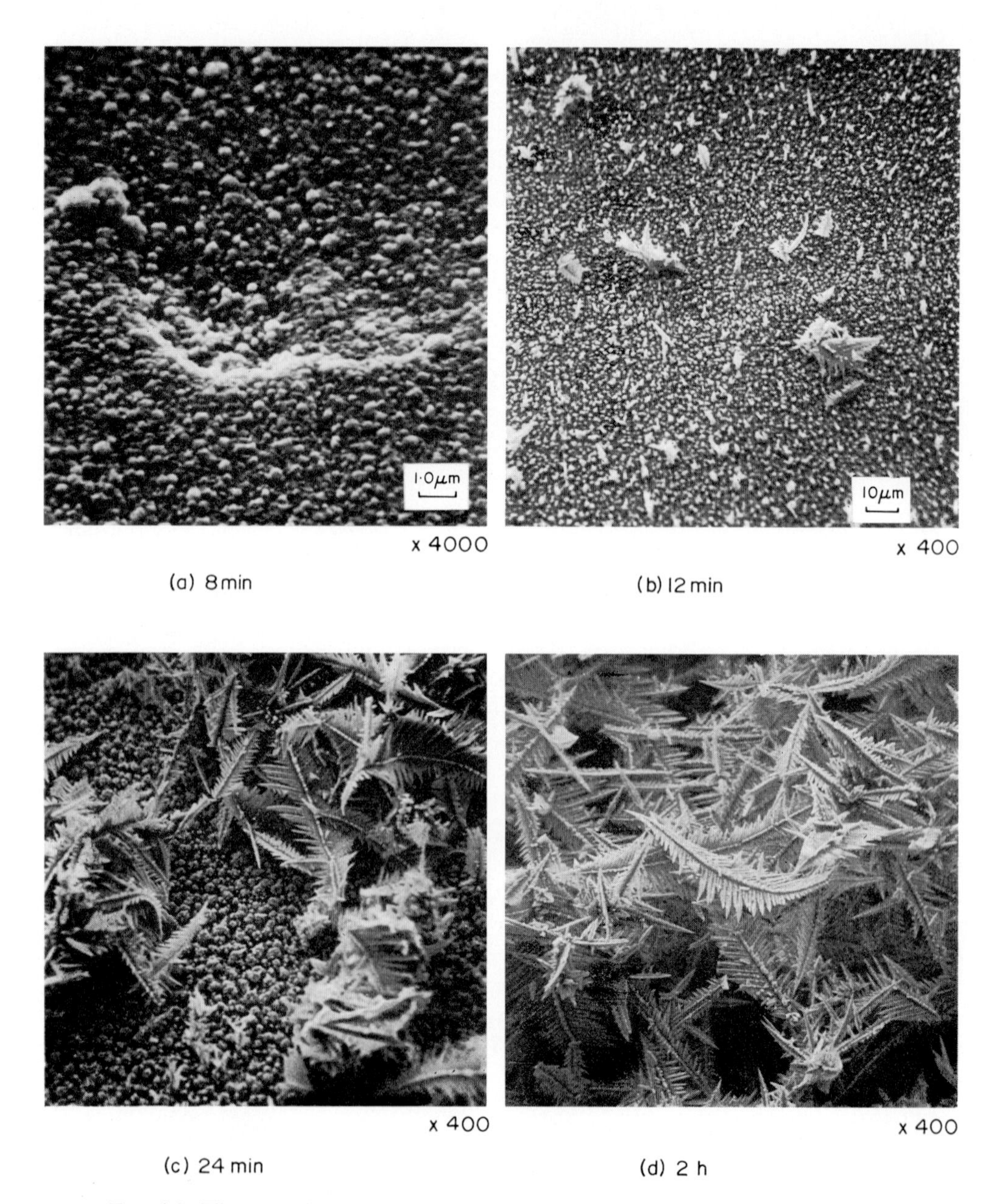

FIG. 6.9. The growth of dendrites in electrodeposits: copper-plated from Watts solution. (Courtesy of Dennis and Fuggle, University of Aston in Birmingham.)

found that porosity in gold electrodeposits was due to the presence of surface contamination from the buffing compound used prior to plating.

6.4.3. Adhesion and orientation of metal films

Several workers have used SEM to look at the structure and orientation of metal films on oxide or polymer substrates. Katz (1970) examined the orientation of copper deposited epitaxially on sapphire by *in situ* ion etching and back reflection diffraction. Rawlins (1970) has demonstrated the possibilities of "Coates" type contrast patterns in the determination of epitaxial relationships in the scanning electron microscope. His studies were of zinc sulphide on silicon substrates but his work is directly relevant to metallurgically oriented studies and his paper is well worth study.

Other published studies, such as that of Bazzarre and Petriello (1970) on vapour-deposited metal permeation barrier or of Milgram (1970) on the adhesion of screen printed silver films, are typical examples of the routine use of the SEM in this field. Arrowsmith (1970), however, has contributed a detailed study of the use of SEM in the investigation of adhesion mechanism between copper and nickel deposits and epoxides. The metals were stripped from the epoxides backing and mating surfaces examined. The epoxides replicated the metal surfaces well but the precise position of the adhesive bond could not be located because of the resolution limit of the microscope.

6.5. CHEMICAL METALLURGY

The fields of study generally associated with chemical metallurgy have warranted a separate section in this review, not because the techniques used are any different to those used in other metallurgical areas, but because the surfaces to be studied must often be regarded as ceramics or semiconductors rather than metals. The products of high-temperature metal oxidation are usually thick oxide scales with complex crystallographic and layer structure (Fig. 6.2), whereas those produced by anodic oxidation may be equally thick yet have a more uniform structure. Studies of mineral deposits on metal in service, or of oxide or other metal ores, are also of major importance in the metallurgical industry.

6.5.1. Products of high-temperature oxidation

The traditional technique for the examination of oxide structure is optical metallography. Generally, however, oxides are not as suitable for metallography as metals; frequently oxides are friable and crack or disintegrate during polishing or else they are so hard that the necessary flatness cannot be preserved at the metal/oxide interface. If it were necessary to select one reason why the SEM comes into its own for oxide studies it must be to study the pore structure of friable oxides and the detail at the metal/oxide interface. Transmission electron microscopy has been more usually applied to examination of oxide films, stripped from the metal surface at an early stage of growth (Field and Holmes, 1965), or to studies of oxide asperities in silhouette (Gulbrausen and Copan, 1960). The advance brought by scanning electron microscopy is illustrated by what is probably the first use of the microscope, to study the growth of an oxide layer (Pease and Ploc, 1965). Pure iron was oxidized in air at temperatures between 500°C and 600°C within the specimen chamber. The progress of oxidation was followed at frequent intervals with particular regard for the growth and distribution of fine whiskers of Fe_2O_3. Whisker density and distribution varied from grain to grain and was shown to be associated with the presence of impurity atoms in the surface of the metal. Pease and Ploc used the whiskers to exemplify the abnormally high brightness associated with the thin surface because of the emission of secondary electrons from their reverse surface.

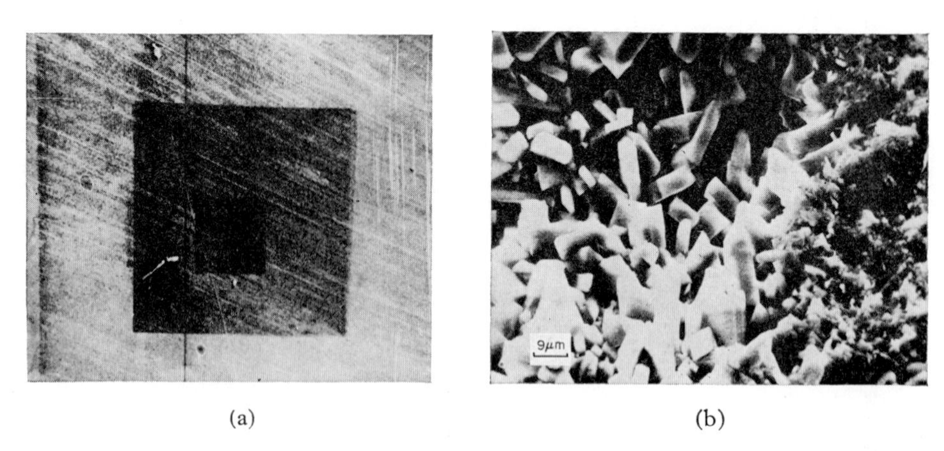

(a) (b)

FIG. 6.10. Example of the good conductivity of corrosion products: the micrograph in (a) of electrolytic condenser grade Al_2O_3 (thickness = 200 Å) was rejected because of persistent charging; the large Al_2O_3 crystals (b) produced by corrosion of aluminium in water at 300°C photographed well without the use of a conducting coating.

By 1966 Castle and Masterson had published representative examples of oxide layers formed on a wide spectrum of metals under diverse oxidation conditions and had concluded that charging, even at 20 kV accelerating voltage, was rarely a problem. This appears to arise from the fact that the oxide layers result, almost by definition as corrosion test pieces, from rapid oxidation processes and contain a non-equilibrium number of atomic and electronic defects (Castle, 1971). We can compare in Fig. 6.10 (a) and (b) the marked charging of the surface of aluminium coated with a very thin, electrical capacitor grade, layer of Al_2O_3 (produced electrolytically) with the acceptable imaging of sapphire (Al_2O_3) crystals produced by oxidation of aluminium in water at 300°C. A less usual charging effect is seen in Fig. 6.2 where the grain boundaries have acquired a high contrast as a result of charging which suggests that impurity segregation to the grain boundary may have occurred. Again the bulk oxide layer, which would normally be considered an insulator, shows no sign of charging. The ease with which oxides can be examined has made the SEM a favoured tool of metal oxidation specialists, particularly for the identification of surface crystal habit (Surman and Castle, 1969; Jones *et al.*, 1970) and correlation of oxide layer structure with the results of electron microprobe analysis (Castle and Masterson, 1966). This may be done relatively easily, not because of the contrast between oxide phases which is often very poor (Castle and Wood, 1968) but by the use of the easily recognizable grain structure. With the more widespread availability of X-ray-analysis attachments individual layers will also be identified by their composition.

Because of the adequate conductivity of the oxide specimen it is generally sufficient to mount the samples on the specimen stubs provided using a conducting glue such as Silver Dag supplied by Acheson Colloids Ltd.

6.5.2. Porosity in oxide layers

The idealized model for the oxidation of metal by diffusion of ions through solid oxide crystal lattice is often complicated by the development of porosity within or between oxide layers. Pores form by a variety of mechanisms, e.g. by coalescence of vacant sites in the oxide lattice or at the scale base, or by mechanical strains resulting from oxide growth. They may range in size from a few angstrom units to many microns and in some cases their presence can be inferred from the oxidation kinetics. Thus when the pores form an almost continuous network across the whole of the oxide layer the rate of gas transport far exceeds the rate of solid-state diffusion and the oxidation rate becomes independent of oxide thickness. In other cases the closed pores progressively reduce the cross-sectional area for solid-state diffusion without actually allowing gaseous oxygen to reach the metal surface. The oxidation rate then decreases logarithmically with time. Generally, however, the presence of pores cannot be confirmed from the oxidation kinetics and some form of metallographic examination is essential. In the conventional metallographic examination the specimen is mounted in resin and polished in cross-section. Very long polishing times may be used to give acceptable results, but there must always be doubts as to whether any pores observed have been introduced, enlarged, or distorted by sectioning and polishing. Alternatively existing pores can be obscured by the accumulation of polishing debris.

In 1965 Castle and Masterson published a study in which the presence of pores in an oxide layer was inferred from the correlation which existed between the oxide grain size as measured in the SEM and that measured by use of a gas absorption technique. The oxide used in this study (see also Castle and Masterson, 1966) was that produced on steel in sodium hydroxide solution at temperatures in the region of 300°C. Porosity cannot be seen in this material with the optical microscope and the conclusion reached by earlier workers (Potter and Mann, 1961) was that it was continuous and impermeable. It is important to note that individual pores could not be seen in the SEM and no claims were made concerning this. Microtomed sections of similar material were, however, observed in the transmission electron microscope (Harrison *et al.*, 1965) and from these it could be seen that the pores did exist and had a diameter of around 200 Å.

The fracture surfaces of fine-grain materials often have a very high roughness factor and when viewed in the scanning electron microscope may appear to be porous. The observation of porosity in such a way must be regarded as quite subjective in the absence of any other confirmatory technique. Porosity is, of course, one of the features of scanning electron micrographs to which authors often refer (e.g. Fyfe *et al.*, 1970;

Menzies *et al.*, 1969). To give more confidence in such observations Castle and Wood (1968) attempted to assess the use of scanning electron microscopy for the detection of porosity in oxide formed under a number of representative conditions. The alloys used for the study—Fe 2% Cr, Ni 2% Cr, Ni 5% Cr, and Ni 7% Cr, Cu 30% Ni—all form complex oxide layers. An example of the oxide layers on iron 3% chromium as seen in an optical micrograph is shown in Fig. 6.11 (a). The difficulty one has in deciding whether the porosity apparent in this micrograph is real or merely an artefact

FIG. 6.11. The identification of porosity. A typical optical metallographic cross-section is shown in (a). Pores may be recognized by their internal structure (b) or their re-entrant shape (c). (After Castle and Wood, 1968.)

of the polishing process is quite apparent. The optical micrograph is useful, however, because there is good colour contrast between the magnetite layer (B) and the Fe_2O_3 layer (A) and spinel layer (C). The dark material (D) is the mount, and thus indicates void, and the bright region (E) is the underlying metal. There is no contrast between the layers (A) and (B) in the SEM. By careful use of fractography one is able with the SEM to trace porosity throughout the oxide layer. Pores at the junction of the layers (A) and (B) are readily identified because of the secondary growth which has occurred within them (Fig. 6.11 (b)). These clearly are not polishing artefacts. Similarly the porosity in the magnetite layer itself appears to be genuine because of the re-entrant nature of some of the pores. The porosity does not, however, penetrate the innermost spinel layer which is revealed as a well crystallized layer growing on a continuous basal film. A further example of a typical pore in the outer oxide layer seen in this work (this time in nickel 7% chromium alloy) is seen in Fig. 6.11 (c). The cleavage patterns in the oxide layer surrounding the pore are noteworthy and may indicate that in this case cleavage initiated from the pore itself. Castle and Wood concluded that the SEM gave a very accurate description in thick layers where the grain size is of the order of 1 micron or greater. Barrier films adjacent to the metal surface, however, frequently have a grain size of the order of 10^3 Å. In this size range the texture of the fractured surface may conceal the true extent of porosity. When the oxide layer has grains in this size range it is essential that the SEM observations are supported by density determinations or surface area measurements as in the work of Castle and Masterson (1966).

The development of macroscopic cracking in oxide layers has been followed by Cox (1969) and by Sharpe (1970). More recently Hunt and Castle (1971) have used a hot stage with a controlled atmosphere to follow the development of cracks in the oxide layers on iron in oxygen by continuous observation.

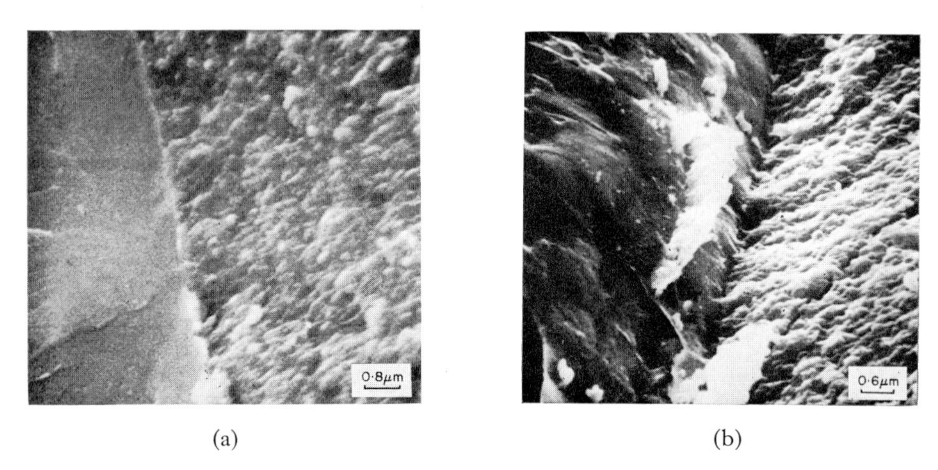

(a) (b)

FIG. 6.12. Examination of the oxide/metal interface. Oxide produced on steel foil in aqueous sodium hydroxide at 300°. The area in (a) shows a cleavage fracture running through the phase boundary which is shown in cross-section. In (b) ductile stretching of the metal has occurred giving distortion in the region of interest.

6.5.3. The oxide/metal interface

In order to study the detail at the oxide/metal interface a prime requirement is a good flat cleavage fracture through oxide and metal in the same field of view. It can be seen that this has been achieved in Fig. 6.12 (a) but not in (b). Figure 6.13 (a, b) is given as a guide to the selection of suitable areas. Howes (1967, 1970) has successfully departed from this requirement in studies of the interface morphology of a range of iron/chromium alloys. In his special case, however, he examines the structure and distribution of cavities which form between the oxide and the metal and uses an oblique viewing angle to achieve it.

6.5.4. Other areas of chemical metallurgy

Scanning electron microscopy has found use in several other areas of chemical metallurgy. For example, Prentice (1970) used it as the microscopic tool in the examination of the effect of water vapour on the combustion of flash-heated beryllium droplets. Samples were withdrawn from the flame by quench sampling and from the morphology of the particles a complete picture of the burning particle geometry was reconstructed. Drew *et al.* (1966) have also reported on the morphology of aluminium particles during combustion. Other workers have used the microscope for studies of chemical transport of metals; for example, Grote (1967) studied the growth of tungsten single crystals at wire tips whilst Gardner and Cahn (1966) examined whisker growth on iron/aluminium alloys. Themelis and Gavin (1963, 1968) have used the SEM in their studies of reduction of iron oxides in smelting technology.

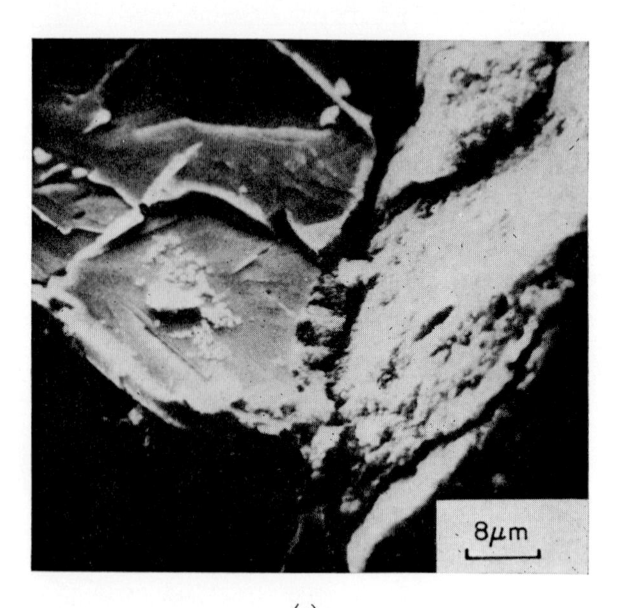

(a)

FIG. 6.13. The selection of an area for interface examination. This area was the only cleavage found to be suitable in the whole fracture cross-section.

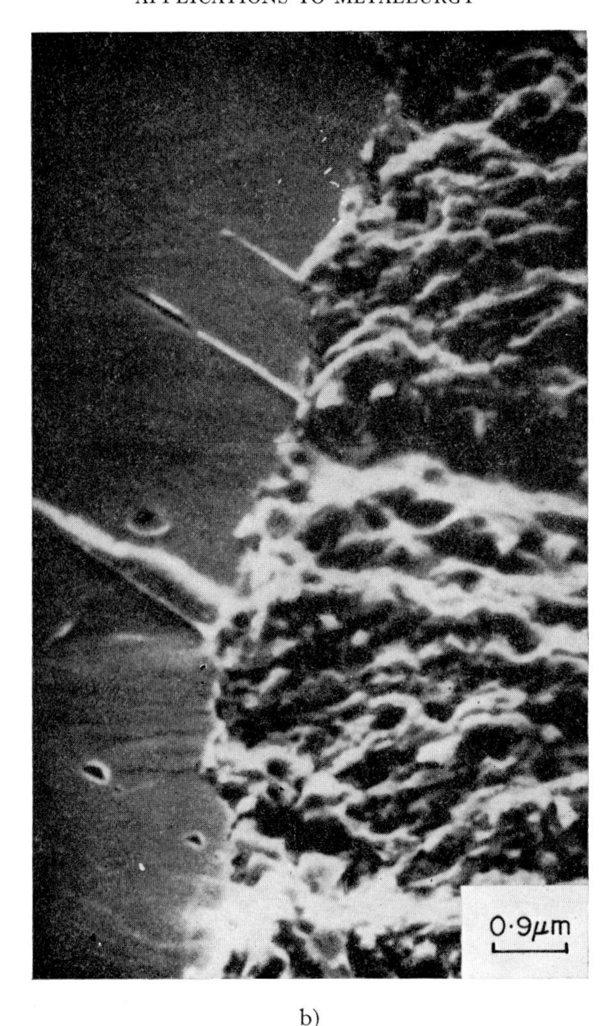

b)

FIG. 6.13 (*cont.*).

6.6. CONCLUSIONS

In this survey of the use of SEM in all areas of metallurgy we have found very few examples indeed where it has been used quantitatively. This clearly is not then its role. On the other hand, more than 100 authors in these related fields adopted scanning electron micrographs for illustration of their work in 1970 alone. Certainly SEM is unchallenged as a means of establishing author/reader rapport in any paper dealing with the complex formations of phases, grain structure, crystal habit, pores, layers or texture of any kind. Its use as an adjunct to concise description of metallurgical phenomena of all kinds will continue to grow. Its use as an adjunct to an analytical facility, whether it be X-ray, Auger, melting point or ion stripping mass-spectrometry, has hardly begun.

ACKNOWLEDGEMENTS

I wish to thank all those who provided examples of their work for this chapter: their names and affiliations are mentioned in the captions to the figures. Without their help examples could not have been drawn from such a wide area of metallurgy. Thanks are also extended to the director of the Central Electricity & Research Laboratories for permission to publish micrographs taken in the course of my research at that establishment, and to Dr. P. Goodhew of the Structural Studies Unit, University of Surrey, for reading and commenting on the draft of this chapter.

REFERENCES

ARROWSMITH, D. J. (1970) *Trans. Inst. Metal Finishing* **48,** 88.
ASBURY, F. F. and BAKER, C. (1967) *Metals and Materials,* 323.
ASHWORTH, V., BODEN, P. J., LEACH, J. S. LI and NEHRU, A. Y. (1970) *Corrosion Sci.* **10,** 481.
ASTM Special Technical Publication No. 430 (1966) *50 Years of Progress in Metallographic Technique.*
BAKER, H. R., BLOOM, M. C., BOLSTER, R. N. and SINGLETERRY, C. R. (1970) *Corrosion* **26,** 420.
BARLOW, R. and GRUNDY, P. G. (1970) *J. Mat. Sci.* **5,** 1005.
BASTON, T. J. and BOWDEN, F. P. (1968) *Nature* **218,** 150.
BAZZARRE, D. F. and PETRIELLO, J. (1970) *Plating,* 1025.
BEACHAM, C. D. (Ed.) (1967) *Electron Fractography,* ASTM Spec. Tech. Pub. No. 436.
BEACHAM, C. D. (1968) in *Fracture: An Advanced Treatise* (ed. H. Liebowitz), **1,** p. 243, Academic Press, New York.
BHATTACHARYYA, S., HOWES, M. A. and JOHARI, O. (1970) *J. Metals* **22,** 49.
BHATTACHARYYA, S. and PARIKH, N. (1970) *Met. Trans.* **1,** 1437.
BJERLE, I. (1970) *A.F.S. Cast Metals Res. J.* **6,** 114.
BRANMER, E. S. and DEWEY, M. A. P. (1966) in *Specimen Preparations for Metallography,* American Edn., Elsevier, New York.
BREINAN, E. M. and KREIDER, K. G. (1970) *Met. Trans.* **1,** 93.
BOWDEN, F. P. and CHILDS, T. H. C. (1968) *Nature* **219,** 1333.
CASTLE, J. E. and MASTERSON, H. G. (1965) *VGB-Speisewassertagung,* p. 7.
CASTLE, J. E. and MASTERSON, H. G. (1966a) *Anticorrosion* **1** (Dec.), 3.
CASTLE, J. E. and MASTERSON, H. G. (1966b) *Corrosion Sci.* **6,** 93.
CASTLE, J. E. and WOOD, C. G. (1968) in *Scanning Electron Microscopy/1968* (ed. O. Johari), IIT Research Institute, Chicago, p. 39.
CASTLE, J.E. (1971) in *Methods of Measurement in Corrosion and Oxidation* (ed. Leach, J. S. Li), Van Nostrand (to be published).
CHAKRAVALI, B., STARKE, E. A. and LEFERNE B. G. (1970) *J. Mat. Sci.* **5,** 394.
CHANCE, D. A. (1970) *Met. Trans.* **1,** 685.
CLARKE, D. R., BREAKWELL, P. R. and SIMMS, G. D. (1970) *J. Mat. Sci.* **5,** 875.
COLE, M., BUCKLOW, I. A. and GRIGSON, C. W. B. (1961) *Brit. J. Appl. Phys.* **12,** 296.
COOKSEY, G. L. and CAMPBELL, H. S. (1970) *Trans. Inst. Met. Finishing* **48,** 93.
COX, B. (1970) *J. Electrochem. Soc.* **117,** 635.
CRAIG, JR., G. E., HARR, R. E., HENRY, J. and TURNER, P. (1970) *J. Electrochem. Soc.* **117,** 1450.
DALAL, H. and GRANT, N. J. (1970) *Met. Trans.* **1,** 536.
DAY, M. G. and HELLAWELL, A. (1967) *J. Inst. Met.* **95,** 377.
DAY, M. G. (1969) *J. Metals.* .
DENNIS, J. K. and FUGGLE, J. J. (1970) *Trans. Inst. Met. Finishing* **48,** 75.
DREW, C. M., KNIPE, R. H. and GORDON, A. S. (1966) *Pyrodynamics* **4,** 325.
EHMAN, M. F. and FAUST, J. W. (1970) *J. Appl. Phys.* **41,** 3170.
EVANS, A. G., GILLING, D. and DAVIDGE, R. W. (1970) *J. Mat. Sci.* **5,** 187.
FAIRING, J. D. (1967) *J. Compos. Mat.* **1,** 208.

FIELD, E. M. and HOLMES, D. R. (1965) *Corrosion Sci.* **5,** 361.
FRIGGENS, H. and HOLMES, D. R. (1968) *Corrosion Sci.* **8,** 871.
FUKUBAYASHI, H., KENWORTHY, H., HIGLEY, L. W. and O'KEEFE, T. F. (1969) in *Scanning Electron Microscopy of 1969* (ed. O. Johari), IIT Research Institute, Chicago, p. 285.
FULMER RESEARCH INSTITUTE (1969) *Composites* **1,** 12.
FYFE, D., SHANAHAN, C. E. A. and SHREIR, L. L. (1970) *Corrosion Sci.* **10,** 817.
GARDNER, G. A. and CAHN, R. W. (1966) *J. Mat. Sci.* **1,** 211.
GERDES, R. J., CHAPMAN, A. T. and CLARK, G. W. (1970) *Science* **167** (Feb.).
GITTINS, A. (1970) *J. Mat. Sci.* **5,** 223.
GROTE, K. H. (1967) *Naturwissenschaften* **21,** 558.
GULBRAUSEN, E. A. and COPAN, T. P. (1960) *Nature* **186,** 959.
HARRISON, P. L., HOLMES, D. R. and TEARE, P. (1965) *VGB-Speisewassertagung*, p. 1.
HARSTON, J. D. and SCULLY, J. C. (1970) *Corrosion* **26,** 387.
HEIDERSBACH, R. H. and VERINK, E. D. (1970) (a) *Corrosion* **26,** 217; (b) *ibid.* 256; (c) *ibid.* 445.
HEYES, A. D. and LANGDON, T. G. (1969) *Nature* **221,** 168.
HOWES, V. R. (1967) *Nature* **216,** 362.
HOWES, V. R. (1970) *Corrosion Sci.* **10,** 99.
HUGHES, A. N., ORMAN, S. and PICTON, C. (1970) *Corrosion Sci.* **10,** 239.
HUNT, M. R. and CASTLE, J. E. (1970) CEGB Report No. RD/L/N 125–70.
INCKLE, A. (1970) *J. Mat. Sci.* **5,** 86.
JACKSON, P. W. and MARJORAM, J. R. (1970) *J. Mat. Sci.* **5,** 9.
JOHARI, O. (1968a) in *Scanning Electron Microscopy/1968* (ed. O. Johari), IIT Research Institute, Chicago, p. 79.
JOHARI, O. (1968b) *J. Metals*, **20,** 26.
JOHARI, O. and BHATTACHARYYA, S. (1968) *Proc. Inst. Conf. on Powder Tech.* ASTM/IITR, Chicago.
JOHARI, O., CORVIN, I., DRAGEN, R. and PARIKH, N. M. (1969) in *Scanning Electron Microscopy/1969* (ed. O. Johari), IIT Research Institute, Chicago, p. 277.
JONES, R. L., STRATTEN, L. W. and OSGOOD, E. D. (1970) *Corrosion* **26,** 399.
JOSHI, A. and STERN, D. F. (1970) *Met. Trans.* **1,** 2543.
KANAI, V. and UCHIBORI, K. (1969) in *Scanning Electron Microscopy/1969* (ed. O. Johari), IIT Research Institute, Chicago, p. 321.
KATZ, G. (1970) *J. Mat. Sci.* **5,** 736.
KLEIN, A. J., HITCHLER, E. W. and LAWSON, G. J. (1969) in *Scanning Electron Microscopy/1969* (ed. O. Johari), IIT Research Institute, Chicago, p. 343.
KOTVAL, P. S. (1969) in *Scanning Electron Microscopy/1969* (ed. O. Johari), IIT Research Institute, Chicago, p. 309.
KOUL, M. K. and BREEDIS, J. F. (1970) *Met. Trans.* **1,** 1451.
KURODA, Y. and TAKADA, H. (1970) *A.F.S. Cast Metal Res. J.* **6,** 63.
LARDNER, E. (1970) *Powder Met.* **138,** 394.
LEE, D. (1970) *Met. Trans.* **1,** 309.
LIFSHIN, E., MORRIS, W. G. and BELON, R. B. (1969) *J. Metals*, 43.
MCGRATH, J. R., BUCHANAN, J. G. and THURSTON, R. C. A. (1962) *J. Inst. Metals* **91,** 34.
MANSFIELD, F. and GILMAN, S. (1970a) *J. Electrochem. Soc.* **117,** 1154.
MANSFIELD, F. and GILMAN, S. (1970b) *J. Electrochem. Soc.* **117,** 1521.
MARANKOWSKI, M. J. and LARSEN, J. (1970) *Met. Trans.* **1,** 1034.
MAREK, M. and HOCHMAN, R. F. (1970) *Corrosion* **26,** 5.
MENY, L., MAILLARD, A., HENRI, G., CHAMPIGNY, M. and LEGRANDE, J. (1970) *Metaux Corr Ind.* **45,** 343.
MENZIES, I. A. and ALDREAD, P. (1969) *J. Electrochem. Soc.* **116,** 1414.
MILGRAM, A. A. (1970) *Met. Trans.* **1,** 695.
MINKOFF, I. and NIXON, W. C. (1966) *J. Appl. Phys.* **37,** 4848.
MURR, L. E. (1970) *J. Mat. Sci.* **5,** 534.
MURRELL, D. L. and ENOCH, R. D. (1970) *J. Mat. Sci.* **5,** 478.
NAYBOUR, R. D. (1969) *J. Electrochem. Soc.* **116,** 520.

Okada, H., Hosoi, Y. and Abi, A. (1970) *Corrosion* **26,** 185.
Orman, S. (1969) *Corrosion Sci.* **9,** 849.
Parker, M. T. and Morrison, J. D. (1969) in *Scanning Electron Microscopy/1969* (ed. O. Johari), IIT Research Institute, Chicago, p. 331.
Pease, R. F. W. and Ploc, A. (1965) *Trans. AIME* **233,** 1949.
Perry, A. J., de la Motte, E. and Phillips, K. (1970) *J. Mat. Sci.* **5,** 945.
Phelps, E. H. and Kemp, M. E. (1970) *Metals Eng. Quart.* **10,** 24.
Potter, E. C. and Mann, G. M. W. (1961) *Proc. 1st Int. Cong. Met. Corrosion,* Butterworth, London, p. 417.
Prentice, J. L. (1970) *J. Electrochem. Soc.* **117,** 385.
Rawlins, T. G. R. (1970) *J. Met. Sci.* **5,** 881.
Reid, F. H. (1970) *Trans. Inst. Metal Finishing* **48,** 115.
Rege, R. A., Szeheres, E. S. and Forceng, J. R. W. D. (1970) *Met. Trans.* **1,** 2652.
Rellick, J. R. and McMahon, C. J. (1970) *Met. Trans.* **1,** 929.
Richardson, J. A. and Wood, C. G. (1970) *Corrosion Sci.* **10,** 297.
Robinson, D. J. and Gabe, D. R. (1970) *Trans. Inst. Met. Finishing* **48,** 35.
Scully, J. C. (1970) in *Scanning Electron Microscopy/1970* (ed. O. Johari), IIT Research Institute, Chicago, p. 313.
Scully, J. C. and Powell, D. T. (1970) *Corrosion Sci.* **10,** 719.
Sharpe, W. B. A. (1970) *Corrosion Sci.* **10,** 283.
Stavros, A. J. and Paxton, H. W. (1970) *Met. Trans.* **1,** 3049.
Stock, T. A. C. and Thompson, K. R. L. (1970) *Met. Trans.* **1,** 219.
Surman, P. L. and Castle, J. E. (1969), *Corrosion Sci.* **9,** 771.
Taplin, D. M. R. (1970) *Metals Eng. Quart.* **10,** 31.
Themelis, N. T. and Gavin, W. H. (1963) *Trans. Met. Soc. AIME* **227,** 290.
Themelis, N. T. and Gavin, W. H. (1968) *A.I.Ch.E.J.* **8,** 437.
Thompson, E. R. and Lemkey, F. D. (1970) *Met. Trans.* **1,** 2799.
Tipper, C. F., Dagg, D. I. and Wells, O. C. (1959) *J. Iron and Steel Inst.* **193,** 133.
Toy, G. M. and Phillips, A. (1970) *Corrosion* **26,** 200.
Tucker, R. C. and Meyerhoff, R. W. (1969) in *Scanning Electron Microscopy/1969* (ed. O. Johari), IIT Research Institute, Chicago, p. 389.
Ugianski, G. M., Skolnich, L. P., Kruger, J. and Stiefel, S. W. (1968) *Nature* **218,** 1156.
Vennett, R. M., Wolf, S. M. and Levitt, A. P. (1970) *Met. Trans.* **1,** 1569.
Verink, E. D. and Parrish, P. A. (1970) *Corrosion* **26,** 214.
Walker, J. L. and Clive, H. E. (1970) *Met. Trans.* **1,** 1221.
Whitehead, R. S. and Noble, F. W. (1970) *J. Mat. Sci.* **5,** 856.
Wilson, D. V. (1970) *J. Inst. Met.* **98,** 133.
Wright, R. N. and Argon, A. S. (1970) *Met. Trans.* **1,** 3065.

CHAPTER 7

Applications to Fibres and Polymers

J. T. SPARROW

7.1. INTRODUCTION

Before scanning electron microscopy the methods of looking at surfaces were restricted to light microscopy and transmission or reflection electron microscopy. This left a magnification gap between 500 × and 5000 ×, which is a very significant range for polymers and fibres. In addition the low depth of field for optical microscopes is a severe disadvantage in the study of fabrics and fibres since the information sought is rarely on a planar surface and a sequence of micrographs taken at various levels of focus is difficult to correlate. Similarly there are restrictions on the full use of the transmission electron microscope (TEM), since both the non-planar nature of the specimens and their fragile construction makes either thin sectioning or replication very difficult. Nevertheless, much information about the external structure of bulk polymers and fibres was gained using optical microscopy and transmission electron microscopy but there was often difficulty in relating the two sets of information because the area viewed in the TEM could not always be identified on the optical micrograph. All these difficulties are resolved with the SEM.

As early as 1958 Sikorski (1958) realized the usefulness of the SEM for the study of polymers and fibres. With the SEM understandable information is yielded immediately, and since the area selected for higher magnification is always at the centre of the low magnification image it is easily identified. A set of related images, with very rarely any part out of focus, is obtainable from 20 × to 50,000 × magnification and can usually be tied into optical and transmission electron micrographs of the same areas. This has meant whole new fields of study have been opened, in particular, studies of bulk polymers and fibres in fracture, fibres damaged by wear and within fabrics. Of course, internal molecular orientation cannot be studied in an SEM as it can with a polarizing microscope, nor the external infrastructure as with a TEM but a useful bridge has been established. A more serious disadvantage is that polymers are not dense nor good electrical conductors, so that techniques of coating and low beam voltage have to be used to reduce beam penetration and charging. There are also some technical difficulties associated with merely handling and mounting specimens and it is worth while offering some solutions to specific problems which may occur.

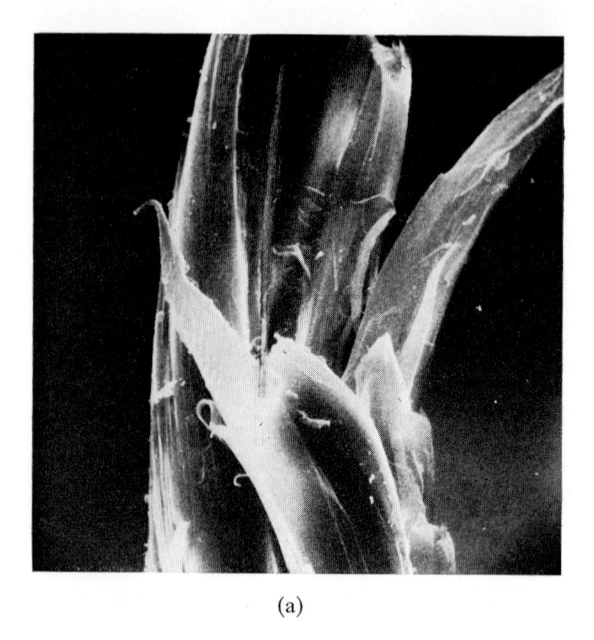

(a)

(b)

FIG. 7.1. Twist break of an Acrilan fibre taken at (a) 20 kV accelerating potential, (b) 5 kV. Increased detail can clearly be seen by using a low accelerating potential. (1000 ×)

7.2. OPERATING CONDITIONS FOR POLYMERS AND FIBRES

It is well known (Boyde and Wood, 1969; Hearle *et al.*, 1970) that more surface detail is revealed in the SEM by using low accelerating potentials. The loss of detail at high voltages is particularly common with materials of low density since the extreme beam penetration resulting from high accelerating potentials produces much unwanted signal coming from the interior of the specimen causing reduced contrast and obscured detail. All polymers and fibres fall into the low density category of materials, and therefore it is recommended that low (around 5 kV) operating voltages be used when examining these types of materials. The advantage of using this low voltage is clearly illustrated in Fig. 7.1 where a twist break of an Acrilan fibre is examined at 20 kV and 5 kV. It is particularly important to use low beam voltages when examining fracture surfaces of this type since all detail must be recorded in order that the fracture process can be interpreted correctly.

The major disadvantage resulting from low beam energy work is a loss of resolution. However, resolution of the order of 50 nm (500 Å) can be achieved with proper operation of the instrument. This means using higher lens currents, a short working distance, careful corrections for astigmatism and a clean instrument. The use of higher lens currents decreases the signal to noise ratio, and forces the operator to make exposures longer than 20 seconds. Some operators have found that if noise is a problem when focusing it is advantageous to make a series of micrographs with the focus altered slightly on each one. Usually one of the micrographs will be in focus.

7.3. COATING MATERIALS

The problems associated with conductive coatings have been discussed in Chapter 4. It will suffice to say here that silver is the easiest metal coating to use for polymers and fibres, but if specimens are to be kept for very long periods of time gold or gold/palladium alloy is recommended. When operating at a low beam voltage, the problems of charging are not as great and a minimum amount of metal is required on the surface of the specimen to suppress charging effects.

Sikorski *et al.* (1968) have recommended the use of an antistatic (Duron) which can be sprayed directly from an aerosol container onto the specimen, or the specimen can be soaked or dipped in the antistatic. Dilute solutions of Duron in isopropyl alcohol can be used and may prove to be more useful than the concentrated solution which comes direct from the container. The advantage of using a dilute solution is that one is more likely to get a monomolecular layer over the surface of the specimen. Using Duron as a spray is particularly useful because it reduces preparation time, although care must be taken that the coating is not too thick or droplets will form, but these are easily recognized in the SEM. It is not recommended to use antistatics in a concentrated form if magnifications greater than 2000 times are required. The use of antistatics plays a very important part when carrying out dynamic experiments in the SEM as will be discussed in Section 7.7.4.

7.4. TEXTILE FIBRES

7.4.1. Introduction

Microscopy has always played an important role in the study of textile materials, because of the fineness of individual fibres. Cotton fibres, for example, are only 15–20 micrometres in diameter and therefore they must be magnified many times in order that surface features and other important characteristics may be examined. Scanning electron microscopy has opened to the textile researcher new fields of investigation which will help to improve fibre and fabric life. Due to the great depth of focus of the SEM more information can be obtained about the nature of fibre fracture, fabric wear, chemical degradation, abrasion and fatigue, and many other aspects of fibre technology.

The mounting and preparation of fibre samples is considerably simpler for the SEM than for other microscopes, and also the amount of correlated, useful information gained is greater. Fibres, even up to bristle size, can be mounted whole in the SEM and in any configuration; by careful mounting and use of the movement of the microscope stage all the external features of a fibre may be studied on one mounted specimen. This has turned out, for instance, to be of particular interest in the study of fibre fracture where the condition of the sides of the fibre a good way from the fracture surface can sometimes add useful information about the strain distribution within the fibre. Due to the very large surface area to volume ratio it is more important than usual to fix the position of small details on fibres and this is easy with the SEM but very complex, and sometimes impossible, with other microscopes. By having SEM images of the fibre at various magnifications it is much easier to correlate the extra information supplied by the optical microscope and the transmission electron microscope.

The mounting techniques for fibres, yarns and fabrics can vary depending upon the nature of the particular study to be undertaken. It will be the object of this section to describe techniques for the preparation of textile materials which will help the investigator to get the most out of his instrument.

7.4.2. Surface studies

A. Hepworth *et al.* (1969b) and Anderson *et al.* (1970, 1971) have shown the usefulness of the SEM for studying the surface topography of chemically treated wool fibres. When examination of surface detail is important the fibre can simply be placed on the specimen holder with the ends attached to the holder with a suitable adhesive (Durofix). This is the easiest mounting technique but it may not provide suitable conduction

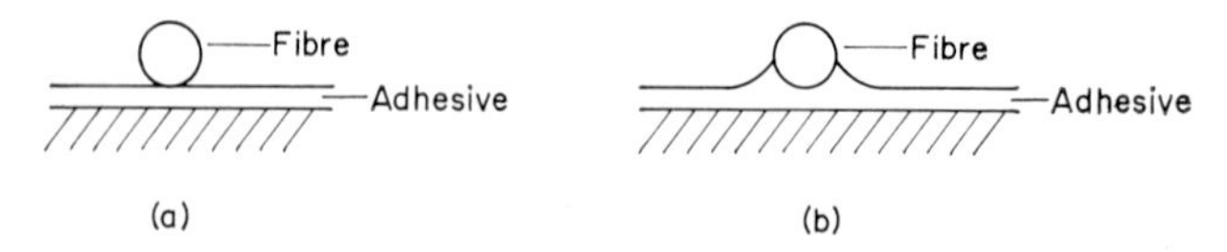

FIG. 7.2. Method of mounting a fibre on the specimen holder. (a) Incorrect. (b) More desirable since metal coating will be continuous from the fibre to the holder.

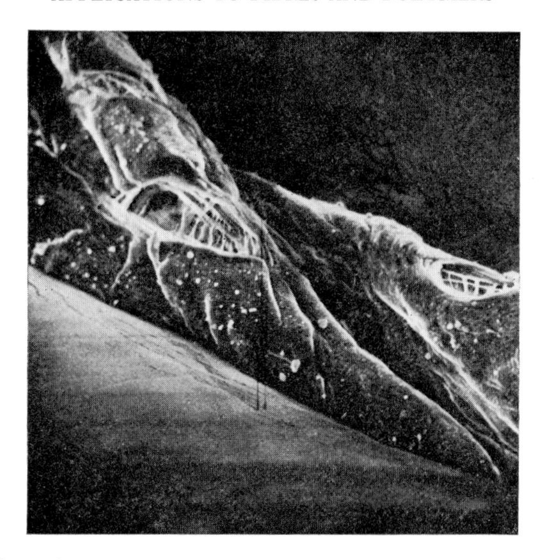

FIG. 7.3. Surface of a fatigued cotton fibre examined after mounting with double-sided adhesive tape. (1700 ×).

and charging may result. Another simple technique is to spread a thin layer of adhesive on the holder and then lay the fibre onto this. It should be noted that better conduction will result if the adhesive is part way up on the sides of the fibre as shown in Fig. 7.2. With the adhesive up the side of the fibre one is sure that there is complete conduction from the fibre to the holder. Another mounting technique which has proven to be successful is to cover the holder with a double-sided adhesive tape (Twinstik) cut to the holder shape. The fibre can now be laid on the tape and is ready for coating in the appropriate manner. Figure 7.3 is a micrograph of the surface of a fatigued cotton fibre which was mounted using the double-sided tape method.

More sophisticated mounting techniques are required if the entire fibre periphery needs to be examined, since by the above methods almost one-half of the fibre surface is covered by the adhesive or is facing the surface of the holder. Sikorski *et al.* (1969) have described a number of specimen holders which can be used for examining the

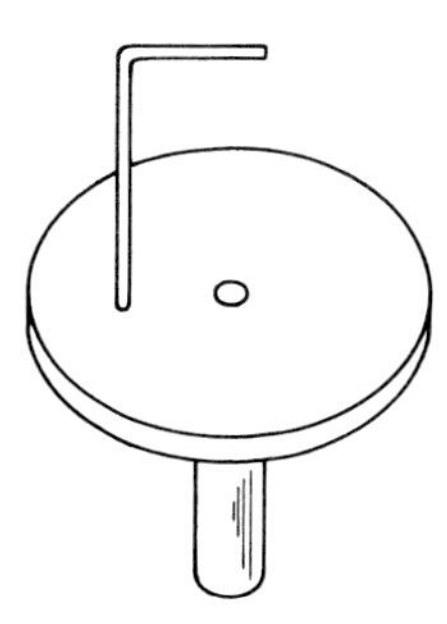

FIG. 7.4. Specimen holder for examining fibre periphery (after Sikorski *et al.*, 1969).

fibre periphery. The simplest is shown in Fig. 7.4. The fibre is threaded through a central hole drilled in the base of the holder and fixed with an adhesive. The other end is attached to the horizontal part of the wire. After coating, the holder is inserted into the specimen stage and tilted 90° to the electron beam. By using the rotation control almost the entire fibre surface can be examined. The amount of image obstruction is negligible since the wire is of small diameter (about 0·8 mm). McKee and Beattie (1970) have also described a more complex device which is a rotation stage that can be used for circumferential examination of fibres.

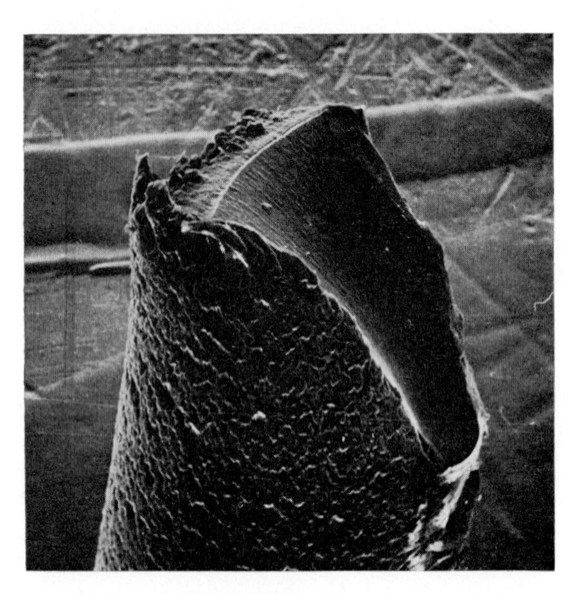

FIG. 7.5. Tensile fracture of nylon bristle. Mounted directly on the holder in a verticle position with Durofix. (70×)

7.4.3. Fracture studies

The SEM, due to its large depth of field has proved extremely useful for studying the fractography of different fibre types as shown by Hearle and Cross (1970), Hearle and Vaughn (1970), Hearle and Sparrow (1971). In this case the mounting techniques are somewhat different from those required for surface studies. The fibre must be held in an upright position so that the fracture surface can be examined from all directions. If the fibre is of sufficiently large diameter it can be cut near the fracture, and by holding it with tweezers the fractured part can be placed in an adhesive in a vertical position on the holder. Figure 7.5 illustrates the tensile fracture of a nylon bristle which was mounted in this manner. With proper use of the tilt and rotation controls the fracture can be examined from all useful directions.

The above mounting technique is suitable for large-diameter fibres which can be handled easily with tweezers. If one is interested in examining the fracture of much smaller fibres, such as cotton or man-made fibres of textile deniers, another technique must be adopted. Again the fibre must be held in an upright position and it is often advantageous to have several fibres mounted on a single specimen holder in order to save time in coating and pump-down time on the instrument. These requirements are achieved by using a simple mounting technique with Sellotape and a special holder. It proves to be extremely successful especially when one is concerned with statistical evaluation on large numbers of fibres.

With the aid of a binocular microscope (preferably a stereozoom binocular microscope) in the low-power range (5 × to 45 ×), the fibres to be examined are placed with tweezers on to the adhesive side of a piece of Sellotape so that their ends are 1 mm or

FIG. 7.6. Specimen holder used for holding fibre/Sellotape arrangement.

less above the edge of the tape. After the desired number have been placed on the tape, it is folded so that the fibres are held between the two adhesive sides. The edge of the tape should be pressed with tweezers so that the adhesive is making good contact with the fibres, thus giving good electrical contact after metal coating. The ends of the fibres protruding above the edge of the tape should be no more than 1 mm long since with a greater length there is more of a tendency for them to move in the electron beam during examination. The fold in the tape can conveniently be used as a reference point so that the order of the fibres will be known. After the tape is folded the part holding the fibres is then cut out so that it can be fitted into the holder ready for coating. The tape should be cut so that once it is fitted into the holder its edge does not protrude too far above the holder surface. The holder mentioned here will be described later.

Fibres mounted between pieces of Sellotape in this manner can be conveniently handled with tweezers, whereas a single fibre is very difficult to manipulate into a specimen holder. No problems are encountered using this mounting technique other than the occasional fibre which may protrude too far above the edge of the tape and thus move in the electron beam.

The specimen holder which is used to hold the fibre/Sellotape arrangement just described is shown in Fig. 7.6. It is simply a specimen holder of normal size for the Stereoscan with the exception of the base which is twice the normal thickness. With a thickened base a step can be cut out to half its thickness. An elliptical hole is drilled

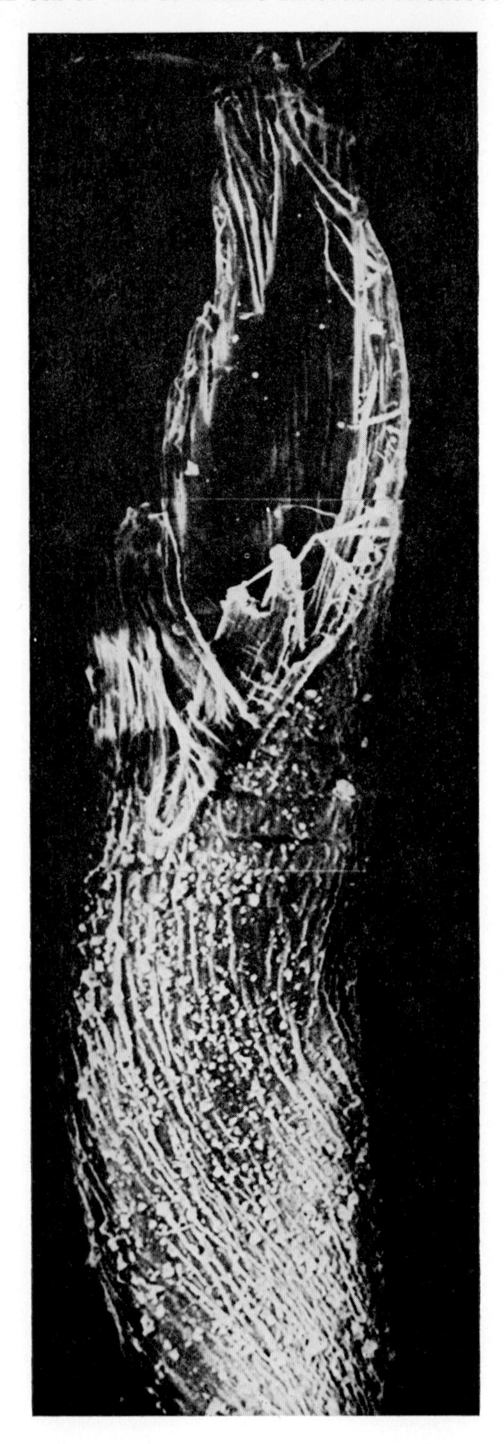

FIG. 7.7. Tensile fracture of a cotton fibre mounted with Sellotape. (1500×)

in the cut-out portion and the assembly fitted with a screw. We now have a holder with a slot down the middle that will, with some variation in construction, accommodate specimens of varying thicknesses and shapes. The fibre samples which are mounted on the Sellotape can be fitted into the holder slot easily with a pair of tweezers. With the mounted fibres in the slot, the movable portion can be pressed tightly onto the rest of the holder and the screw tightened. Figure 7.7 illustrates a fractured end of a cotton fibre which was examined using the above mounting technique. The photograph was made by taking overlapping micrographs and fitting them together after printing. A

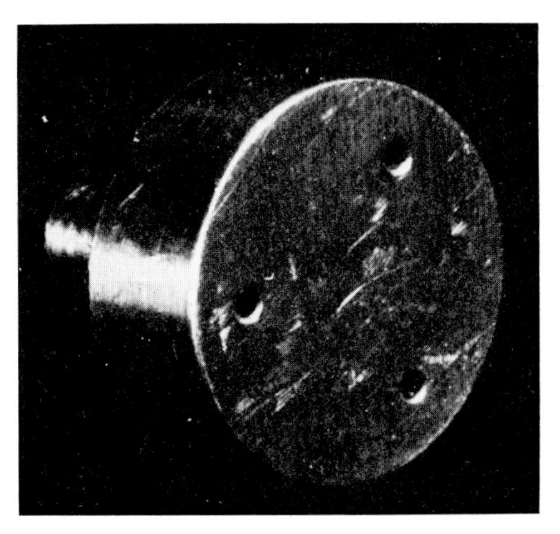

FIG. 7.8. Specimen holder used for making and examining fibre and yarn cross-sections.

scan rotation unit is sometimes useful when making montages since by using it the direction of the scan in relation to the specimen can be changed enabling the operator to always have the fibre in a vertical or horizontal position on the screen.

7.4.4. Studies of cross-sections

In some instances it may be necessary to examine cross-sections of fibres or yarns using the SEM. The fibre sections can be prepared similar to the method described by Ford and Simmens (1958) and with a specially prepared specimen holder one can prepare the cross-section directly on the holder. Small holes are drilled around the outside edge of a standard holder as shown in Fig. 7.8. If these holes are drilled in a circle concentric with the centre of the holder, each cross-section when prepared can be brought into view with a simple movement of the rotation control on the specimen stage. The cross-sections are made by pulling a bundle of fibres through one of the holes and sectioning with a sharp razor blade by moving it across the surface of the holder.

The fibres are inserted in the holes in the following way. A thread is looped through one of the holes and a tuft of fibre is then placed through the loop. The bundle must be large enough so that the fibres are held tightly when pulled into the hole by the thread. After the desired number of holes are filled, the fibres are sectioned by a sharp razor

FIG. 7.9. Method used for making fibre and yarn cross-sections.

blade and are ready for coating after cutting the fibres at the bottom surface of the holder. The process is illustrated in Fig. 7.9. Figure 7.10 illustrates the use of this sectioning method for determining the distribution of sizing material between the fibres of a wool yarn.

Normally, unmodified fibre cross-sections when examined in the SEM simply reveal a record of the blade cut, which would be very smooth under ideal conditions. However, if some components have been removed or separated by chemical treatment, mechanical processing or enzyme digestion, then details are revealed in the cross-sections. Anderson and Lipson (1970) have shown that some histological components of chemically treated wool fibres are visible by scanning electron microscopy when cross-sections are examined.

FIG. 7.10. Part of a cross-section of wool yarn illustrating how sectioning can be used for determining the distribution of sizing material within the yarn. (385 ×)

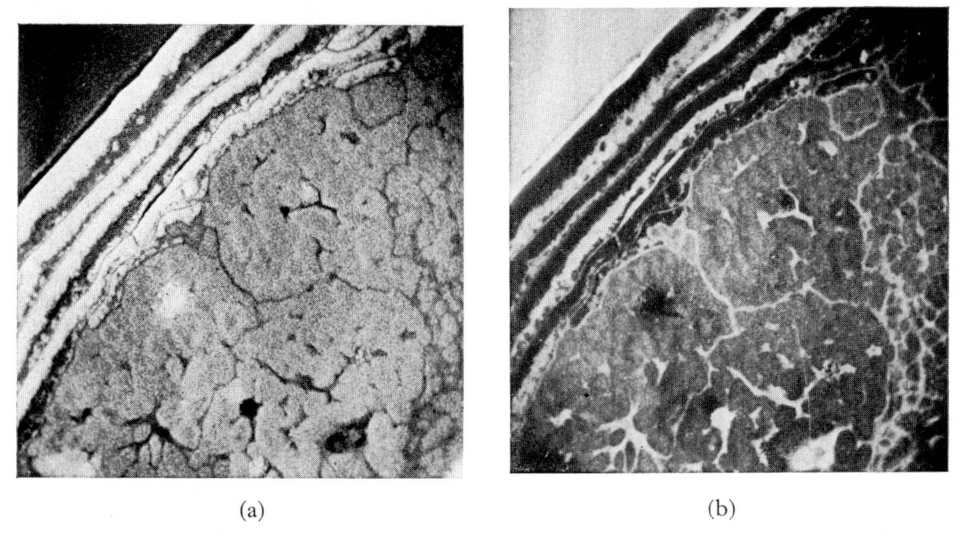

(a) (b)

FIG. 7.11. (a) Micrograph obtained by secondary electron emission of a 1600 Å section of human hair stained with silver (6600 ×). (b) Same section as in (a) but by transmission scanning electron microscopy (6600 ×). (Brown and Swift, 1970, Swift 1971.) (Compliments of J. A. Swift, Unilever Research Laboratory, Isleworth, Middlesex.)

Brown and Swift (1970) and Swift (1969, 1971) have successfully stained keratin fibre cross-sections with Gomori's silver-methenamine reagent and been able to produce sufficient image contrast with the flat fibre sections to enable them to identify many of the histological components of the fibres (see Fig. 7.11 (a)). The image contrast observed here was attributed to the differential deposition of silver in the

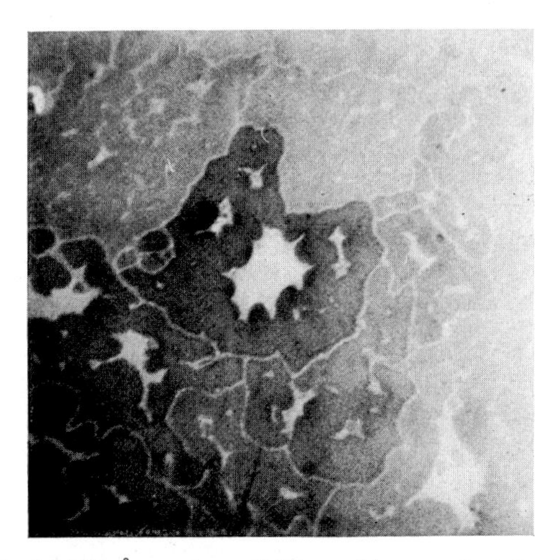

FIG. 7.12. TSEM of 1600 Å section of Shropshire wool stained with silver and showing ortho- and para-cortex (7200 ×) (Swift, 1972a). (Compliments of J. A. Swift, Unilever Research Laboratory, Isleworth, Middlesex.)

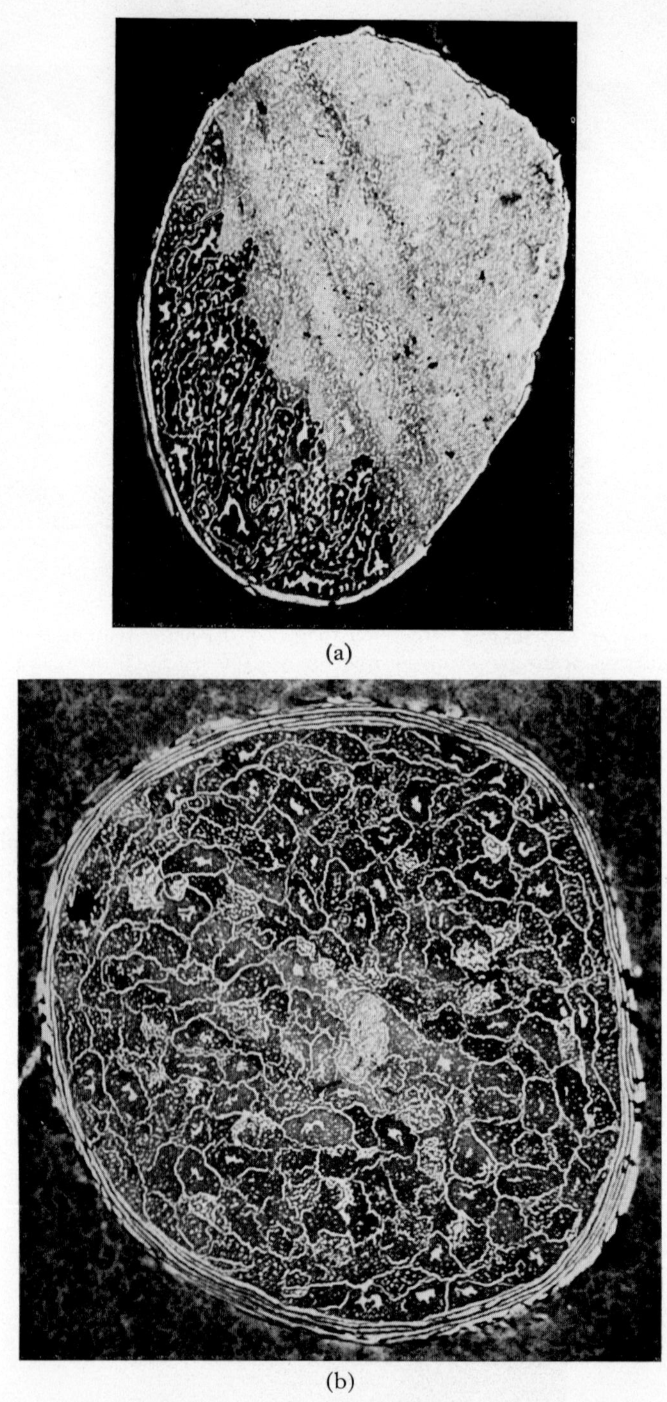

(a)

(b)

FIG. 7.13. (a) Transmission scanning electron micrograph of a complete 1600 Å section of Shropshire wool digested with papain-bisulphite and with *no* metal staining (1620×). (b) Same as (a) but of Yak hair (1200×) (Swift, 1972b). (Compliments of J. A. Swift, Unilever Research Laboratory, Isleworth, Middlesex.)

various histological components. This technique or a similar one could prove to be useful for examining cross-sections of other natural fibres with the SEM.

Brown and Swift (1970) and Swift (1972a, b) have also used the SEM in the transmissive mode to examine cross-sections of keratin fibre (see Figs. 7.11 (b) and 7.12). As was pointed out, using the SEM in transmission has advantages in some circumstances over direct transmission electron microscopy. With the transmission scanning electron microscope (TSEM) one can obtain a single micrograph of an entire fibre cross-section (see Fig. 7.13), which is not always possible with some TEMs when working with large

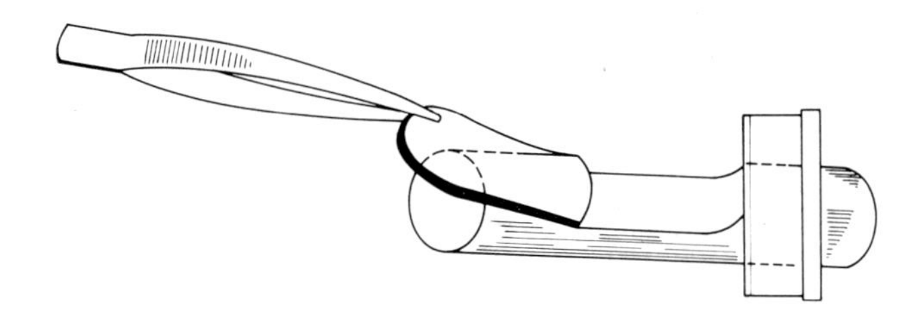

FIG. 7.14. Illustration of peeling technique (after Scott, 1959).

diameter fibres since the minimum magnification of the TEM is not low enough to do this. Another disadvantage of using the TEM is that relatively thin sections (up to 100 nm thick) are required while with the TSEM sections up to 1 μm thickness can be examined with a resolution of 50 nm (500 Å). Resolution for the TSEM can be as good as 6·5–7·0 nm (65–70 Å) with thinner sections. Figure 5.3 is a micrograph of a $\frac{1}{2}$-μm-thick section of human hair examined in transmission.

Although at present the literature on TSEM techniques is scarce, the next few years promise an increase in the use of transmission scanning electron microscopy for examining fibre cross-sections.

7.4.5. Studies of internal structure

Occasionally it may be necessary to study the internal structure of fibres for detection of defects in manufacturing, e.g. inclusions, or structural morphology. A method is needed whereby a fibre can be opened up without disrupting polymer morphology so that it can be examined in the SEM or replicated and examined in the TEM or TSEM. Scott (1959) has developed two types of splitting or peeling of fibres for examination in the TEM which reveal internal morphology of the fibre. This technique can be used with the SEM also. One type of splitting is called orientation splitting and the other cleavage plane splitting. For examination in the SEM we are mainly concerned with cleavage plane splitting which is illustrated in Fig. 7.14. In this technique the fibre is fixed to a glass slide or to the SEM holder and placed on the stage of a stereomicroscope. An oblique cut is made partway through the fibre with a razor blade. The cut polymer is pulled back parallel to the fibre axis with a pair of tweezers. The split will start down the fibre and work its way to the outside of the specimen where it will then propagate parallel to the fibre axis.

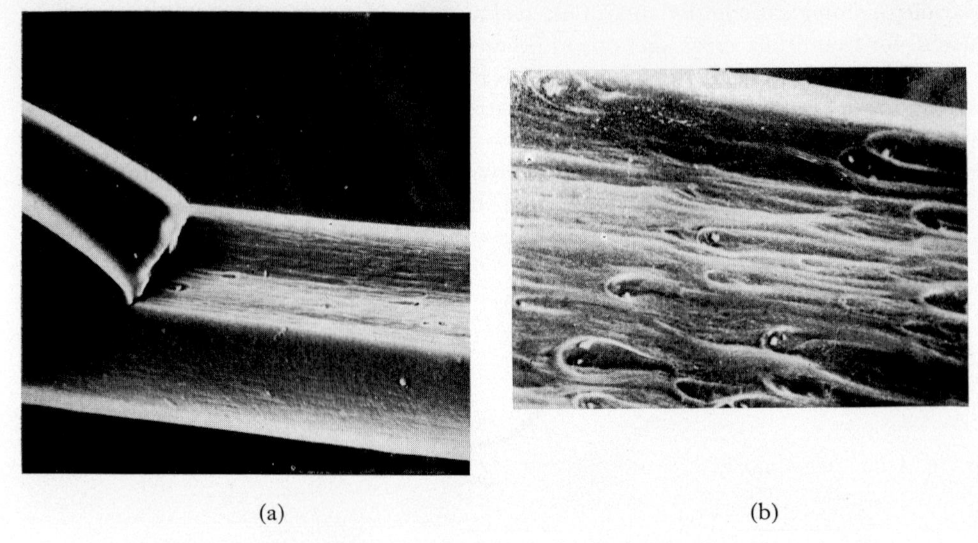

(a) (b)

FIG. 7.15. (a) A peeled nylon fibre (1300 ×). (b) Peeled surface of a nylon fibre at high magnification showing titanium dioxide inclusions (3240 ×). (Compliments of W. D. Emery, Unilever Research Laboratory, Port Sunlight, Wirral, Cheshire.)

FIG. 7.16. Wash/wear nylon sock illustrating the peeling behaviour of nylon (1640 ×). (Compliments of W. D. Emery, Unilever Research Laboratory, Port Sunlight, Wirral, Cheshire.)

For examination in the SEM the split fibre can be mounted by one of the methods described earlier if the splitting was not carried out on the holder. After examination it may be desirable to remove successive layers of the fibre and this can be done by repeating the same procedure. Van Veld *et al.* (1968) have used the technique with success to illustrate the fibrillar nature of some synthetic fibres and to reveal inclusions within the fibres themselves. It was found that each type of fibre had a characteristic peeling behaviour which could be affected by changes in fibre nature or production conditions. Figure 7.15 illustrates a nylon fibre which has been peeled to reveal inclusions within the fibre. It is interesting to note that during wear nylon fibres peel in layers as illustrated in Fig. 7.16. Haly *et al.* (1970) have used this peeling technique to examine the cortex of wool fibres and it may be a useful technique for examining other natural fibres as well.

7.5. TEXTILE YARNS AND FABRICS

7.5.1. Introduction

When studying textile materials it is particularly useful to examine fabrics for different types of wear, chemical attack, defects in processing, results of washing, drying, chemical finishing, etc. In this area the SEM again proves useful because of its great depth of field and the ease of sample preparation. Sikorski and Hepworth (1969) and Goynes (1969) have illustrated the importance of the SEM for studying fabric abrasion. The object of this section is to explain preparation techniques which are useful when studying fabrics in the SEM. The same mounting techniques that are employed for single fibres can be used when textile yarns are to be examined.

7.5.2. Preparation techniques

As with fibres and yarns the fabric may be placed on to the holder on a layer of liquid adhesive, though the adhesive may run in between the yarns or between individual fibres obscuring information which is present. On the other hand, the edges of the fabric may be attached to the holder with the adhesive but this may not provide sufficient electrical conduction. A mounting method like the one used for single fibres or yarns is successful for fabrics. The holder is covered with a double-sided adhesive tape and the fabric placed on this. The edges are trimmed to the dimensions of the holder and the specimen is ready for coating. It may be necessary to paint the edge of the fabric and holder with Aquadag to ensure that after coating there is sufficient conduction from the fabric to the specimen holder.

Occasionally when examining worn areas of fabric mounted with adhesive tape charging will develop along the worn edge as in Fig. 7.17. This can be suppressed or eliminated (see Fig. 7.18) by first coating the adhesive tape, while on the holder, with metal. The adhesive will still retain its holding properties so that the fabric can be placed on the metal-coated adhesive and then coated in the normal manner. By using this preparation technique individual fibres in the yarns can also be examined at high magnifications.

Cross-sections of fabrics can be examined by using the specimen holder described in Section 7.4.3. The fabric is sectioned with a sharp razor blade and the sectioned

FIG. 7.17. Worn cotton fabric mounted with uncoated Twinstik showing charging behaviour along the worn edge. (13×)

FIG. 7.18. Worn cotton fabric mounted on Twinstik that has been coated with silver. (40×)

area cut to an appropriate size to fit the holder, or the fabric can be fitted into the holder and sectioned there. It may be useful to paint with Aquadag the edges of the fabric which are in contact with the holder so that there is good conduction.

7.5.3. Use of X-ray microanalysis

X-ray microanalysis is useful at times when particular studies are being carried out on textile materials. Unilever Research Laboratories at Port Sunlight, Cheshire, have used X-ray analysis during wash/wear trials to determine the chemical nature of particulate matter remaining on fabrics after washing. Figure 7.19 illustrates material remaining on a yarn in a nylon fabric after washing. A particle was singled out on a fibre, as in Fig. 7.20, to determine its elemental composition by X-ray analysis. Analysis was carried out on the EDAX System (supplied by Nuclear Diodes, Inc.) which can be fitted to the SEM.

To carry out the analysis the scanning beam was switched to spot and the beam moved to the particle in Fig. 7.20. This is easily done since the image remains on the visual screen (due to the long-persistence phosphor) for a time long enough to position the beam directly on the particle. (It should be remembered that when the scanning beam is switched to spot for point analysis of particles the brightness on the record screen should be turned completely off, and should be lowered on the visual screen so that the phosphor on the screens will not be burnt.) The whole range of X-rays is collected and a spectrum shown on the display screen of the EDAX system (see Fig. 7.21). It was found that the particle contained elements which are components of clay.

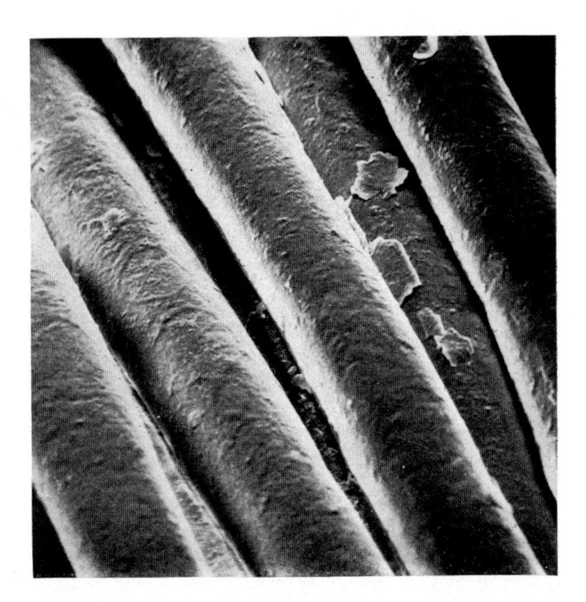

FIG. 7.19. Particulate matter remaining on a used nylon fabric (720 ×). (Compliments of W. D. Emery, Unilever Research Laboratory, Port Sunlight, Wirral, Cheshire.)

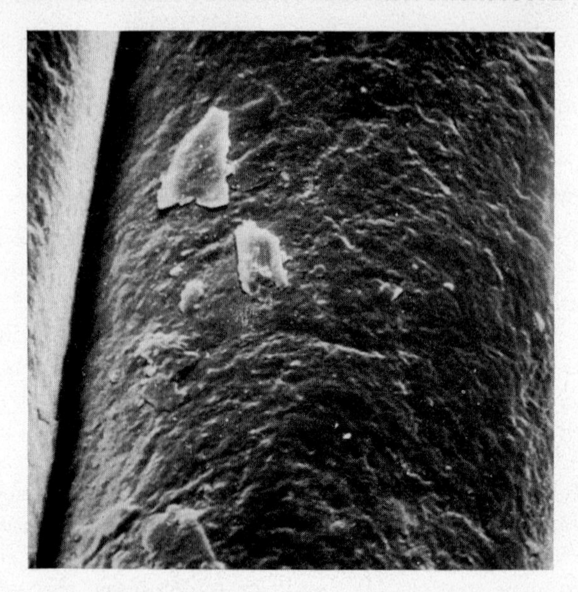

FIG. 7.20. Higher magnification micrograph of a particle on a nylon fibre. For X-ray analysis the scanning electron beam was switched to spot and positioned on the particle. (Compliments of W. D. Emery, Unilever Research Laboratory, Port Sunlight, Wirral, Cheshire.)

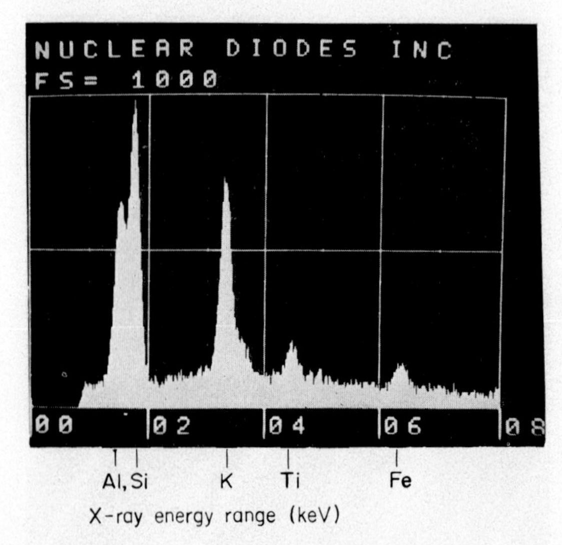

FIG. 7.21. Photograph of the display screen of the X-ray analyser showing the elements which are present in the particle in Fig. 7.20. The elements detected indicate that the particle is clay. (Compliments of W. D. Emery, Unilever Research Laboratory, Port Sunlight, Wirral, Cheshire.)

Coating of the fabric should be done with carbon as described in Chapter 4 since carbon cannot be detected by the analyser. Gold, silver and gold/palladium are detected by the system and may confuse matters if they are used for coating. High accelerating potentials must be used for the excitation of X-rays.

7.6. BULK POLYMERS

7.6.1. Introduction

Although the SEM does not have the resolving power to reveal morphological details of spherulitic crystallites in polymers it does prove to be useful for examination of fracture surfaces, gross morphological details, inclusions in thin films, etc. This section will describe some techniques which will be helpful if bulk polymers are to be examined. Princen (1971) has illustrated many applications of the SEM to the study of polymers.

7.6.2. Powders

Polymer powders can sometimes be difficult to examine due to charging effects, and because of their delicate nature they must be sprinkled onto a holder which is coated with an adhesive. The powders cannot be pressed into the adhesive or surface detail will be destroyed and if the viscosity of the adhesive is too high the powder will sink into it completely. Since the powder is sprinkled onto the holder good electrical conduction is not insured after coating and charging problems may result.

7.6.3. Large specimens

As has been mentioned earlier the standard SEM stage will accommodate specimens up to 1 cm^3 without having to remount the specimen for complete examination of the surface. If large pieces of polymer are to be examined they must be cut to an appropriate size for the specimen stage being used. After the polymer is cut, the piece to be examined is simply glued to the holder and is ready for coating. After coating, the sides and any area not to be examined should be painted with Aquadag, since, as was pointed out in Section 7.5.2, the coating material probably is not thick enough down the sides. If an almost perfect cube has been coated one should paint a very small area of the top surface and then carry the conducting paint down the sides to the holder.

7.6.4. Thin films

Thin polymer films can be mounted by sticking the film to the holder with a thin layer of Durofix or Aquadag and then painting the edges with Aquadag. It has been noted that if the film is not in good contact with the holder, i.e. a bubble in the adhesive between the film and holder, charging will result in the area over the bubble. Precautions should be taken to ensure that this does not happen. Mounting with double-sided adhesive tape is also useful if the edges of the film are painted with Aquadag.

If extremely smooth films are being examined it may be necessary to tilt the specimen to an angle much steeper than normal before any detail can be recognized.

Another alternative is to try collecting only primary electrons, but increased noise adds limitations here.

Cross-sections of films can be examined for defects in manufacturing, i.e. inclusions or voids within the film. These are prepared by sectioning with a sharp razor blade and cutting to the appropriate size to fit the holder described in Section 7.4.3, or sectioning while in the holder. Again the edges are painted with Aquadag (as in Section 7.5.2) before coating.

7.7. DYNAMIC AND SPECIAL EXPERIMENTS

7.7.1. Introduction

Recently it has become more common to use the SEM for observing experiments *in situ*. Sikorski *et al.* (1969) first carried out a number of experiments in which they observed wool fibres in the SEM during different types of deformation. Wool fibres were stretched and crack development followed in the surfaces of cuticular cells. Other experiments were carried out in the SEM in which they observed the breakdown of the cuticular cells due to friction and the removal of the cells with a razor blade. Since 1970 the introduction of TV scanning has made possible the continuous recording of experiments carried out in the SEM, although the cheaper and somewhat time-consuming method of time-lapse photography has been used for a number of years (Hearle, J. W. S., Reeves, D. A. and Sparrow, J. T., unpublished) to study deformation of adhesive bonded non-woven fabrics.

It will be the object of this section to describe special stages and give reference to a number of dynamic experiments on textile materials which have been carried out in the SEM. It is also worth mentioning here that Weitzenkamp (1969) carried out a special experiment in the SEM in which he measured fibre potentials and calculated charge densities for organic polymer fibres.

7.7.2. Special stages

In order to carry out dynamic experiments in the SEM special specimen stages or straining devices need to be constructed or the standard stage modified slightly. Sikorski *et al.* (1969) and Hepworth *et al.* (1969a) have described a relatively easy and simple method to observe straining processes in the SEM. One end of the sample is attached to a special holder operated by the rotation control of the specimen stage while the other end is attached to a cantilever on which are mounted two strain gauges. By simple operation of the rotation control the sample can be extended and the stress recorded. This arrangement gave excellent results and it was possible to follow changes in the surface topography of fibres and polymer tapes while the sample was under a known stress. Holliday and Newman (1969) describe a simple specimen straining device in which the specimen is wound around a rotating pillar which is fixed into the holder insert hole on the specimen stage. The sample is wound around the pillar by use of the rotation control on the specimen stage. Using this device, knots of synthetic fibre were tightened and broken as well as loops of fibres.

McKee and Beattie (1970) have successfully used a more sophisticated extension/compression stage and a rotation/twisting stage for examination of fibres, yarns,

FIG. 7.22. Extensometer stage for use in the Cambridge Stereoscan (Cross *et al.*, 1971).

fabrics, films, etc., in the Cambridge Stereoscan. These stages replaced the specimen stage proper and required translation of the drive to provide the desired movements. With the stages the following types of deformation were carried out: tensile fractures, knot fractures, loop tensile fractures, twisting and compression. A circumferential viewing of surfaces was also carried out on the rotation stage.

Cross *et al.* (1970, 1971) have described an extensometer stage, shown in Fig. 7.22, which replaces the complete standard stage. The straining part of the stage is shown

FIG. 7.23. Straining part of stage with ruptured non-woven fabric.

in Fig. 7.23. X and Y movements are possible but Z movement, rotation and tilt movements are not available. Strains can be measured on the external micrometer which is used to separate the clamping jaws. This stage has been used with great success for the examination of the deformation of adhesive bonded non-woven fabrics as shown in Fig. 7.24.

If special stages cannot be constructed by the user a tensile stage is available as an accessory from Japan Electron Optics Laboratory Co., for their JSM-U3 SEM, or, for the Cambridge Stereoscan, Applied Research and Engineering Ltd., is prepared to construct special stages on request. Cambridge Instrument Co. Ltd. has also announced the availability of an extensometer stage.

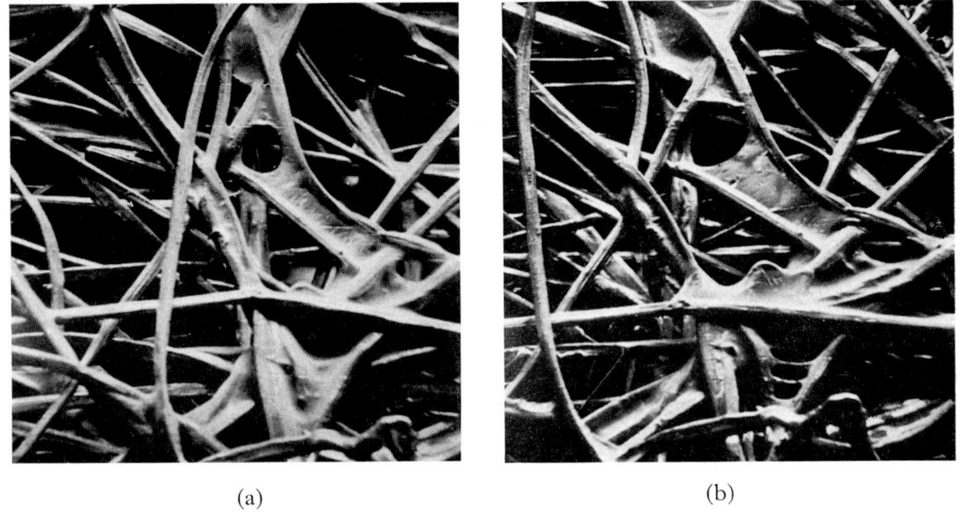

(a) (b)

FIG. 7.24. (a) Adhesive bonded non-woven fabric before straining. (b) Fabric after 13% extension. (90×)

7.7.3. Recording of information

As a result of the increasing demand for using the SEM to observe dynamic experiments the need has arisen for some method of recording the experiments as they are performed. Some instrument manufacturers have recently developed TV scanning systems in which the specimen is scanned by the electron beam at rates similar to those used in ordinary televisions with the information being recorded on video-tape or onto ciné film from the TV monitor. An alternative recording technique is to use time-lapse photography.

A time-lapse photographic recording technique is simple but very time consuming. The specimen is moved or acted upon and the appearance of the specimen after this action is recorded in the usual manner on Polariod film or 35-mm negative film. The micrographs are then rephotographed onto ciné film in the correct sequence. The instrument magnification used will determine how much extension or movement is required for each micrograph. If this movement is too great a jerky motion will result when the film is projected.

Time-lapse photography was first mentioned by McKee and Beattie (1970). Van Veld *et al.* (1970) produced a motion picture film from a sequence of still Polaroid photographs in which they examined the motion of individual filaments and yarns in a fabric under strain. If this method is used great care must be taken to position each photograph properly when recording on the ciné film or a jerky motion will result.

Hearle *et al.* (1971) have described a similar method of recording in which a ciné camera is mounted in front of the record CRT and single exposures are made direct onto the 16-mm ciné film. If the video amplifier has a provision for reversing contrast a positive film can be made in one step thus cutting down on the film-processing time.

Before making a choice as to which system is best one must consider the advantages and disadvantages of both systems.

TV scanning and fast scan monitors. These two systems are expensive and in some cases require modifications to the SEM. They are clearly necessary for direct viewing and recording of continuous events. Specimen manipulation is easy as it can be viewed directly. TV monitors are available for lecture and demonstration purposes. Other than the high cost of equipment the main disadvantage to these systems is the loss of resolution due to the fast scan times and the difficulty in recording extremely slow changes to the specimen.

Time-lapse photography. The main advantages of this recording technique are the low cost of equipment and the good resolution due to recording direct from the record CRT onto ciné film. It cannot be used for recording continuous changes to the specimen and is very time consuming.

7.7.4. Coating for dynamic experiments

Usually polymers and fibres are coated with silver, gold, or gold/palladium alloy so that they will be electrically conductive for examination in the SEM. Such a coating is successful if a straightforward examination of the surface is to be carried out but the inelasticity of a metal coating makes it impractical for use with dynamic experiments. The metal coating will crack and charging will result. One must turn to antistatic materials for use as a coating when dynamic experiments are to be carried out. Duron is a very good antistatic to use; it comes in an aerosol container and fabrics can be easily sprayed or materials can be dipped or soaked in concentrated or diluted solutions. It might prove useful to try using Duron at different concentrations to see if one concentration works better than another with the particular material being examined. Duron is usually diluted in isopropyl alcohol. Magnification will probably be limited to a maximum of about 2000 times.

7.8. CONCLUSIONS

It has been shown that there is a wide range of uses for the SEM in the field of polymers and fibres. Since its commercial availability new areas of research have been opened which will enable investigators to gain more knowledge about these materials. The techniques which have been described have proven to be particularly useful. Careful use of these preparation techniques will result in good-quality micrographs. It is again important to stress the need for using low beam voltages especially if fracture surfaces are being examined. Investigators should at least try different voltages with the particular specimen and investigation to see if one voltage is better than another. Dynamic experiments are very useful for studying deformation processes of textile materials. Hopefully in the future better SEM techniques will be discovered which will help to reveal even more information about polymers and fibres.

REFERENCES

Anderson, C. A., Katz, H. J., Lipson, M. and Wood, G. F. (1970) *Text. Res. J.* **40,** 29.
Anderson, C. A. and Lipson, M. (1970) *Text. Res. J.* **40,** 88.
Anderson, C. A., Goldsmith, M. T., Katz, H. J. and Wood, G. F. (1971) *Applied Polymer Symposia No. 18*, Interscience Publishers, New York, N.Y. p. 715.
Boyde, A. and Wood, C. (1969) *J. Micro.* **90,** 221.

BROWN, A. C. and SWIFT, J. A. (1970) *Beitr. electronenmikroskop Direkstabb. Oberfl.* **3,** 299.
CROSS, P. M., HEARLE, J. W. S., LOMAS, B. and SPARROW, J. T. (1970) *Proc. 3rd Annual SEM Symposium*, IIT Research Institute, Chicago, Ill., p. 81.
CROSS, P. M., HEARLE, J. W. S., CROSS, J. D. and SANDS, A. (1971) *Text. Res. J.* **41,** 629.
FORD, J. E. and SIMMENS, S. C. (1958) *Shirley Institute Memoirs* **31,** 289.
GOYNES, W. R. (1969) *Proc. 2nd Stereoscan Collog., Engis Equipment Co.*, Morton Grove, Ill., p. 149.
HALY, A. R., SNAITH, J. W. and ANDERSON, C. A. (1970) *Text. Res. J.* **40,** 1128.
HEARLE, J. W. S. and CROSS, P. M. (1970) *J. Mat. Sci.* **5,** 507.
HEARLE, J. W. S. and VAUGHN, E. A. (1970) *Rheologica Acta* **9,** 76.
HEARLE, J. W. S., LOMAS, B. and SPARROW, J. T. (1970) *J. Micro.* **92,** Pt. 3, 205.
HEARLE, J. W. S. and SPARROW, J. T. (1971) *Text. Res. J.* **41,** 726.
HEARLE, J. W. S., CLARKE, D. J., LOMAS, B., REEVES, D. A. and SPARROW, J. T. (1971) *Proc. 25th Anniversary Meeting of EMAG, Inst. Physics*, Institute of Physics, London, p. 210.
HEPWORTH, A., BUCKLEY, T. and SIKORSKI, J. (1969a) *J. Sci. Inst.*, series 2, **2,** 789.
HEPWORTH, A., SIKORSKI, J., TUCKER, D. J. and WHEWELL, C. S. (1969b) *J. Text. Inst.* **60,** 513.
HOLLIDAY, P. and NEWMAN, P. H. (1969) *J. Sci. Inst.*, series 2, **2,** 444.
MCKEE, A. R. and BEATTIE, C. L. (1970) *Text. Res. J.*, **40,** 1006.
PRINCEN, L. H. (1971) editor, *Applied Polymer Symposia No. 16*, Interscience Publishers, New York, N.Y.
SCOTT, R. G. (1959) *Symposium on Microscopy*, ASTM Special Technical Publication No. 257, p. 121.
SIKORSKI, J. (1958) *Fourth Intern. Conf. Electron Micro., Berlin*, **1,** 686.
SIKORSKI, J., MOSS, J. S., NEWMAN, P. H. and BUCKLEY, T. (1968) *J. Sci. Inst.*, series 2, **1,** 29.
SIKORSKI, J., MOSS, J. S., HEPWORTH, A. and BUCKLEY, T. (1969) *Proc. 2nd Stereoscan Collog., Engis Equipment Co.*, Morton Grove, Ill., p. 25.
SIKORSKI, J. and HEPWORTH, A. (1969) *Proc. 2nd Annual SEM Symposium*, IIT Research Institute, Chicago, Ill., p. 249.
SWIFT, J. A. (1969) *J. Text. Inst.* **60,** 30.
SWIFT, J. A. (1971) *Applied Polymer Symposia No. 18*, Interscience Publishers, New York, N.Y., p. 185.
SWIFT, J. A. (1972a) *J. Text. Inst.* **63,** 64.
SWIFT, J. A. (1972b) *J. Text. Inst.* **63,** 129.
VAN VELD, R. D., MORRIS, G. and BILLICA, H. R. (1968) *J. Appl. Poly. Sci.* **12,** 2709.
VAN VELD, R. D., BRYANT, F. D. and BILLICA, H. R. (1970) *Text. Res. J.*, **40,** 1138.
WEITZENKAMP, L. A. (1969) *J. Sci. Inst.*, series 2, **2,** 561.

CHAPTER 8

Applications to Solid-state Electronics

D. C. NORTHROP

8.1. INTRODUCTION

The applications of scanning electron microscopy to solid-state electronics cover a very broad spectrum of materials and of contrast modes. All that can be done in the space available here is to categorize the various applications and give some specific examples of each. When read in conjunction with the remainder of the book this will serve as a guide to what can (and to what cannot) be achieved with the scanning electron microscope.

We can first distinguish two broad areas of application:

(i) the investigation of homogeneous and nearly homogeneous materials,
(ii) the investigation of semiconductor device structures.

In the first category fall the study of materials for variations in composition, in electrical properties and in crystal perfection. A wide range of contrast mechanisms is applicable here, including secondary electron emission, X-ray emission, Auger electron emission, cathodoluminescence, and internal electric effects.

It should be emphasized from the onset, however, that not all techniques are suitable for all materials. Cathodoluminescence has application only for light-emitting materials like ZnS and GaP, but it is not useful for silicon and germanium (some materials like GaAs and InSb exhibit cathodoluminescence in the infrared). Internal electrical effects in the specimen can only be observed where their magnitude exceeds the current noise in the specimen, implying an upper limit on the conductivity of the specimen (modulation of the primary beam can improve the sensitivity of the method by allowing the use of a tuned or phase-sensitive detector, but this facility is not available on all scanning electron microscopes). Specimen analysis is also difficult to achieve in semiconductors for two reasons:

(i) impurity levels of interest are well below the detection limit of X-ray and reflected primary modes of contrast;
(ii) many semiconductors and almost all the important impurities have low atomic numbers, making them unsuitable for measurement by X-ray emission.

8.2. INVESTIGATION OF MATERIALS

There are a number of uses for the scanning electron microscope which parallel the uses of the optical microscope in the study of surfaces. The advantages lie in the greater magnification and depth of field of the scanning electron microscope, coupled with the fact that the various "analysis" modes can often be used to identify foreign particles on the surface. Figure 8.1 shows thermal etch pits on silicon, illustrating greater detail than would be obtainable optically. Exactly what kind of detail can be seen is not easily predictable, but it is often rewarding to examine specimens just to see what turns up.

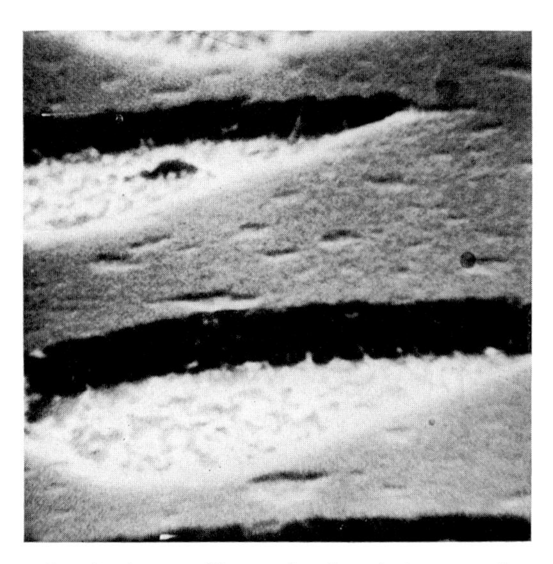

FIG. 8.1. Thermal etch pits on silicon, showing their exact alignment, and flat bottoms.

8.2.1. Internal specimen currents

When electrodes are connected to the specimen, as shown in Fig. 8.2, a small potential difference will cause an induced current to flow due to the excess carriers produced by the primary beam. When the applied voltage is low enough to ensure that the electrons and holes produced by the primary beam recombine before reaching the electrodes the current, i, flowing in the external circuit is given by

$$i = eE(n\,\mu_n + p\,\mu_p)$$

where

$$n = p = \frac{i_0}{e} \cdot \frac{W_0}{\omega} \cdot \tau.$$

In this expression n and p are the excess electron and hole numbers produced by the primary beam, μ_n and μ_p are the electron and hole mobilities, e the electronic charge, E is the local electric field at the point where the primary beam strikes the specimen, W_0

and ω are the primary electron energy and the specimen ionization energy and τ is the minority carrier lifetime.

Variations in specimen current from point to point on the specimen can therefore reflect variations in

(a) resistivity (through variations in E),
(b) carrier mobility (μ_n and μ_p),
(c) minority carrier lifetime (τ).

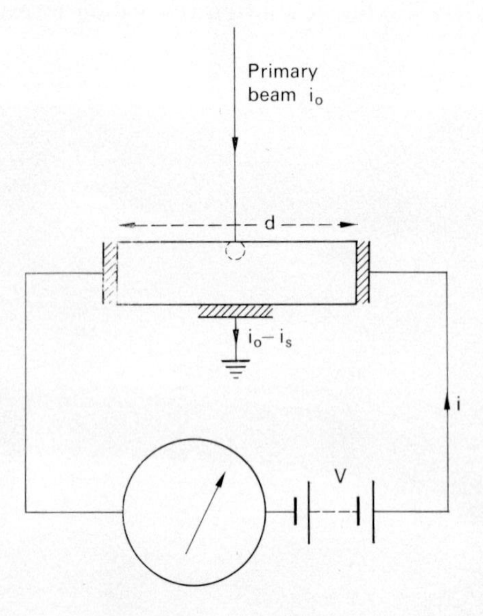

FIG. 8.2. Circuit arrangement for specimen current measurement.

The situation is further complicated by the fact that the value of lifetime is a compound of surface and volume recombination processes which vary in relative importance as the primary beam energy is varied to change the penetration depth for the primary electrons. The carrier generation rates ($i_0/e \times W_0/\omega$) also vary linearly with primary beam energy. Variations of picture brightness may therefore show one or more of a number of specimen parameters, with the possibility of varying the primary voltage to distinguish between them. Another possible aid is that of varying the potential difference V applied to the specimen until it is large enough for the current i to be independent of V. This happens when all the carriers excited by the primary beam reach the electrodes before recombination. The current then becomes

$$i = i_0 \times (W_0/\omega)$$

and will be independent of variations of mobility and lifetime and resistivity. Variation of W_0 would then give a proportional change in current provided that the surface recombination velocity is not too high. If surface recombination is important then ω will in effect depend on the distribution of secondaries in the specimen, and increasing W_0 will give a more than linear increase in current by reducing the effect of the surface.

In both high field and low field conditions the photoelectric response time is τ, the minority carrier lifetime, and if this is long its effect on the resolution must be considered.

The magnitude of the specimen current is small: typically the value of ω for semiconductors is between 3 and 4 eV, so that when W_0 is 20 keV the specimen current cannot exceed 5000 times the primary current. Usually the primary current is about 10^{-10} amps, giving a *maximum* specimen current of 5×10^{-7} amps. This maximum is achieved only for complete collection of the specimen current at the electrodes, and clearly the resolution of the instrument for variations of resistivity and lifetime will be lost under these conditions. The most important limitation on resolution in this mode is the carrier drift length, $\mu E\tau$, the distance travelled by the carriers (on average) between ionization and recombination. When the field E is reduced to obtain good resolution the specimen current is reduced by the factor $\mu E\tau/d$, where d is the distance between the electrodes. This factor can be as small as 10^{-3}.

Basically then this is a low resolution mode of operation of somewhat limited usefulness.

8.2.2. Specimen voltages

When the beam scans the specimen without an applied field the excess charge carriers will move by diffusion down the concentration gradient. In doing so they will produce voltages whenever there are specimen inhomogeneities resulting in small variations in the positions of the band edges relative to the Fermi level. The mechanism by which these voltages are established is identical with that which operates when a *p–i–n* junction is used as a photovoltaic cell.

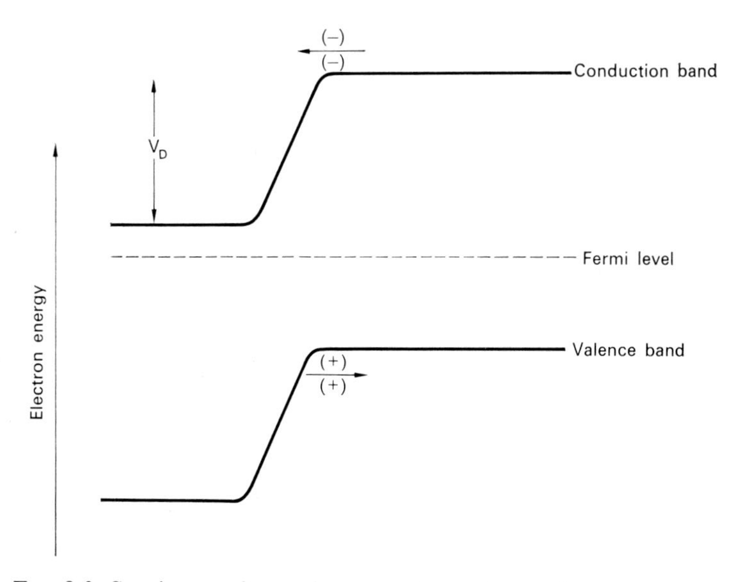

FIG. 8.3. Specimen voltages due to gradient of impurity concentration.

Figure 8.3 shows schematically a somewhat exaggerated amount of band bending, such as that which might occur at a discontinuity in the impurity concentration. When electron–hole pairs are created on the right-hand side (p-type) those electrons which diffuse to the potential discontinuity are attracted to the left by the built-in field in that region, whereas holes reaching the discontinuity are prevented from crossing it by the same field. The argument is inverted for electron–hole pairs approaching the discontinuity from the left, so that both sets of carriers charge the material positively on the right-hand side, building up a potential difference which opposes the built-in field in the specimen. The additional induced field gives a potential difference which can be observed in the external circuit. Clearly the greatest voltage which can be developed in this case is V_D, the diffusion potential of the discontinuity, because at this value the local field in the specimen becomes zero and no further charge separation takes place. This saturation will be reached when the charge collected from that volume of the specimen within a carrier diffusion length $(D\tau)^{\frac{1}{2}}$ is enough to cancel the internal field. It is difficult in practice to know whether saturation is reached.

This contrast mode has a resolution which will normally be determined by the carrier diffusion length $(D\tau)^{\frac{1}{2}}$, where D is the diffusion constant $\mu e/kT$ and τ the minority carrier lifetime. Variations in voltage from point to point on the specimen may reflect changes in mobility, lifetime (including surface recombination) and impurity concentration. Detailed quantitative analysis of the results is difficult in most cases, but it is comparatively easy to detect output voltages as low as 10^{-6} V from most semiconductor specimens without the complication of current noise.

8.2.3. Total specimen current

The current flowing to earth through the specimen is $(i_0 - i_s)$ (see Fig. 8.2). The primary current can be held accurately constant by most microscopes, so that this current shows variations in the true secondary emission (work function variation) and in the reflection coefficient (atomic number). The current to be measured is low (less than 10^{-10} amps normally) and requires additional amplification, comparable with that in the scintillator and photomultiplier of the secondary electron collector. One of the main differences between this mode and the normal secondary emission mode is that effects due to the specimen–collector geometry are eliminated.

8.2.4. Cathodoluminescence

Cathodoluminescence can usefully be observed in many semiconductors of the III–V and II–VI classes. Variations in intensity and colour of luminescence reflects variations in impurity concentration, and this may be studied as a function of depth by varying the primary beam voltage. The resolution obtainable is usually better than 1000 Å, depending on the mechanism of cathodoluminescence emission, which may vary from one material to another. Usually it is necessary to have a very good understanding of the materials under investigation to design a good cathodoluminescence experiment.

8.3. INVESTIGATION OF SEMICONDUCTOR DEVICES

The commonest use of the SEM in semiconductor technology is in the investigation of devices. These are now so small that important features cannot be resolved in a good optical microscope, so that the SEM has an important role to play simply in the observation of surface topography using the conventional secondary emission mode of operation.

Figure 8.4 shows a number of photographs of silicon microcircuits taken with the secondary emission mode. They show a variety of faults none of which would have been detectable using the optical microscope. The important thing to be said about these photographs (and those which follow illustrating other modes of operation) is that they must be studied in relation to the device manufacturing processes and the subsequent history of the device in use. The point is that scanning electron microscopy has an important part to play in semiconductor technology in two ways:

(i) in device manufacture it helps to optimize processing with respect to yield and performance;
(ii) examination of devices which have failed in operation can yield evidence about failure mechanisms in what were initially successful devices. This can also react in an important way on manufacturing processes.

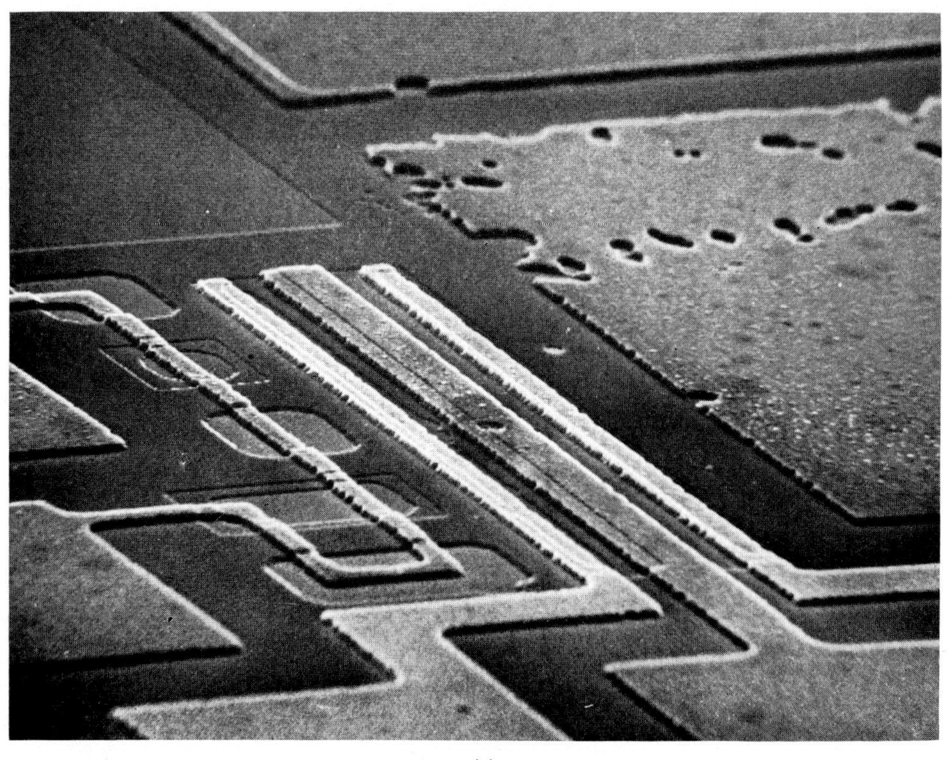

(a)

FIG. 8.4. (a) Part of silicon microcircuit.

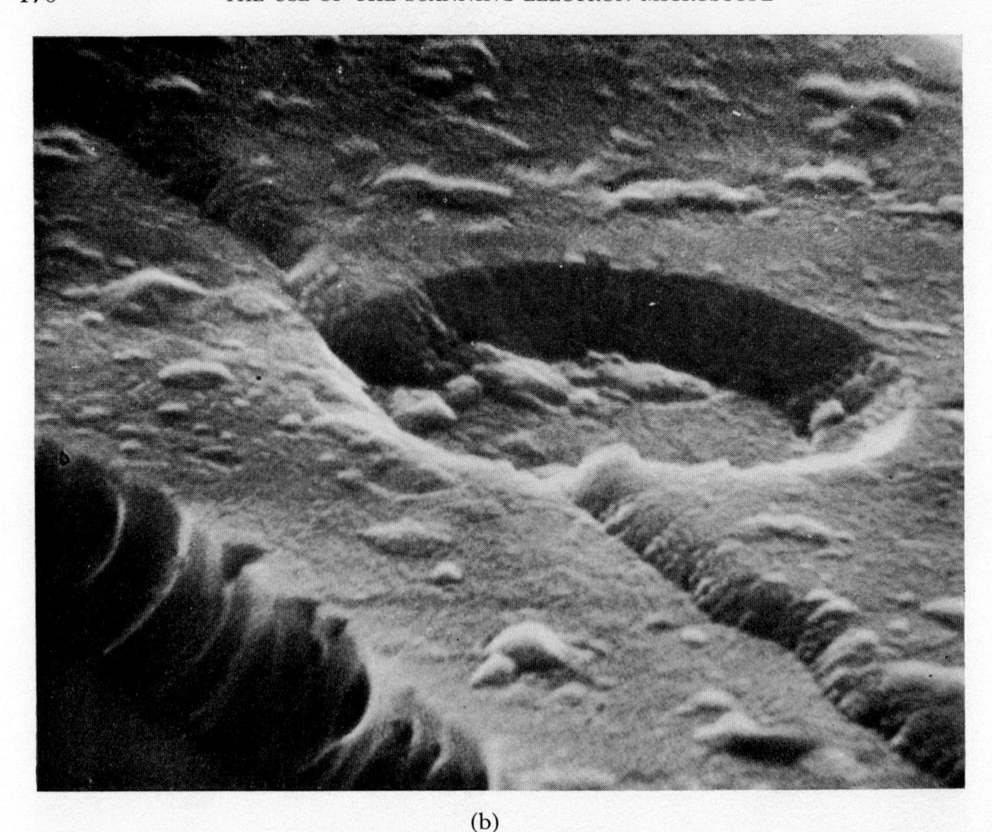

(b)

FIG. 8.4. (b) Enlargement of part of (a).

To achieve results it is therefore best to compile a whole series of photographs following the complete history of a device through every stage of manufacture and testing to eventual failure, preferably using more than one contrast mechanism at each stage to obtain the maximum amount of information.

Figure 8.4 shows a transistor (a) in a silicon microcircuit and (b) an enlargement of part of one of the aluminium interconnection stripes in which the aluminium has penetrated a hole in the oxide mask. This is a clear case of a fault being identified in a particular stage of manufacture. Figure 8.5 shows part of another silicon microcircuit in which the aluminium film shows signs of crystalline structure and is lifting away from the silicon. This particular device has passed electrical tests and was within specification but it is not known when this malformation occurred, nor what effect it might have on the useful life of the device; only a careful stage-by-stage examination can answer these questions. Figure 8.6 shows another fault in an integrated circuit where there is an open circuit in an aluminium interconnection. Here the fault is obvious but the reason for it is not obvious; it would be interesting to know when this fissure developed and whether it is connected with the somewhat crystalline structure of the aluminium film. Examples of this kind can be multiplied almost indefinitely, but it is probably obvious by now that the high magnification and depth of field of the

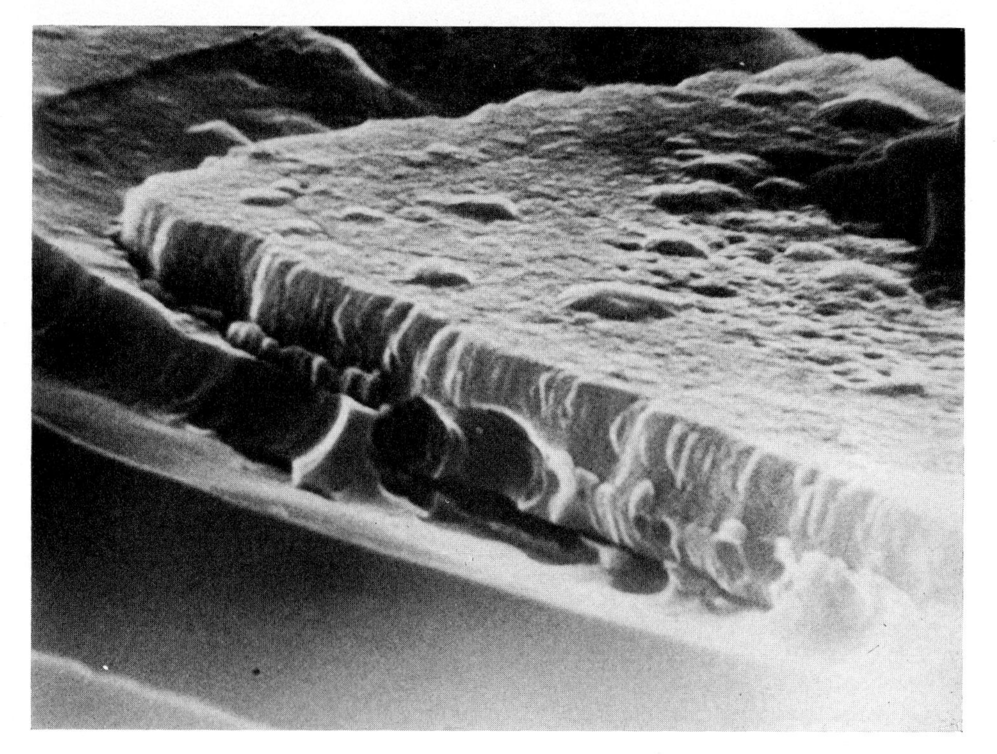

FIG. 8.5. Aluminium films lifting from microcircuit.

microscope allow direct observation of details not visible through an optical microscope, and that the pictures obtained are immediately recognizable and easy to interpret.

8.3.1. Measurements on devices

A variety of measurements can be made on *p–n* junctions, on discrete devices and on integrated circuits. Indeed the whole range of contrast mechanisms can be used either singly or in combination to monitor manufacturing processes, to investigate device operation, or to study the mechanisms of device failure. Clearly it is impractical to catalogue all the possible variations of technique in a book of this nature, but a number of examples will serve to illustrate the versatility and potential importance of scanning electron microscopy in the field of industrial semiconductor electronics.

Figure 8.7 shows an MOS transistor at comparatively low magnification ($\times 110$) with the microscope used in the conductive mode. In Fig. 8.7 (a) the charge depletion region between source and drain is comparatively wide, whereas in Fig. 8.7 (b) the drain electrode is disconnected and the charge depletion region is narrower and more sharply defined. Without going into details it is possible to see how the state of the device changes the picture, allowing an inspection to be made of its functioning.

The photographs of Fig. 8.8 show an MOS sixteen-bit, parallel access, four-phase shift register. Figure 8.8 (a) shows the device with -30 V connected to the no. 1 pulse

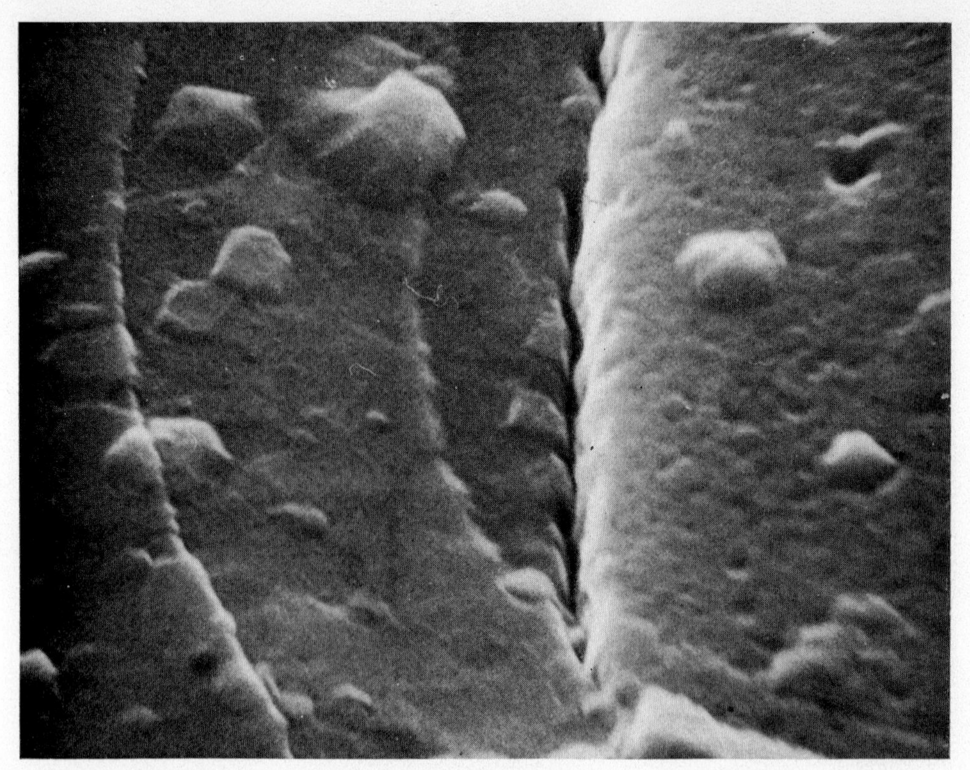

FIG. 8.6. Broken aluminium interconnections at oxide step in planar micro-circuit.

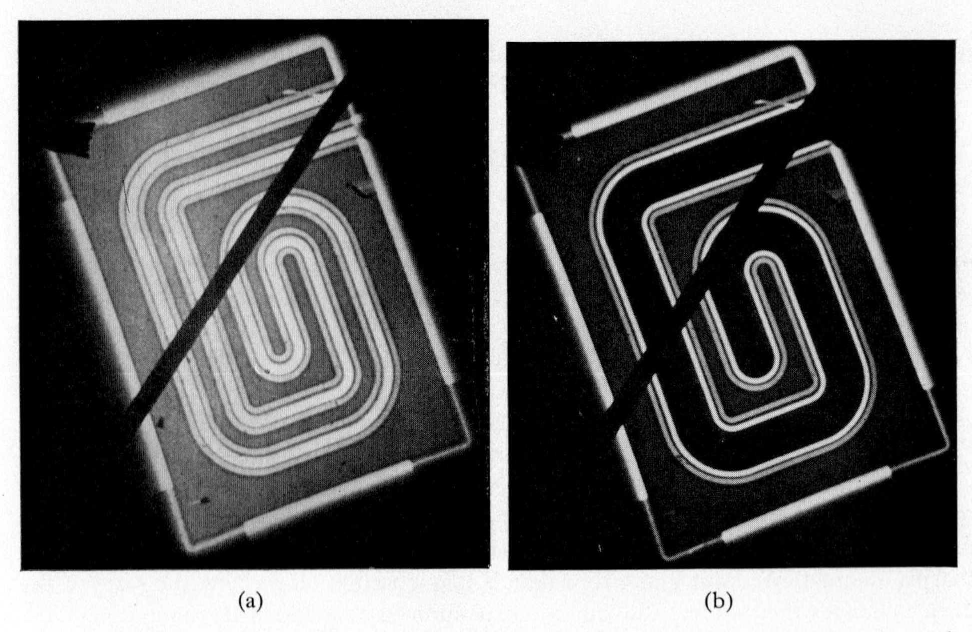

(a) (b)

FIG. 8.7. An MOS switch at 250°C showing the charge depletion layer (a) with the drain connected and (b) with the drain electrode disconnected, conductive mode (×110). (Photographs are reproduced by permission of Marconi-Elliott Microelectronics, Witham, Essex.)

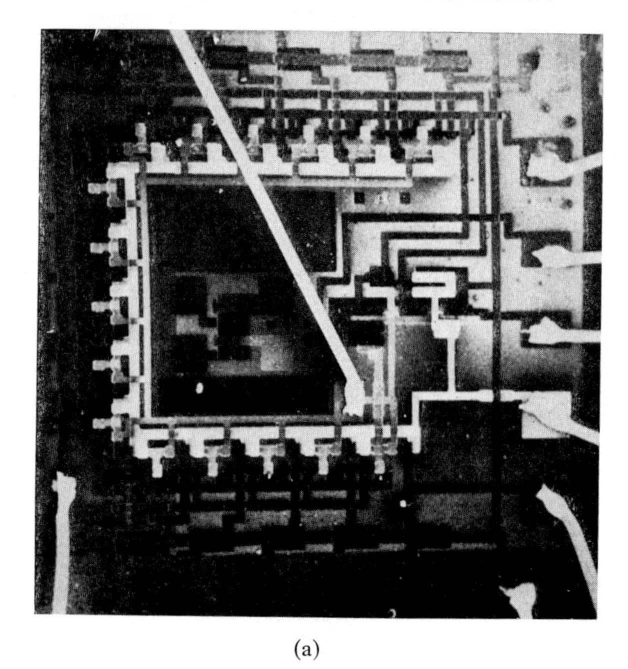

(a)

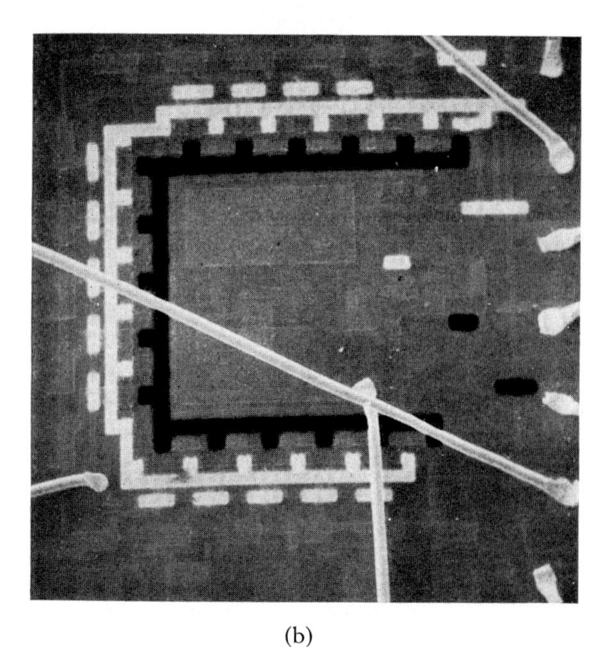

(b)

FIG. 8.8. An MOS sixteen-bit, parallel access, four-phase shift register photographed at ×40 magnification using various combinations of contrast mechanisms. (a) −30 V on no. 1 clock pulse line, specimen current amplifier connected to inner clock pulse line, mixed emissive and conductive mode signals. (b) As (a) but emissive mode only, showing surface topography and voltage contrast.

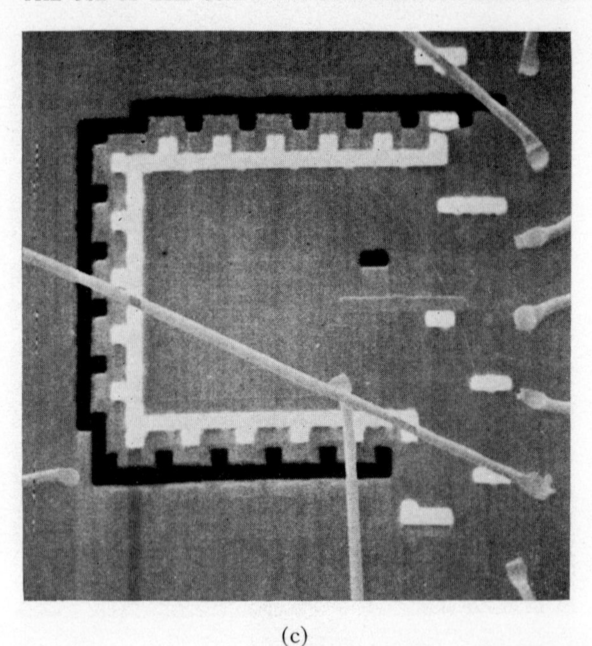

(c)

FIG. 8.8. (c) −30 V connected to inner clock pulse line, specimen current amplifier connected to no. 1 clock pulse line mixed emissive and conductive mode signals. (These photographs are reproduced by permission of Marconi-Elliott Microelectronics, Witham, Essex.)

line, and the specimen current amplifier connected to the inner clock pulse line with all other connections earthed. The picture is produced by mixing the signal from the specimen current amplifier with the normal secondary emission signal. Figure 8.8 (b) shows the same device under the same conditions, but using the secondary emission signal only. It gives topological detail and the electric field patterns produced by the applied voltage, but lacks the detail given in Fig. 8.8 (a) of the conducting regions in the circuit. Finally, Fig. 8.8 (c) reverts to a mixture of emissive and conductive mode signals, but with −30 V connected to the inner clock pulse line and the specimen current amplifier connected to no. 2 clock pulse line. These photographs show very clearly the way in which the conducting regions of the shift register are controlled by the clock pulse signals. They also show how important it is to be able to combine signals due to different contrast mechanisms: in this particular case the addition of the emissive mode signal shows where the conducting regions lie relative to the device structure.

Where semiconductor devices are made from cathodoluminescent materials it is possible that the cathodoluminescent contrast mechanism of the scanning electron microscope will provide a useful tool for material assessment and device investigation. Figure 8.9 shows photographs obtained using the infrared cathodoluminescence of gallium arsenide to investigate the structure of injection lasers. Figures 8.9 (a) and (b) are photographs of the same device taken at different magnifications. They show striations of cathodoluminescent intensity, which proved to be associated with the

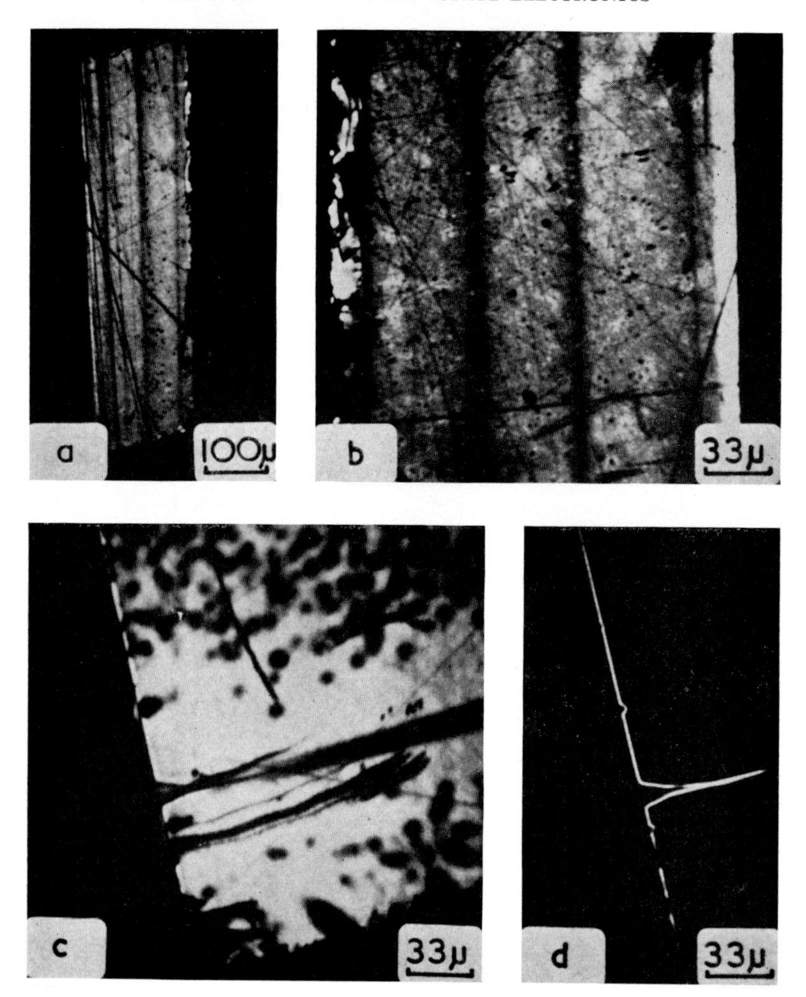

FIG. 8.9. Gallium arsenide *p–n* junction lasers examined by cathodoluminescence. (a) and (b) are the same device examined at different magnifications with the electron beam incident on one of the polished faces of the laser; (c) and (d) are photographs of another device and compare cathodoluminescence with conductive mode signals at the same magnification. (These photographs are reproduced by permission of Dr. D. A. Shaw, Communications Research Center, Ottawa.)

diffusion schedule used in the manufacturing process. They also show what is almost certainly a dislocation pattern. The second two photographs are on a second device. Figure 8.9 (c) shows the cathodoluminescence pattern near the laser junction and Fig. 8.9 (d) is a conductive mode signal showing clear evidence of a diffusion spike.

It remains to make two important points about semiconductor measurements. Firstly, the interpretation of the results often depends upon a detailed knowledge of the physics of the devices being measured. It is not appropriate to deal with such a detailed and lengthy topic here, and normally individual investigators will know all that is necessary about their own specimens. The unwary should be warned, however, and should note that careful design of experiments can often aid the interpretation of results. The second point is a prediction for the future. Most of the applications of scanning electron microscopy to semiconductors are qualitative in nature, but they could be made quantitative. Much is yet needed before this will be possible, but the information which can be obtained is such that the effort to obtain it will probably be made. If the electron probe could be used for analysis and for field and potential measurements on a quantitative basis it would add greatly to out knowledge of semiconductor materials and devices.

REFERENCES

The reader is referred to the following books and conference reports, all of which contain extensive bibliographies.

THORNTON, P. R. *Scanning Electron Microscopy*, Chapman & Hall, 1968.

NIXON, W. C. Scanning electron microscopy, *Contemporary Physics* **10,** 71, 1969.

Proceedings of the Annual Scanning Electron Microscope Symposia. Sponsored and Published by IIT Research Institute, Chicago, Ill.

CHAPTER 9

Applications to Biological Materials

P. ECHLIN

9.1. INTRODUCTION

Following the introduction of commercially available scanning electron microscopes in 1965 a large number of papers have appeared which attest to the value of this instrumentation in biological research. A reasonably complete list of specific applications of electron probe instrumentation to biological problems is given by Johnson (1969), Wells (1969, 1970) and in the bibliography compiled by Rossi (1968). Further details may be found in the five annual scanning electron microscope symposia held at IIT Research Institute since 1968. The limitations of space only allow passing reference to the applications of the scanning electron energy analysing microscope to biological studies. Crewe and his co-workers have constructed and recently described (Crewe, 1970; Crewe and Wall, 1970) a high-resolution SEM which uses a field emission source and is capable of at least 5 Å point resolution using transmitted electrons in thin specimens. Using this form of instrument they have elegantly shown that it is possible to determine atomic or molecular species on the basis of the unique energy loss suffered by the electrons as they pass through material. These characteristic energy losses are then used to provide picture contrast. Stroud *et al.* (1969) have applied these techniques to cells and were able to demonstrate that at different levels of energy loss, particular structures are distinguished with high contrast in unstained sections. These and associated techniques will have important applications in the study of cell and molecular structure. Although this form of instrumentation is not yet commercially available, it promises to open a new and exciting tool for use in biological research.

Many of the applications of scanning electron microscopy to biological samples are dependent on adequate preservation of the material prior to its examination in the microscope column. Some of the newer methods will be described in this present chapter, and more detailed preparative techniques are given by Echlin (1968), Boyde and Wood (1969), and Marszalek and Small (1969). Some general reviews on the application of scanning microscopy to biological systems are presently available (Hayes and Pease, 1968); this present chapter will concentrate on some of the recent advances in tissue preparation and specimen signal acquisition.

9.2. MODES OF EXAMINATION

9.2.1. Topographical detail

The bulk of the information available from scanning electron microscopy falls within this category and it is this mode of operation which has, up to the present, been most useful in biological research. There is some question, however, as to how much new information has been revealed by this technique, which could otherwise have been shown by more conventional methods. The principal advantage of this mode of scanning microscopy is the speed and ease of obtaining information. The interaction

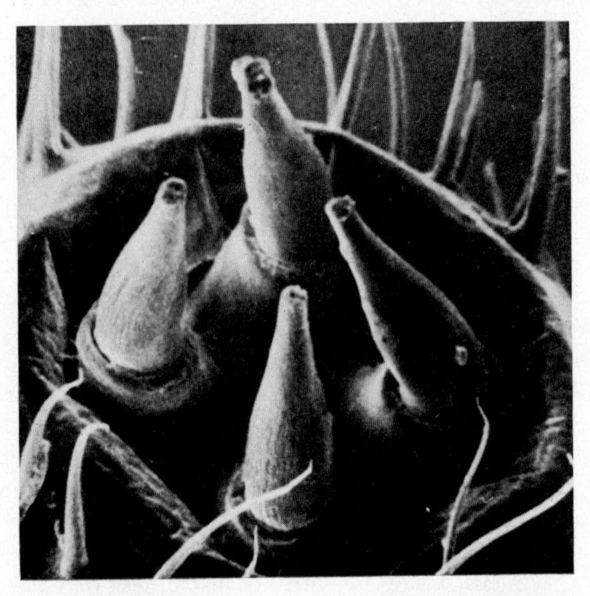

FIG. 9.1. Micrograph of a group of basiconic sensilla at the tip of the antenna of *Polymicrodon polydesmoides* illustrating the great depth of focus of the SEM. (Compliments of N. Sakwa, Zoology Dept., University of Manchester.)

of the high-voltage primary electron beam with the specimen gives rise to low-voltage secondary electron emission which may be collected and the information displayed on a cathode-ray tube. As far as topographical detail is concerned, the images are closely similar to those formed by reflected light optical microscopy, but with a much greater depth of focus and at a considerably increased resolution (see Figs. 9.1 and 9.2). In practice the depth of focus is better than for light microscopy at similar magnifications by a factor of at least 300. It is worthwhile noting that the typical probe currents of 10^{-2} to 10^{-3} nA are far below the 10 to 10^{-1} μA of a conventional transmission electron microscope or the 10^{-1} to 10^{-3} μA of a conventional electron microprobe. This is of great importance in reducing specimen contamination and specimen damage due to heating. The converse situation is also useful to biologists. By utilizing the beam-blanking units available on some instruments, it is possible to identify and selectively destroy a particular point on biological tissue by intense spot irradiation.

The secondary electrons which form the signal have low energies of the order of 4–6 eV. Some of the primary electrons may penetrate the specimen and, following scattering through large angles, may re-emerge from the surface as back-scattered electrons. Such electrons have undergone little energy loss and may in turn generate further low-energy secondaries which may lessen the overall quality of the picture.

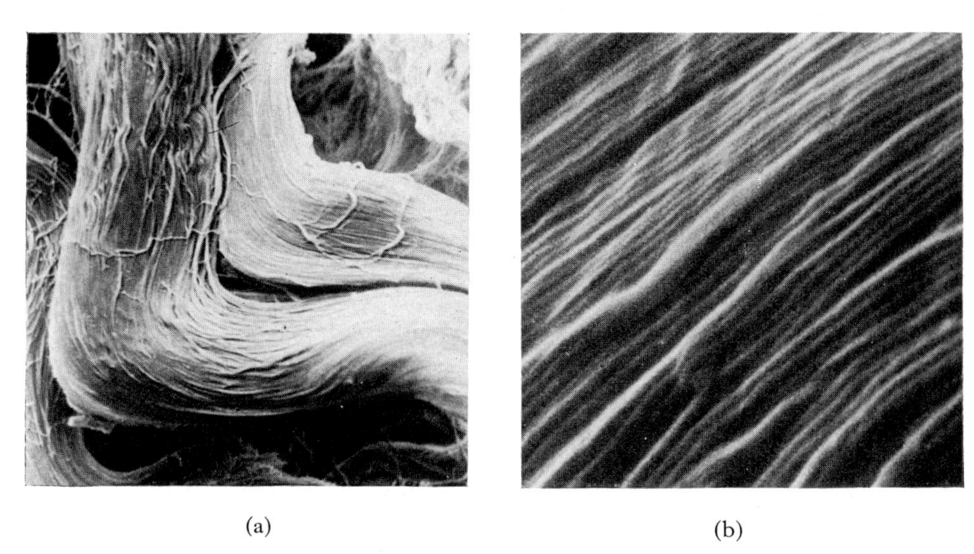

(a) (b)

FIG. 9.2. Formazin fixed adult human skin prepared by the method of Finlay, Hunter and Steven, *J. Microscopy* **93,** 73–76 (1971). Dense fibrous tissues such as this are an exception to the rule of examining biological materials at low kV. (a) Collagen fibres taken at 5 kV. The constituent fibrils can be resolved only poorly. (b) Resolution of the fibrils within the fibre made possible by working at 20 kV. (Compliments of J. Hunter, Rheumatism Research Centre, Manchester Royal Infirmary, Manchester.)

However, the primary excited secondary electrons form a large proportion of the total signal, and since they are derived from an area only marginally wider than the irradiating electron probe, fine detail is obtained from the surface of specimens. The resolution is considerably improved if the specimen is coated with some suitable conducting material such as evaporated carbon or a gold/palladium alloy. At the accelerating voltages normally used in the examination of biological material most of the secondary electrons are derived from within a few hundred angstroms of the surface of the specimen. The amount and nature of the secondary electrons depends on a number of factors including the angle of the incident beam, the atomic number of the specimen, and the surface properties of the specimen. Some degree of topographic information may also be obtained utilizing the primary electrons which are back-scattered from the specimen. Such electrons show little change in energy from the primary beam. The characteristics of the back-scattered primaries are dependent on the nature of the specimen. The lower the atomic number of the specimen so the greater the absorbance of the primary electrons. On a flat surface this provides a measure of element discrimination as elements of low atomic number appear lighter in contrast than those of high atomic number.

On some commercially available scanning electron microscopes, it is possible to modulate the amplitude of the scanning wave forms, as opposed to the normal brightness modulation, and provide micrographs from which quantitative measurements of signal intensity may be made. By measuring the amplitude of the scan lines at given points, accurate measurements may be made of the signal intensity. The immediate application of this technique, known as "Y-modulation", to biologists is that it provides a sensitive measure of any irregularities in surfaces which appear uniformly smooth by other methods.

9.2.2. Cathodoluminescence

On removal of the scintillator from the electron-collector system or by adding an additional photomultiplier and light guide, it is possible to detect luminescence emitted naturally from the specimen or as a result of excitation by the primary beam. The light can arise either from the natural luminescence of the material or from artificially introduced phosphors. Unfortunately the efficiency of production and collection of photons is low and the generation of light quanta takes place in a volume considerably bigger than indicated by the cross-sectional parameters of the primary beam. This generation of light quanta is lowered even further if thin sections are used, and it is necessary to have long exposure times in order to build up an acceptable picture image. Many of the dyes which have been used so successfully in fluorescent light microscopy are readily quenched when used in a cathodoluminescence system which adds to the difficulties of obtaining a picture. Cathodoluminescence closely resembles fluorescence optical microscopy and although very little of the former has been carried out on biological systems, it would appear that the same parameters and characteristics may be applied to both systems. Thus low concentrations of excitable material are needed to provide visible light which means that specific flurochromes may be added to living cells in quantities too small to cause toxic damage. Although the system has a high contrast the resolution is only marginally better than 100 nm and it is difficult to see how any accurate quantitation may be achieved. De Mets and Lagasse (1971) have investigated the use of some organic chemicals as cathodoluminescent dyes for the SEM.

In the absence of but a few definitive studies in the organic field (see, for example, Pease and Hayes, 1966; Pease and Hayes, 1967, 1969) it is difficult to predict all the applications of cathodoluminescence. On the instrumentation side we presently have a system with a resolving power of between 80–100 nm and a depth of focus which is better than that for reflected light optical microscopy at comparable magnifications by a factor of at least 300. Provided a number of specific resistant dyes are found the potential uses of cathodoluminescence would appear to fall into two main divisions. Firstly, the observation of naturally occurring luminescent structural material in plant and animal tissues and the localization of any luminescent by-products of their metabolism. Secondly, where fluorescent stains which have a specificity for a particular region of the cell or for a particular molecular species are added to living or fixed cells. These secondary luminescent studies may take the form of staining well-defined cellular entities or end-products of reactions in tracer techniques or for detecting foreign substances and used in conjunction with labelled antibody methods. The only experimental work which has been carried out on the cathodoluminescence of biological material is the preliminary study by Muir (1970) on modern and fossil wood (see Fig. 9.3), and although this work is being extended, nothing further has yet been published.

(a)

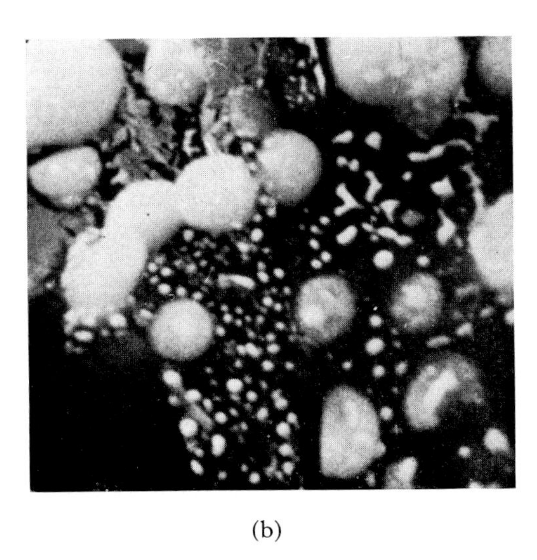

(b)

FIG. 9.3. (a) Tracheids of fossil wood showing a torn wall (centre) with sphaeroids on the secondary wall on either side. (b) Same field as (a) in the cathodoluminescence mode. Note the small granules of zinc sulphide in the wall (from Muir, 1970). (Compliments of M. Muir, Geology Dept., Royal School of Mines, Imperial College, London.)

9.2.3. X-ray microanalysis

Significant advances have been made in the X-ray microanalysis of some of the low atomic number elements, such as Na, K, Mg, Ca, which are commonly found in biological material. For an up-to-date and authoritative account of progress in this field reference should be made to the paper by Hall (in press). X-rays emitted from the specimen as a result of excitation by the electron probe are analysed to determine their chemical composition. The spectrum of the X-rays produced is characteristic of the elements in the target, and by monitoring the spectra of the X-rays produced as the specimen is scanned by the electron beam it is possible to build up a picture of the elemental distribution in the specimen. Suitable crystals in spectrometers are used for the dispersive analysis of the longer wavelength X-rays, and there are now suitable solid-state and gas-flow proportional counters for measuring the energy of X-rays with a resolution approaching 150 eV. The non-dispersive systems are more sensitive at very low X-ray generation rates as they can be made to intercept a much greater solid angle. The advantage in biological studies of the non-dispersive detectors is that it is possible to use much lower beam currents to detect elements, thus avoiding the possibility of engendering thermal and radiation damage within the specimen.

One limitation to the X-ray mode of operation is that since the X-ray production is a bulk effect the resolution is limited to about 200 nm, which is within the range of optical light microscopy. Attempts to improve the resolution by using a finer electron probe or thinner specimens have not met with much success because of the lower rate of X-ray production. In a recent paper which made a comparison of X-ray analysis in the scanning electron microscope, Sutfin and Ogilvie (1970) stress that one of the limitations of non-dispersive techniques is their inability to analyse in the ultra-soft region. They consider that the best possibility for solving this is in the development of gas-flow proportional counters to analyse in the region below 1 keV.

Nevertheless, the X-ray microanalysis mode of operation does provide a sensitive device for detecting elements. It is difficult to give meaningful figures to illustrate the sensitivity of the system since detection and measurement depend on the other elements which can give a confusing background signal. Hayes and Pease (1968) give approximate figures for detection limits using X-ray detection, of about 10^{-12} g of copper and 10^{-10} g of boron within a volume of about 10 μm^3. The appropriate instrumentation is available which permits quantitative analysis of single points on a specimen surface (see Figs. 7.20 and 7.21), semi-quantitative analysis of single line scans across the specimen, and for qualitative distribution of selected elements over a scanned raster. The selected elemental distribution may be displayed on a second video display unit, and can be compared directly with the signal obtained from secondary electron emission which provides topographical information concerning the specimen.

The full potential of X-ray microanalysis of biological material has yet to be realized. The biologist is primarily interested in the low atomic number elements such as sodium and potassium, and in most plant and animal systems these elements are present in the tissue in very small amounts and frequently masked by the main constituents of the cytoplasm. In order to accurately localize elements/ions in the cell fluids it is necessary to immobilize the tissue in order to prevent excessive migration of the elements being analysed. Very little work has been done in this field, but it would appear that 2–5 μm sections of quench frozen material cut on a cryomicrotome may be the best approach. These frozen sections could be kept at low temperatures inside the microscope column and the microanalyses carried out *in situ*. Such an approach would effectively immobilize the elements and, together with the use of efficient cryoprotective substances, should cause a very minimum of damage to the specimen. In spite of the relatively low accelerating voltage employed in scanning electron microscopy such thin specimens may be damaged unless very low beam currents are employed. An alternative approach would be to use thinner (*ca.* 500 Å) sections cut on a cryo-ultramicrotome which should lessen the beam damage. Unfortunately as the section becomes thinner, so there is less material to give rise to X-rays, and this in turn creates serious problems in being able to detect the emissions. It would seem that the only way out of this impasse would be to use inordinately long recording times, coupled with an electronic sorting of signal from noise; by using even more sensitive solid-state or gas-flow detectors, or by using more highly refined fully focusing or semi-focusing spectrometers.

The detection and localization of artificially introduced specific heavy metal stains may be a useful adjunct to X-ray microanalysis. The use of X-ray microanalysis may be particularly useful in the detection of the end product of enzyme cytochemistry where this is a heavy metal or a suitable ligand which may be bound to a heavy metal. The specificity of silver salts for halides, the stochiometric binding of uranium to DNA, the affinity of lead salts for RNA containing nucleoproteins, and the availability of ferretin-coupled antibodies, to name but a few, show that this aspect of the technique may provide yet another tool for microanalytical studies.

9.2.4. Transmissive operating mode

Additional information may be obtained on the internal structure of specimens which are thin enough to permit the penetration of the incident electrons. By placing a scintillator and photomultiplier below the specimen, such a transmitted electron signal may be detected and displayed on the cathode-ray tube. An aperture is placed in the electron beam axis and rejects any diffracted electrons and admits only those which penetrate in a direct line. By off-centring the aperture it is possible to carry out dark-field studies.

The images that one is able to obtain by this technique are similar in most respects to the image obtained by the more conventional transmission electron microscope (see Fig. 7.13). One advantage of the scanning transmission electron image is that it has a much higher inherent contrast and as the final image is electronically amplified this contrast may be controlled and enhanced. This feature is most useful in the examination of biological material, for it is here that the natural contrast is low. Another advantage is that the very low beam current 10^{-2} to 10^{-3} nA produces only negligible specimen damage.

The SEM transmission mode is able to produce images of specimens which are 3–4 times thicker than those examined in the conventional transmission microscope running at the same accelerating voltage (see Fig. 5.3). In practice, this means it is possible to examine specimens of between 0·5 and 1·0 μm. This is because the detector in the SEM acts as an image intensifier, and such a system in a transmission microscope will only operate on specimens thin enough to allow the passage of a beam of electrons which are nearly monochromatic.

The resolution in the transmission mode of scanning electron microscopes is presently between 5 and 7 nm, but as gun brightness increases and probe diameter decreases this will be improved. The recent paper by Swift and Brown (1970) admirably demonstrates the range of sectioned biological material which may be examined in the SEM operated in the transmission mode. These workers have shown that it is possible to examine quite large specimens up to 5 mm^2 of resin-embedded material at resolutions of between 15 and 20 nm. This large specimen size and intermediate resolution conveniently fills a gap which exists between the capability of the light microscope and the transmission electron microscope. This mode of operation provides a useful adjunct to the SEM which may become more pertinent with the advent of microanalysis of thin specimens. It cannot, however, in its present form compete with the excellence of information which may be obtained from conventional transmission electron microscopes which is usually better than 0.5 nm. It should be noted that the type of transmission scanning electron microscopy described in this section is normally possible in the commercially available scanning electron microscopes. It should not be confused with the high-resolution scanning transmission electron microscopy pioneered by Crewe and his co-workers, which employs field emission sources operating at ultra-high vacuum pressures.

A useful technique which has been applied in transmission electron microscopy is of examining the remains of the specimen after accurately controlled microincineration. There seems to be no reason why the same technique should not be applied to scanning electron microscopy and microanalysis carried out on the incinerated remains. Such incineration may either be by heat on one of the newly designed hot and cold stages, or by a cold ashing process in which oxidation is carried out using radio-frequency excited oxygen.

9.3. ANCILLARY TECHNIQUES

9.3.1. Introduction

In the past few months a number of ancillary techniques and smaller pieces of instrumentation have become available for use with the SEM which are of considerable use to the biologists. Three of these devices and techniques will be described, and although they will be dealt with separately, it will become clear that they may most profitably be used together.

9.3.2. Low-voltage operation

Boyde and Wood (1969) emphasize that the scanning microscope should not operate at as high an accelerating voltage as possible when examining biological material. A complex interplay of factors determine the operating conditions and, more importantly, the optimum amount of information which may be obtained from the specimen. By operating the primary beam at high accelerating voltages this permits a reduction in probe diameter with a consequent increase in resolution. However, this increase in beam voltage also leads to deeper penetration by the beam, which in turn may on a highly ornamental sample give rise to secondary electron emission from a depth of several micrometres instead of 20–30 nm (see Figs. 3.7, 3.8, 4.5, 7.1 and 10.1). These secondaries will contribute spurious brightness variations in the image which can give rise to errors in interpretation. Another consequence of deeper penetration is that it may give rise to uneven specimen charging which will either emit more secondaries or will deflect the primary beam. The net effect of this uneven charging is uneven contrast and problems of astigmatism at higher magnifications. The contrast variations, like the brightness variations, make meaningful interpretations more difficult.

However, these are extreme examples, for it is usual to apply a conductive coating of carbon and some heavy metal to the specimen prior to its examination in the microscope. The conductive coating of metal serves a number of functions, all of which improve the specimen image. The metal layer allows any electrical or thermal build up to leak away and is in itself a better emitter of secondaries than the underlying lower atomic number of the biological material. But the application of the metal coat, although improving the information in the emissive mode, precludes obtaining meaningful data using the cathodoluminescent and X-ray microanalysis mode of operation.

Boyde and Wood (1969) maintain that for examining biological material the accelerating voltage should not exceed 10 kV. At this voltage the resolution is in the 20–30-nm range, but this is accompanied by a dramatic decrease in artefacts.

Much has been talked and written about improving the resolution of electron optical images, and only passing reference has been made concerning the information content of the specimen. One of the outstanding advantages of scanning electron image information is that it provides a much wider range of information transfer systems than is available from transmission eleetron optical systems, albeit at a reduced resolution. In some instances, as will be seen later, it is not practicable to apply a surface coating, and in order to avoid the adverse charging effects, it is necessary to operate at considerably reduced accelerating voltages. We have carried out a series of experiments in which fresh material was examined uncoated at an accelerating voltage of between 1 and 2 kV at a resolution of between 80–90 nm. There is considerably less beam penetration and, more importantly on biological material, a considerable reduction in the radiation flux impinging on the specimen.

As well as examining fairly hard biological material, such as pollen grains and insect cuticle, it has also been possible to view hydrated materials such as mesophyll leaves. Heslop-Harrison and Heslop-Harrison (1969) were able to observe considerable detail in an untreated leaf of the xerophyte *Dianthus plumarius*, and attributed some of their success to the ability of the vacuum system in the Stereoscan scanning microscope to take up any water which is evaporating from the specimen. With mesophytes and hydrophytes, although there was considerable cellular collapse due to the loss of water, the presence of dissolved minerals and ions in the cell sap are thought to have contributed to the generation of secondary electrons, and provided a conductive pathway. The advantage of the low-voltage operation is that it allows a rapid examination of material at medium resolution without the disadvantage of the time-consuming preparative and coating techniques. Even fairly labile specimens may be examined provided this is done within a few minutes of their being introduced into the microscope column.

9.3.3. Ion beam etching

It is now possible to selectively remove parts of specimens inside the microscope column. An ion source has been designed to operate in the specimen chamber of the Stereoscan SEM (Echlin *et al.*, 1969). The ion source is a demountable cold cathode argon discharge unit which may be focused onto any area within the dimensions of the specimen stub. The ion beam has been used in two ways. Firstly, the beam was broadened and the whole of the specimen stub irradiated for known time intervals. Following each irradiation the specimens were examined and if necessary photographed. Secondly, the beam was carefully directed onto the centre of the specimen and irradiation continued until a small part of the specimen was removed. In practice it was found possible to traverse the specimen at known spatial increments across the etching profile, and in homogeneous material it is possible to successively examine and photograph non-etched through to etched material. Following the ion beam etching, the time of which varies with the nature of the material, the specimens were examined in the emissive mode of operation at between 1 and 10 kV accelerating voltage. In spite of the absence of a surface coating it has been possible to obtain a resolution of better than 100 nm. Using this ion source the internal morphology of a number of complex resistant biological objects has been examined. An examination of ion-beam-etched Ambrosia pollen grains (see Echlin *et al.*, 1970a) show that the first signs of etching are in the small openings of the tectum. As the etching proceeds, these holes are progressively enlarged and within a short time the tectum is entirely removed. The underlying bacula are revealed not as solid pillars, but as short spikes, and it is thought that some etching of these structures takes place before the tectum has finally disappeared. Once the exinous pollen grain wall has been removed by the ion beam, the cellulosic intine is rapidly disintegrated, revealing the underlying cytoplasm which is even more quickly removed. Figure 9.4 is an example of the effects of ion etching on a leaf.

The advantage of having the ion beam source within the microscope is that it is possible to carry out a controlled etching process and immediately examine the uncoated specimen at low accelerating voltages. Should high-resolution results be required, the specimen may be removed from the column, given a metal coat, and re-examined. Further etching may then be carried out as it is a simple process to remove the metal coating. This work has been carried further, and preliminary studies (Echlin *et al.*, 1970b) have shown that it is possible to ion etch soft biological material maintained at low temperatures. The etching rates appear to be decreased, which may mean either a reduction in the heating effect and/or the thermal effect has little significance in the etching process.

Mention should be made of two other methods of ion-beam etching which, although they suffer from the disadvantage that they may not be carried out *in situ* in the microscope column, may provide a way around the problems of interpretation of the final image.

Boyde and Stewart (1962) employed a glow-discharge sputtering technique to etch polished surfaces of dental tissue. Ions extracted from the glow discharge are attracted towards the surface to be etched. Each ion is thought to knock off one atom from the bombarded surface, but this "one for one" is not always maintained because

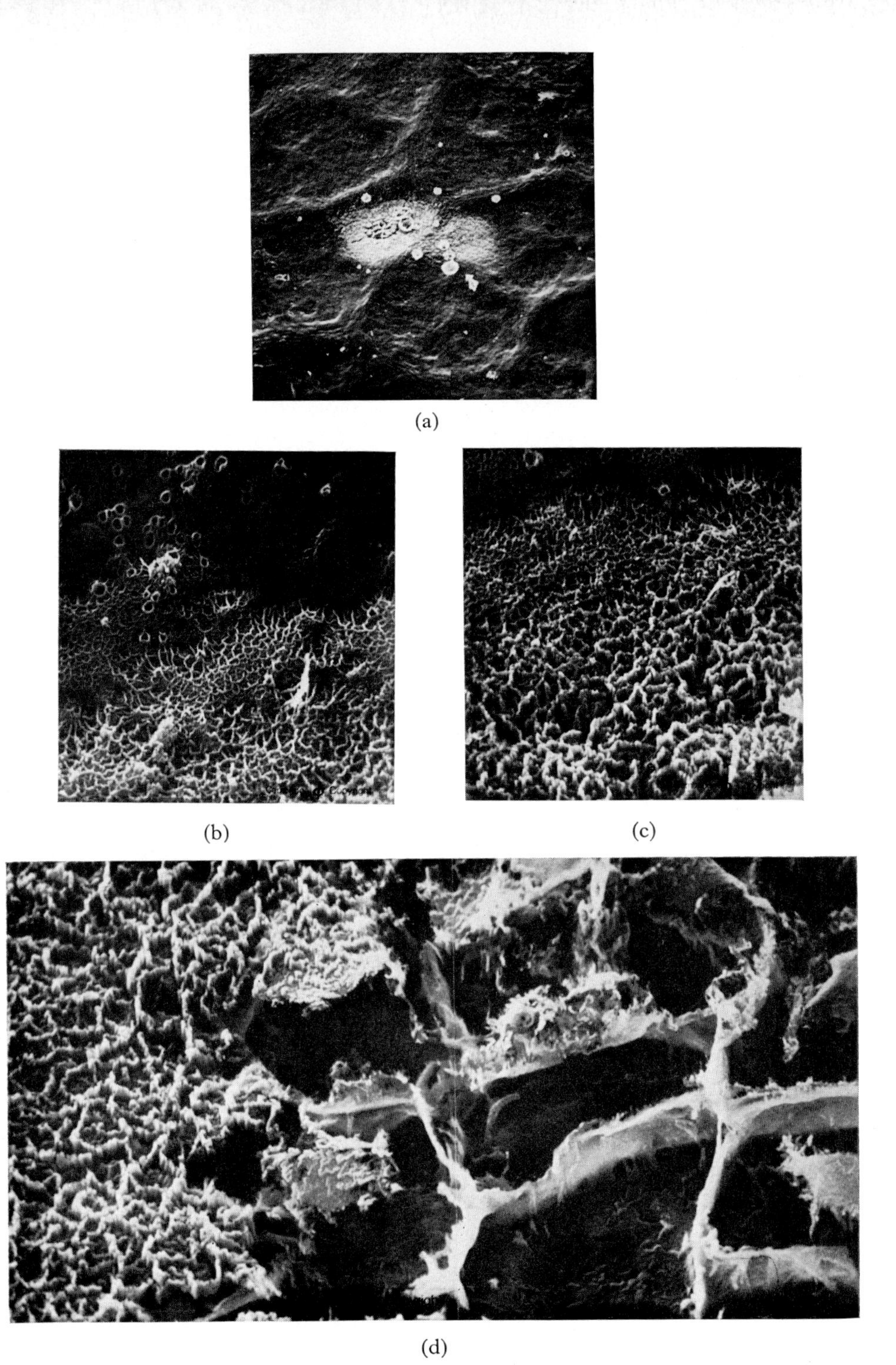

(a)

(b)

(c)

(d)

FIG. 9.4. The effects of ion etching on an oak leaf. (a) The surface of a leaf showing the etched region in the centre ×60. (b, c, d) Areas from the outer perimeter of etching to the centre, respectively ×1500. Note that in (d) the underlying structure can be seen where the heaviest etching has occurred. (Compliments of D. Kynaston and P. Echlin, Cambridge Scientific Instruments Ltd., Cambridge.)

of the direction of entry of the ions and the inadequate control of the energy distribution of the ions. The rate of sputtering is determined by the ion mass and energy, and the atomic weight of the target atoms and their binding energy. These workers considered that although differences in specimen density may be related to specimen topography, some of the differences in etching rates were related to differences in mineral content and crystallite orientation. The observed effects were unlikely to be chemical as an inert gas (argon) was employed at a low pressure. However, the heating effect was not entirely negligible and may have contributed to the removal of material.

Another approach, which shows considerable promise, is seen in the recent paper by Stuart *et al.* (1969). These workers employed a radio-frequency sputtering using a voltage alternating with a frequency of 10 MHz. This alternately in half-cycles induces bombardment of the specimen with ions and neutralization with electrons. In practice, it was found that after a few cycles the specimen builds up a negative charge because the lighter weight electrons are more mobile than the heavier ions. Once the negative charge had built up it meant that there was a positive ion bombardment during all but a small fraction of the cycle, and with an increased efficiency. Using this technique Stuart *et al.* (1969) found that the etching profiles were dependent on the gas used, the pressure of the system and the duration and power of the etching. They used a variety of gases, some, i.e. hydrogen, which were chemically more reactive than others (argon).

There are a few problems which still remain, particularly regarding the correct interpretation of the image (see Fig. 9.5). It is not entirely clear whether the ion beam removes material by a charring and sputtering process, or whether the mechanism is one of true vaporization. When the ion beam hits the specimen it may cause it to accumulate a positive charge. This results in a deflection of further ions and a consequent diminished etching in this region, whereas other non-charged regions will continue to etch. One way to prevent this would be to alternately pulse the ion beam and the electron beam to give a neutralizing effect, but this would also give rise to considerable contamination of the scintillator. It is also clear that two quite different etching profiles may be obtained if the etching is carried out at two different angles. These differences are thought to be related to the crystallinity and sub-structure of the specimen. Crystalline materials have differential rates of etching along different crystal planes.

Finally, it is obvious that soft tissue etches much more rapidly than hard tissue. It is possible to have the situation where a small piece of resistant material protects an underlying layer of soft tissue, which everywhere else is being rapidly eroded away. This gives rise to the "mushroom effect" with a small piece of material perched on top of a long column. The interpretation of these spikes, which always point in the direction of the ion beam, is problematic. They may represent either a genuine difference in sub-structure, crystallinity or hardness, or may represent a charged area which had deflected further ions.

In order to interpret the appearance of ion-etched material, it is necessary to distinguish between the differences due to selective etching based on tissue resistance, the difference arising from variations in the crystallinity and sub-structure of apparently homogeneous materials, and the uneven charging effects of the ion beam. It is likely that more meaningful information may be obtained from ion-etched material if, as

Boyde and Stewart (1962) suggested, a fine metal grid be placed over the specimen. This would protect some of the tissue, and it is the profiles of this protected material which may contain information free of artefact. Another approach suggested by Boyde and Wood (1969) would be to infiltrate soft biological material with a suitable embedding material and prepare a cut or smooth surface. The etching rate might then be related to compositional difference reflected in the degree to which infiltration had taken place.

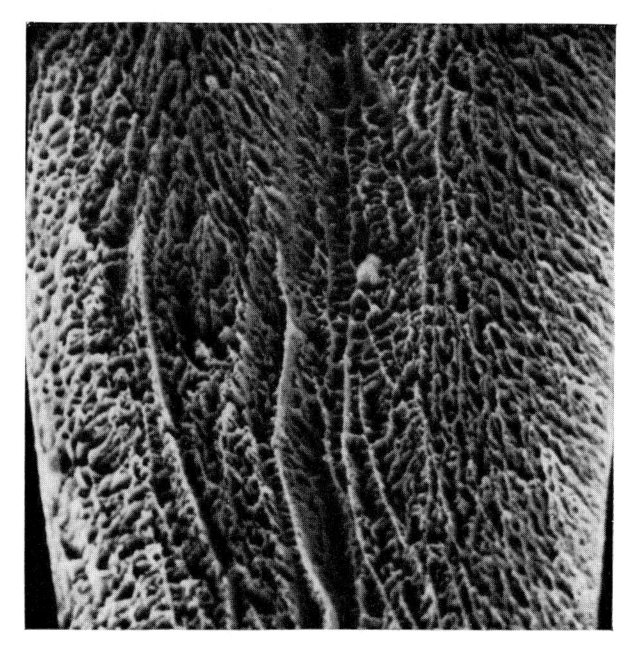

FIG. 9.5. A cotton fibre which has been ion etched. Note the complexity of the etched surface. (Compliments of J. T. Sparrow, Dept. of Textile Technology, UMIST.)

9.3.4. Examination of specimens at low temperatures

Preliminary studies have been made on a versatile temperature-controlled stage module which has been developed for the Stereoscan SEM. The preliminary design aspects of the stage are given in the paper by Echlin *et al.* (1970a).

A wide range of labile biological tissue has been examined in the microscope at temperatures ranging from +25°C to −180°C. The two principal methods of specimen preparation were as follows. Initially it was found that biological material could be quench-frozen by plunging specimen stubs with the material mounted on the surface into liquid nitrogen at −196°C. With plant material such as leaves of *Lagarosiphon major* (an aquatic angiosperm) and *Tradescantia bracteata* (a terrestrial angiosperm) and root hairs of *Sinapis alba* and *Zea mays*, adequate preservation was obtained with little or no evidence of ice-crystal damage. It is clear that the cellulose wall of plant cells is sufficiently robust to prevent or retard extensive ice-crystal formation and growth.

These procedures were unsatisfactory for optimal preservation of microalgae, protozoa and animal tissue and cells. Such extremely hydrated and delicate tissues were drained or damp-dried and either placed directly onto cooled aluminium specimen stubs or onto clean platinum discs which in turn were fixed to cooled specimen stubs by a spot of silver dag. The specimen stubs were then rapidly plunged into liquid Freon 22 (monochlorodifluoromethane) maintained at its melting point of —140°C by liquid nitrogen. The Freon 22 quench-freezing invariably gave specimens free of ice-crystal damage. A mixture of iso-pentane and methylcyclohexane, although an adequate quench coolant, was unfortunately transferred via the specimen stub into the microscope and it was impossible to maintain an adequate vacuum in the instrument.

It has been found to be undesirable to infiltrate specimens with cryoprotective substances such as dimethyl sulphoxide, polyvinylpyrrolidone and glycerol which have been shown to substantially lower the nucleation rate of water. Although these chemicals are incorporated into material prior to their preparatory cryofracture techniques, their low vapour pressure precludes their presence in specimens which are going to be introduced into the microscope column operated at room temperatures because they are a source of contamination and may obscure specimen detail. However, although their mobility will be considerably reduced at low temperatures, we have been unable to demonstrate any advantage to using them prior to the quench-freezing. Heat is only extracted from samples through the surface and as a consequence one only obtains vitrification within the upper 1–2 μm. Below this depth there is always the danger of ice-crystal damage.

Moor (1970) has shown that there are alternative ways to reduce the critical freezing rate other than by using cryoprotective substances. Such methods include ultra-rapid quench-freezing at a rate of at least 10^4 °C/sec, and the application of high pressure during the freezing process which lowers the rate of ice-crystal growth. It is now possible to obtain water vitrification to a depth of about 200 μm from the surface of a specimen, and as the emissive mode of scanning microscopy only obtains meaningful information from within a few hundred angstroms of the surface, the low-temperature preparative techniques would appear to be more than adequate for the available instrumentation technology.

Following the quench-freezing the specimen stubs were quickly (within 4–5 seconds) transferred to the cold stage maintained at —180°C with liquid nitrogen and the microscope column pumped down to 1×10^{-4} torr its working vacuum. An immediate examination of the specimens revealed that they were covered with ice. This water is derived from two sources. In the case of aquatic organisms or cells maintained in an aqueous medium, the water was part of their immediate environment. Such organisms were either examined in their natural state, or carefully washed in isotonic salts solution to remove any surface debris. A certain amount of atmospheric water will condense onto the specimen during the transfer from the quench-coolant to the specimen stage. We were initially concerned at the presence of this water, and elaborate steps were taken to construct a dry-environment specimen transfer chamber. But as will be shown it is relatively easy to remove the surface water within the microscope so these cumbersome procedures were abandoned. The temperature of the stage was raised to between —100°C and —90°C whereupon the high vacuum within the microscope column caused the surface water to sublime revealing the tissue surface below. The removal of water had to be carefully monitored, and the temperature was not allowed to rise above —85°C. If too much water was removed, by allowing too great a rise in temperature, then the tissue collapsed. The same tissue showed considerable collapse and surface artefacts when fixed in buffered glutaraldehyde and then taken through a freeze-substitution procedure. Examination of material on the cold stage dramatically demonstrates that water is a vital structural component in many plant and animal cells.

The high stability of the specimen stage and the fine-temperature-control mechanism allowed the removal of ice to be accurately monitored either by direct observation or by following the temporary increase in pressure in the specimen chamber. Surprisingly large amounts of water could be removed from the specimen and in no instance did we observe or experience inoperatively high pressures. When the removal of the surface water was completed, and this usually took place within 3–5 minutes of placing the specimen on the cold stage, the temperature of the stage was maintained at between —140°C and —100°C during the examination of the specimen.

All the specimens were examined in the emissive mode of operation over a wide range of accelerating voltages without the benefit of surface coating. Attempts were made to spray "Duron" antistatic spray onto the surface of the *Tradescantia* leaves prior to quench-freezing. This procedure was abandoned as it gave such consistently poor results because the micro-droplets of the antistatic agent obscured any surface detail. In spite of the absence of a surface coating it has been possible to obtain a resolution of the order of 1000 Å. Although accelerating voltages as high as 30 kV were employed, it was usual to take pictures at 20 kV or lower as this reduces the energy impinging on the specimen. It was possible to measure the radiation dosage per unit area falling on the specimen—a feature of particular importance when examining biological material.

Under typical operating conditions the total power dissipated in the specimen within the volume occupied in the electron range is of the order of a microwatt. For biological materials this can correspond to the power density of the order of one watt per gram dry weight, or 1 joule/sec/g dry weight.

The cold stage has been particularly useful in examining microalgae and protozoa, because they retained their three-dimensional shape while in the frozen state. It was possible to see details and the orientation of the pellicular warts on the myonemes of *Euglena* and the array of cilia on *Paramecium*. Preparation of these objects by more conventional means invariably resulted in serious collapse. The cold stage has been particularly useful in examining tissue culture cells. Mouse L. cells, which are sensitive to adverse conditions, were grown on platinum discs, washed several times with a buffered isotonic salts solution, before being quench-frozen. The cells retain their natural configuration, and show such features as the nuclear bulge and the delicate protoplasmic protruberances which are believed to be associated with the attachment of the cells to the substrate. It is also possible to see the fine filamentary appendages at the tips of the cells. Examination of the sample at different tilt angles relative to the primary electron beam revealed further information about the specimen. These preliminary studies clearly demonstrate the usefulness of the cold stage in the examination of biological material and the instrumentation and procedures employed ensure that the specimens faithfully retain their natural three-dimensional configuration. The only treatment which has been applied to the specimens is a quench-freezing technique which has already proven satisfactory in the preservation of cells and tissues. The results so far obtained are often better than results obtained by fixation and freeze-drying techniques. An advantage of the latter technique is that specimens may be coated before examination, thus improving both resolution and ease of interpretation. A distinct advantage of the cold stage is that images may be obtained within a few minutes of the specimen being removed from its natural environment. Future applications of the cold stage would lie in the direct examination of material prepared by cryofracture techniques (see Haggis, 1970) and in the X-ray microanalysis of cryocut thin sections at low temperatures. Preliminary experiments by Hall (personal communication) have shown that this is possible using an adapted Stereoscan cold stage.

9.4. PREPARATION OF LABILE BIOLOGICAL MATERIAL

9.4.1. Introduction

One of the main disadvantages to the examination of biological material at high resolution in both the TSEM and SEM is that the observations usually have to be made on material in a highly dehydrated state. Water is an important structural component of most biological material, and elaborate steps are taken to ensure that it is removed from plant and animal material prior to their examination in the high vacuum of electron probe instruments. Ideally, this dehydration should be accomplished without causing any specimen damage.

Certain biological materials, such as wood, the hard outer covering of seeds and pollen grains, diatoms and bone are readily examined in the scanning electron microscope. The very fact that they are hard means that they are able to resist collapse following the removal of the "structural water". Fortunately these tough resistant structures frequently contain the morphological characters which are of vital importance in taxonomy and systematics, and already we have a considerable corpus of knowledge gleaned from such features. Plant cells are generally easier to examine in the scanning microscope simply because they possess a rigid cellulosic cell wall, which may be marked or impregnated and encrusted with such resistant material as lignin, suberin and sporopollenin. Difficulties arise when hydrophytes are examined, for although they have cellulosic cell walls, their internal water is of vital structural importance. With the exception of hard animal tissue, such as teeth and bone, which are strengthened with calcium salts, most animal cells are extremely prone to collapse in a high vacuum. This same sensitivity is shared by many microbes and protozoa.

It is extremely difficult to know where the boundary exists between "structural" or "bound" water and the watery environment which bathes most cells. There would appear to be a continuous gradient of water in living tissue, the "bound" water being more difficult to remove than environmental water. The problems arise during removal of water by one means or another, for although it is desirable to remove "environmental" water, removal of "bound" water could so alter the natural configuration of the specimen as to introduce serious artefacts.

It is thus usually necessary to pretreat delicate biological material before it is examined in the scanning microscope. Unless adequate care is taken to preserve the three-dimensional topography of delicate specimens, the taxonomic use of any distinctive characters will be obliterated by the artefact of collapsed cells. The adequate preservation of the surface is vital to the meaningful interpretation of any image obtained by electron probe instrumentation. Yet it is not always necessary to preserve the surface of the entire structure, because, depending on the level at which detail is being sought, the collapse of the whole cell or tissue may not be manifest. If, however, the interrelationship of prominent features over a wide area is important, then although this large area must be adequately preserved, the intervening details are of little consequence. Although it may be dangerous to make "general rules" for any biologic preparative technique, it is probably only necessary to ensure that the preparative technique maintains preservation to about one-fifth of the size of the features being examined. Thus, there seems to be little point in ensuring faithful preservation of material down to the 10 Å level when the feature to be examined is of the order of 100 nm.

A lot has already been written on the various preparative techniques, and it is impossible to cover all the methods which have ever been used. For those readers who are interested in a technique as applied to specific plant, animal or microbial material it is suggested that they refer to the excellent bibliographies referred to earlier. More general references to methods which have been applied to larger groups of organisms are given by Echlin (1968), Boyde and Barber (1969), Boyde and Wood (1969), Heywood (1969), Marszalek and Small (1969). The paper by Boyde and Wood (1969) is particularly useful, as it contains an excellent review of the techniques for preserving soft animal tissue, most of which have been pioneered and tested by Boyde and his co-workers.

The purpose of this section can be no more than to describe in outline the general type of methods which have been utilized in the preservation of labile biological material, and this will be followed by a brief discussion of some newer techniques which are becoming available. The methodology which follows is only really applicable to the examination of material which is to be examined in the emissive mode of operation. When the cathodoluminescent and X-ray microanalysis modes are to be used, it may be necessary to modify these methods.

9.4.2. Minimal preparation

Before any preparative technique is initiated, it is necessary to ensure that the surface of the specimen is free of any extraneous and foreign material. With dry resistant specimens this may be accomplished by a jet of clean air across the specimens once they have been firmly affixed to the specimen stub. The small cans of ultra-clean Freon gas which are commercially available are ideally suited for this purpose. With wet specimens it is necessary to wash the sample in a suitable ionically balanced solution, which is then followed by a brief rinse in distilled water. The latter treatment will remove any salt crystals which may obscure surface detail. For strongly adherent material it may be necessary to carry out a more turbulent washing procedure, or even resort to a brief exposure to ultrasonic waves.

As previously indicated, most plant material can be examined with no more treatment than controlled air drying followed by the use of chemical desiccants. Even some aquatic plants have been examined by these methods and as was shown earlier Heslop-Harrison and Heslop-Harrison (1969) were able to observe considerable detail in an untreated leaf of the xerophyte *Dianthus plumarius*.

9.4.3. Fixation and dehydration techniques

A measure of success has been obtained using some of the preparative methods devised for the examination of material at the level of the light microscope. Such procedures are adequately described in books on preparative microscopy and will not be dealt with here. Following the fixation, it is necessary to dehydrate the specimens, and this is accomplished by passage through graded ethanol or acetone. Better results have been obtained using the preparative techniques which have been applied to material prior to its examination in the transmission microscope. There are many recipes available, but the general concensus of opinion is that a fixative containing an organic aldehyde such as formaldehyde or glutaraldehyde probably gives the best results. The organic aldehydes tend to cross-link free amino groups and toughen tissue and this lessens the degree of collapse which usually occurs in the ensuing dehydration procedures. A number of workers recommend the use of hardening agents such as Parduczs fixatives which tend to harden biological tissue. It is clear, however, that the dehydration procedure is where the damage occurs in spite of the seemingly judicious application of fixatives. The answer would seem to lie in adopting dehydration techniques which do not involve distortion of the material by surface tension forces which occur as the receding liquid evaporates.

9.4.4. Sectioned material

Internal details of specimens may be revealed by subjecting the material to mild disruptive techniques such as sonication and alternate freezing and thawing or even gentle mechanical teasing apart of tissues. Although these methods disrupt the specimens and show internal structure, only a limited value may be placed on the authenticity of the structures revealed because of the ever-present danger of introducing artefacts. For fuller details of this approach reference should be made to the paper by Boyde and Wood (1969). Following fixation and dehydration, considerable success has been obtained from an examination of wax-embedded material. The samples are prepared in the usual way, embedded in wax and thin sections cut using a glass knife to lessen the chance of scratch marks. The sections may then be dewaxed in xylol, washed in ethanol and examined in the microscope. A limited amount of success has been obtained using epoxy-resin embedded material from which the resin is removed using sodium methoxide.

Cleveland (1970) has recently described a plastic infiltration technique as a means of preserving tissue for the SEM. The method is based on the theory that plastic monomers are able to pass into the cell, but upon partial polymerization the short-chain polymer is too large to diffuse back. The extracellular polymer is washed off the surface of the specimen without removing the intracellular polymer, and on full polymerization the specimens are well preserved. An added advantage of this technique is that it allows suitably sectioned parts of the same material to be examined in the conventional transmission microscope as well as the scanning microscope, thus providing the valuable interface between the two techniques.

Haggis (1970) has demonstrated that it is possible to examine the internal structure of complex tissues such as mouse kidney and heart by the technique of cryofracture. In

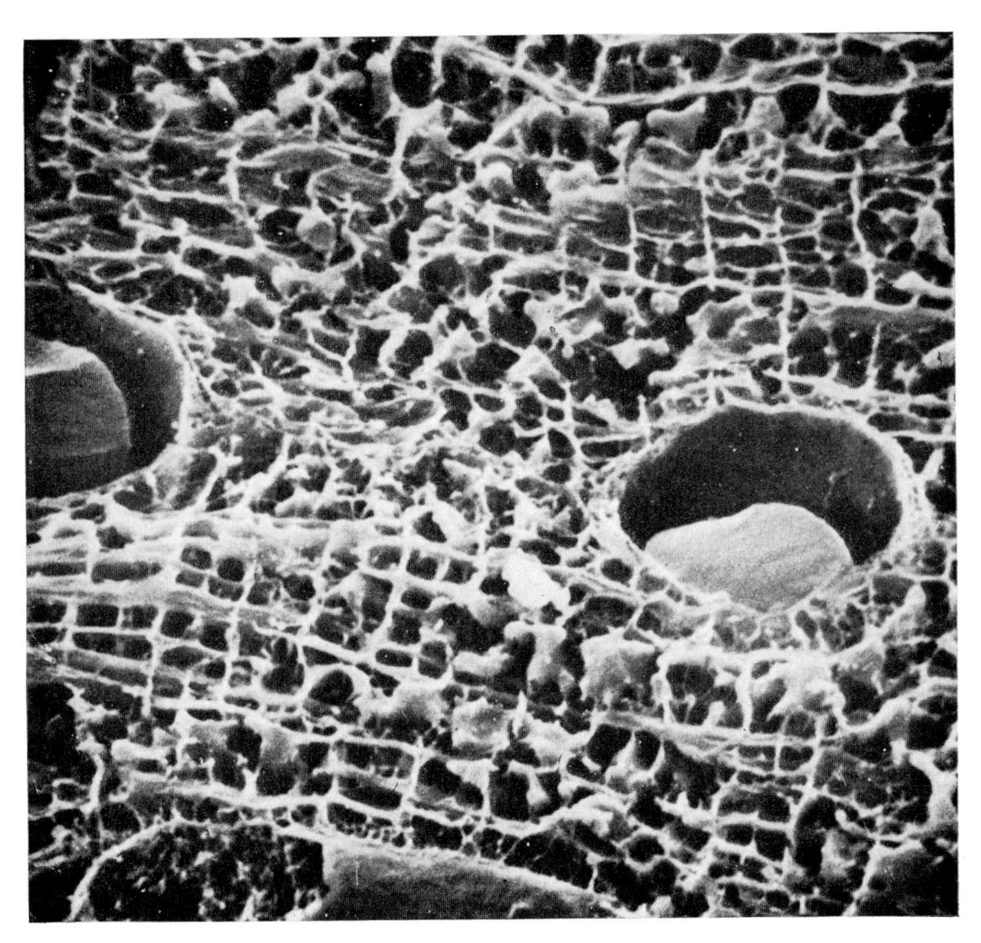

FIG. 9.6. The internal structure of heart muscle cells as revealed by cryofracture. This is a longitudinal fracture through heart muscle cells which includes two capillaries (with a fractured red cell in each) and a muscle cell nucleus with its membrane partly torn away (extreme bottom) (from Haggis, 1970). (Compliments of G. H. Haggis, Electron Microscope Centre, Canada Dept. of Agriculture, Ottawa, Canada.)

this method small pieces of tissue are quench-frozen and then fractured to reveal internal detail. The advantage of this method is that the fracture will tend to follow planes of structural weakness and thus reveal more than a cut in a single plane (see Fig. 9.6). Cryofracture takes place at a much lower temperature than that employed on a cryomicrotome and thus avoids or lessens the problems of ice crystal growth or recrystallization. The fractured surface is identical to that revealed by the now familiar freeze-etching technique, and provided the temperature is maintained at below −150°C, the sample may be examined direct in the microscope. If, however, such a facility is not available, then it is necessary to take the specimen through a freeze-drying

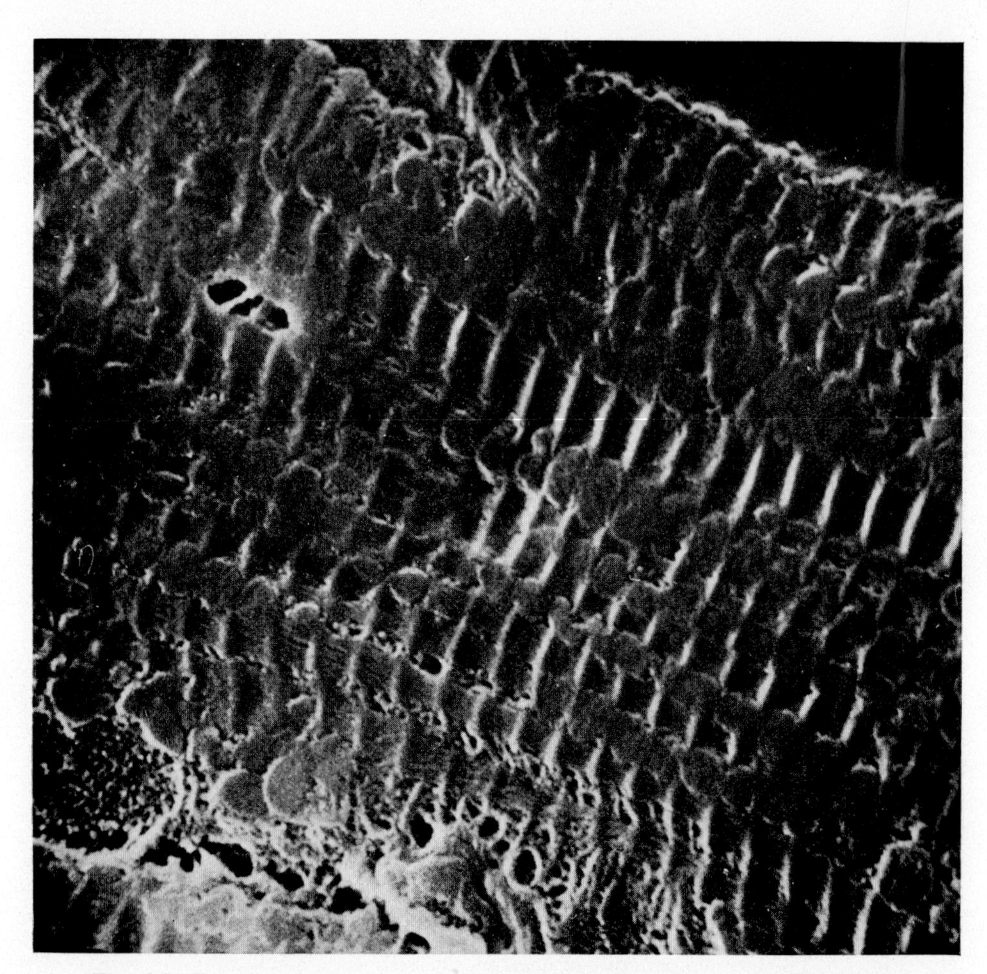

FIG. 9.7. Freezing damage reduced by replacing the tissue water with other solvent. In Fig. 9.6. the muscle cytoplasm shows extensive ice-crystal damage with fragments of Z-lines, myofibrils and mitochondria barely identifiable. The arrangements of the myofibril Z-lines and mitochondria are much better preserved as shown in this micrograph by replacing the tissue water with other solvent prior to freezing. (Compliments of G. H. Haggis, Electron Microscope Centre, Canada Department of Agriculture, Ottawa, Canada.)

procedure in order to adequately preserve the surface detail. It has recently been shown by Haggis (private communication) that if, after fixation and prior to freezing, the tissue water is wholly or partly replaced by another solvent, e.g. alcohols, amyl-acetate, dioxane, etc., then freezing damage can be much reduced as shown in Fig. 9.7.

9.4.5. Freeze-drying

In this procedure the sample, which may or may not have been prefixed, is rapidly frozen, and while still frozen, the water removed by sublimation at a high vacuum. In order to minimize ice-crystal damage it is necessary to lower the nucleation rate of water, and this may be achieved by a number of methods. It is possible to infiltrate the tissue with physiological amounts of cryoprotective substances such as glycerol, dimethyl sulphoxide or polyvinylpyrrolidone, all of which substantially lower the rate of ice-crystal formation. Unfortunately, the low vapour pressure of the chemical precludes their presence in samples which are going to be introduced into the microscope column operated at room temperatures because they are a source of contamination and may seriously obscure specimen detail. Boyde and Wood (1969) claim to have achieved a lower nucleation rate by using a 2% chloroform solution, the advantage being that the chloroform would be more easily removed in the high vacuum. Reference has already been made to Moor (1970) who has demonstrated that it is possible to lower the critical freezing rate other than by using cryoprotective substances. Boyde and Wood (1969) have described in detail the procedures involved in freeze-drying, especially as applied to the preparation of material for examination in the SEM, and reference should be made to this paper. Figure 9.8 is an example of a freeze-dried specimen.

9.4.6. Critical point drying

Horridge and Tamm (1969) have applied the technique originally devised by Anderson (1951) to the preparation of the protozoan *Opalina*. The organisms were fixed and then dehydrated in alcohol. The alcohol was replaced by some non-polar solvent such as amylacetate and the amylacetate and organism placed in a critical point apparatus. The bomb was flushed out with liquid carbon dioxide at room temperature to replace the amylacetate, and once the amylacetate had been removed the bomb was sealed and the temperature raised to 50°C. During the heating the carbon dioxide goes through its critical point and forms a gas without the formation of a phase boundary. The pressure was then released, the CO_2 allowed to escape and the dried organisms were placed on the specimen stub. The preparative procedure preserves the fine structural detail as well as the metachronal waves of the ciliary action. Boyde and Wood (1969) have also used this procedure and, while they attest to its usefulness, point out that if not carefully applied may easily give rise to artefacts and result in bulk shrinkage. At present, too little work has been carried out using the critical point drying method to come to any firm conclusion regarding its usefulness. The principal advantage of the procedure would appear to be that it is easy to assemble the necessary apparatus and it is a relatively quick procedure.

All the methods so far described suffer from the disadvantage in that the tissue being examined must be sufficiently robust to retain its natural configuration in the absence of structural water. This appears to be the case with plant cells which, as has been mentioned previously, are constrained by a cellulosic wall. Many animal cells may also retain their natural tridimensional configuration, but for those that do not it is necessary to allow the water to remain inside. To examine such wet specimens

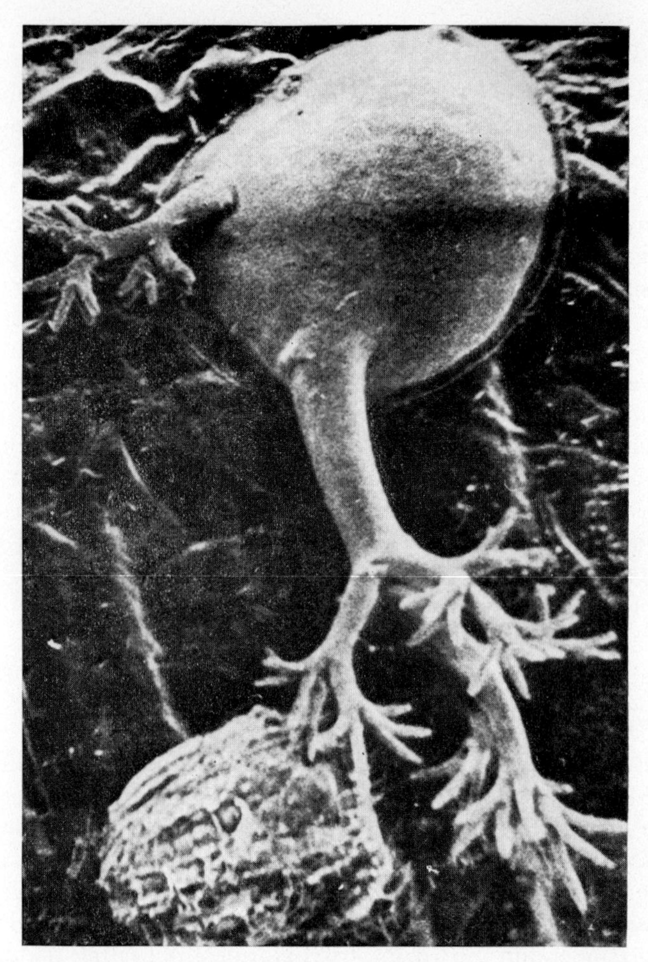

FIG. 9.8. A protozoam, *Dendrocometes*, which is found on the gill plates of *Gammerus*, a freshwater shrimp. *Dendrocometes* is a protozoan predator which sucks out the body contents of other protozoan. The gill plates of the shrimps were fixed in glutaraldehyde followed by Parducz's fixative (osmium tetroxide and mercuric chloride), briefly washed, and quickly frozen in liquid nitrogen. They were then freeze dried at 60°C and 2×10^{-4} torr for 5–6 hours. *Dendrocometes* is shown with a protozoan prey caught on the lower tentacle. (Compliments of T. Allan, Dept. of Zoology, University of Manchester.)

it is necessary to use specially designed wet chambers or to cool the specimen to a point where the vapour pressure of water is so low that it defies rapid sublimation in the relatively high vacuum in the SEM.

9.4.7. Examination of living material

The SEM has several advantages over transmission electron microscopy as far as examining living specimens. Firstly, it is not necessary to cut sections as the specimen may be examined in its entirety, and secondly, the beam current is several orders of magnitude lower than in the transmission microscope. Concomitant to this there is a considerable reduction in the amount of energy impinging on the specimen. The only organisms which have successfully been examined in the scanning microscope are specimens which can either retain their water against a high vacuum, or can undergo a considerable dehydration/hydration cycle and still remain viable. Before discussing the attempts which have been made to examine hydrated material and living organisms, it should be remembered that Thornley (1960) calculated that the lethal dose for most robust bacteria was certainly less than that given by a beam current of 10^{-2} nA scanning a 100 μm^2 raster in 1 second. Hayes and Pease (1968) were able to demonstrate that all stages of development of *Tribolium confusum* were able to survive within a vacuum of 10^{-3} torr, and Humphreys *et al.* (1967) were able to examine untreated *Artemia* cysts in the scanning microscope and demonstrated that the organisms resumed activity following hydration. For the examination of wet and labile material it is necessary to use a wet stage in which the organisms remain in an environment of relative high humidity.

Another possible approach to the problem has been outlined in the recent paper by Cowley *et al.* (1970). These workers, who have built a high-voltage SEM, suggest that it should be possible to examine material in air. One advantage of their approach is that the radiation effects of fast electrons will be considerably less than those employed in conventional scanning electron microscopes.

Lane (1970) has recently described an environmental control stage for operation in the SEM. Although the design parameters provide a high water-vapour pressure at room temperature for viewing wet specimens, it is not at all clear whether this does do what the designer intended. It was hoped to provide a sufficiently high gas-flow rate so that a cloud would surround the specimen. The gas in this cloud is continuously diffused away into the vacuum and resupplied from the stage. From the preliminary results it would appear that the evaporation of the water vapour lowered the temperature sufficiently to freeze a film of water onto the specimen, thus making it extremely difficult to obtain meaningful information.

An earlier approach by Thornley (1960) was to make a small totally sealed wet chamber covered with a thin film through which the incident beam passed. A 50-μm aperture is sealed on top of the wet cell and this is covered with a plastic film only a few hundred angstroms thick. Such a film can withstand the pressure differences between the interior of the cell and the interior of the microscope column, while still allowing free entry of the primary electron beam. If the vacuum pumping system in the microscope is sufficiently efficient, the plastic film may be dispensed with, as the relatively slow passage of water vapour through the small aperture could be dealt with by the pumps or trapped onto a suitably chilled surface. This would not, however, avoid excessive scattering of the primary beam by water molecules. This wet-cell approach may prove to be useful if sufficiently small and sensitive scintillators and detectors may be constructed so as to operate inside the specimen chamber.

Swift (unpublished data) has constructed a moist chamber for use with scanning transmission electrons and preliminary results indicate that it is possible to visualize microalgae at a resolution approaching that obtained by light microscopy.

9.4.8. Examination of materials at low temperatures

Details of this method of specimen preparation have been given earlier in the chapter under the heading of specialized techniques for use in the SEM. Ideally one would like to examine untreated quench-frozen specimens, but as indicated earlier, this may not give completely adequate preservation unless very specialized and expensive instrumentation is employed. For most work it should be possible to employ partially preserved or cryoprotected material, which at some stage in its preparative procedure should be infiltrated with a dilute solution of one of the commercial antistatic agents. Sikorski (personal communication) has found that a one-millionth dilution of Duron in water considerably improves the conducting capability of biological specimens.

9.4.9. Surface coating

Following the adequate preparation of the specimens by whichever means is considered best, it is usual to coat the specimen with a layer of suitable conducting material prior to its examination in the microscope. This layer, which is usually in the form of evaporated carbon and/or a heavy metal such as gold, provides a route for the transferences of excess thermal and electrical build up, both of which may considerably distort the final image. A number of surface coatings and the techniques for their application are now available, and for further detail reference should be made to the publications at the beginning of this section. In some instances it is not practicable to apply a surface coating, and in order to avoid the adverse charging effects it is necessary to operate at considerably reduced accelerating voltages. We have carried out a series of experiments in which fresh material was examined uncoated at an accelerating voltage of between 1 and 2 kV at a resolution of between 80–90 nm. There is considerably less beam penetration and, more importantly on biological material, a considerable reduction in the radiation flux impinging on the specimen.

The advantage of the low-voltage operation is that it allows a rapid examination of material at medium resolution without the disadvantage of the time-consuming preparative and coating techniques. Even fairly labile specimens may be examined provided this is done within a few minutes of their being introduced into the microscope column. Another way around the problem of surface charging would be to adopt the procedure suggested by Pfefferkorn (1970), in which the specimens are exposed to the vapour of OsO_4 before being placed in the microscope.

9.4.10. Surface replication

Following preparation by one or more of the general methods already described, it may be desirable to make a surface replica to examine in the microscope. The techniques for this procedure are adequately described in modern manuals of electron microscopy techniques and will not be dealt with further at this point. The one advantage of this approach is that it does provide a cast of the surface which may also be examined in the TEM. The surface-replication techniques are more useful in scanning microscopy where they are taken from tissues which are difficult to examine in the SEM.

This is an extension of a technique utilizing silicone rubbers to make stomatal counts on angiosperm leaves, and the replicating agents which may be employed include polystyrene (Chapman, 1967), celluloid (Fujita *et al.*, 1969) and synthetic elastomers (Carteaud, 1970). A disadvantage of such substances is that they employ non-polar lipidophillic solvents, and it may be worth trying to make replicas from such substances as methylated celluloses and water-soluble epoxy, acrylic, and polyvinyl resins.

REFERENCES

Anderson, T. F. (1951) *Trans. N.Y. Acad. Sci.*, Ser. 11, **13,** 130.

Boyde, A. (1970) *Proc. 3rd Annual Scanning Electron Microscopy Symposium* (ed. O. Johari), IIT Research Institute, p. 105.

Boyde, A. and Stewart, A. D. G. (1962) *J. Ultrastruct. Research* **7,** 159.

Boyde, A. and Barber, V. C. (1969) *J. Cell Sci.* **4,** 223.

Boyde, A. and Wood, C. (1969) *J. Micros.* **90,** 221.

Carteaud, A. J. P. (1970) *Proc. 3rd Annual Scanning Electron Microscopy Symposium* (ed. O. Johari), IIT Research Institute, p. 217.

Chapman, B. (1967) *Nature* (*Lond.*) **216,** 1347.

Cleveland, P. H., Branson, D. R. and Schneider, C. W. (1970) *Third Annual Stereoscan Colloquium*, presented by Engis Equipment Corp., Chicago.

Cowley, J. M., Smith, D. J. and Sussex, G. A., (1970) *Proc. 3rd Annual Scanning Electron Microscopy Symposium* (ed. O. Johari), IIT Research Institute.

Crewe, A. V. (1970) *Proc. Roy. Soc.* (*London*) (in press).

Crewe, A. V. and Wall, J. (1970) *J. Mol. Biol.* **48,** 375.

De Mets, M. and Lagasse, A. (1971) *J. Micros.* **94,** 151.

Echlin, P. (1970) *Proc. Roy. Soc.* (*London*) (in press).

Echlin, P. (1968) *J. Royal Micros. Soc.* **88,** 407.

Echlin, P., Kynaston, D. and Knights, D. (1969) *Proc. 27th Ann. Meeting EMSA, Minneapolis* (ed. C. J. Arceneaux), p. 10.

Echlin, P., Paden, R., Dronzek, B. and Wayte, R. (1970a) *Proc. 3rd Annual Scanning Electron Microscopy Symposium* (ed. O. Johari), IIT Research Institute, p. 49.

Echlin, P., Wayte, D., Dronzek, B. and Kynaston, D. (1970b) *28th Annual Meeting EMSA, Houston* (in press).

Fujita, T., Tokunaga, J. and Inoue, H. (1969) *Archvm Histol. Jap.* **30,** 321.

Haggis, G. H. (1970) *Proc. 3rd Annual Scanning Electron Microscopy Symposium* (ed. O. Johari), IIT Research Institute, p. 99.

Hall, T. *Physical Techniques in Biological Research*, Vol. I, 2nd ed. (ed. G. Oster), Academic Press (in press).

Hayes, T. L. and Pease, R. F. W. (1968) *Adv. Biol. Med. Phys.* **12,** 85.

Heslop-Harrison, Y. and Heslop-Harrison, J. (1969) *Proc. 3rd Annual Scanning Electron Microscopy Symposium* (ed. O. Johari), IIT Research Institute, p. 119.

Heywood, V. H. (1969) *Micron* **1,** 1.

Horridge, G. A. and Tamm, S. L. (1969) *Science* **163,** 817.

Humphreys, W. J., Hayes, T. L. and Pease, R. F. W. (1967) *Proc. 25th EMSA Meeting, Chicago* (ed. C. J. Arceneaux), p. 50.

Johnson, V. (1969) *Proc. 2nd Annual Scanning Electron Microscopy Symposium* (ed. O. Johari), IIT Research Institute, Chicago, p. 483.

Lane, W. C. (1970) *Proc. 3rd Annual Scanning Electron Microscopy Symposium* (ed. O. Johari), IIT Research Institute, Chicago, p. 43.

Marszalek, D. S. and Small, E. B. (1969) *Proc. 2nd Annual Scanning Electron Microscopy Symposium* (ed. O. Johari), IIT Research Institute, Chicago, p. 231.

Moor, H. (1970) *Proc. Roy. Soc.* (*London*) (in press).

MUIR, M. (1970) *Proc. 3rd Annual Scanning Electron Microscopy Symposium* (ed. O. Johari), IIT Research Institute, Chicago, p. 129.

PEASE, R. F. W. and HAYES, T. L. (1966) *Nature* **210,** 1049.

PEASE, R. F. W. and HAYES, T. L. (1967) *Proc. 25th EMSA Meeting, Chicago* (ed. C. J. Arceneaux), p. 122.

PEASE, R. F. W. and HAYES, T. L. (1969) *Ann. N.Y. Acad. Sci.* **157,** 497.

PFEFFERKORN, G. E. (1970) *Proc. 3rd Annual Scanning Electron Microscopy Symposium* (ed. O. Johari), IIT Research Institute, p. 89.

ROSSI, F. (1968) Published by Engis Equipment Co., Morton Grove, Illinois.

STROUD, A., WALTER, L. M., RESH, D. A., MALBECK, D. A., CREWE, A. V. and WALL, J. (1969) *Science* **164,** 830.

STUART, P. R., OSBORNE, J. S. and LEVIS, S. M. (1969) *Proc. 2nd Annual Scanning Electron Microscopy Symposium* (ed. O. Johari), IIT Research Institute, p. 243.

SUTFIN, L. V. and OGILVIE, R. E. (1970) *Proc. 3rd Annual Scanning Electron Microscopy Symposium* (ed. O. Johari), IIT Research Institute, p. 17.

SWIFT, J. A. and BROWN, A. C. (1970) *Proc. 3rd Annual Scanning Electron Microscopy Symposium* (ed. O. Johari), IIT Research Institute, p. 113.

THORNLEY, R. F. M. (1960) Ph.D. Dissertation, Cambridge University.

WELLS, O. C. (1969) *Proc. 10th IEEE Annual Symposium on Electro Ion and Laser Beam Technology* (ed. L. Marton), San Francisco Press.

WELLS, O. C. (1970) *Proc. 3rd Annual Scanning Electron Microscopy Symposium* (ed. O. Johari), IIT Research Institute, p. 509.

CHAPTER 10

Faults

MRS. P. M. CROSS

10.1. INTRODUCTION

In this chapter it is proposed to discuss the recognition of faults which may occur on the instrument or in photomicrographs, and the first steps to locating the source of the faults. Some of these faults can, and indeed must, be found and corrected by the operator; others need the attention of a trained microscope engineer, but it will save time and trouble if the likely area of the fault is pin-pointed before his arrival. Not all faults which give false micrographs are due to defects in the performance of the instrument; it often happens that the condition and position of the sample itself can cause some of these faults, but if no picture at all can be obtained it can usually be assumed to be an instrument fault.

The following general comments should be read in conjunction with information in the operating manual of a particular instrument.

10.2. FAULTS WHICH RESULT IN NO IMAGE ON THE DISPLAY TUBE

10.2.1. Electronic faults

These are the class of faults which most often require the aid of an engineer, however it is worth checking one or two points before calling him. First check all the available fuses. If no gun voltage is being supplied, check then for a break in the cables from the machine to the gun. Some care must be taken in this operation since the gun operates at very high voltages. All the high-voltage power must be switched off and each piece of the apparatus to be tested must be very carefully earthed before touching it. With intermittent faults the cables and conduction paths inside the gun should be carefully checked. If no break can be found it is worth looking at the power-supply pack (with the same precautions) to see if a fault is visually obvious, though this will probably be the point at which the engineer is called.

If the first and second lenses are not working, the water pressure and safety relay on the water line should be checked before calling the engineer since if the water pressure drops the lenses are automatically cut off.

If all the components of the column are apparently working but no signal is received the position of the final aperture and the condition of the collector should be examined.

If the final aperture has slipped it will cut off the beam; this fault may be caused either by replacing the aperture carelessly or by rough handling of the aperture holder, and can be found visually by dismantling the column, putting a light in the specimen chamber and looking down onto the final lens. Collector faults do not usually completely cut out the image but give a weak and noisy picture; however, if the aluminium coating on the scintillator is badly holed or the high-tension connection is completely broken, the image may be fairly indistinguishable from noise. Other machine faults in the pre-amplifier, amplifier or scanning systems are best found by an engineer.

10.2.2. Vacuum faults

Poor vacuum is usually caused by leaks, mishaps with the baffle valves or contamination from the specimen. If the trouble starts immediately after the column has been dismantled, a failure due to a nick, misplacement or dirt in an O-ring seal should be suspected; or a piece of contamination may have been accidentally left in the system. The column should be dismantled again, and the whole system, especially the O-rings, carefully examined. Any O-ring with the slightest sign of damage should be replaced. If this does not stop the trouble, the position of the leak can sometimes be detected by spraying the suspected region on the outside of the column or elsewhere (while the system is running) with a volatile liquid, that is recommended by the manufacturer. If the pressure suddenly changes, either up or down, the position of one leak is in the region wetted by the liquid. Although any volatile liquid will show the presence of a leak the anaesthetic and fire risks associated with ether or alcohol make them unwise choices, the safest liquid is inhibited trichloroethane (Inhibisol). If the leak occurs at the junction of two demountable parts of the system it can probably be cured by thorough cleaning of the surface or replacement of the O-ring. If, however, the leak occurs in a component (for instance, in the bellows to the diffusion pump) a temporary repair may be made by sealing the area with Apiezon compound, or a more permanent repair with a vacuum resin such as Tor-seal or a silicon varnish such as GE-Vac; the component will probably have to be replaced eventually.

If part of the vacuum cycle is completed and then there is a stoppage, a fault in the relays or sticking of the baffle-valves may be suspected. The latter is fairly unlikely and the former may simply be caused by dirty relays, especially in areas with high industrial pollution. Sometimes the relays can be induced to work by shutting them manually; eventually, however, the whole vacuum system will have to be shut down and the relays cleaned. Remember to leave the backing lines running for half an hour after the diffusion pumps have been shut off, in order to give the pumps time to cool down before they come to atmospheric pressure. The relays should be cleaned with fine emery paper and wiped with alcohol.

With the more frequent use of controlled environment cells and frozen but slowly evaporating specimens, as well as other samples with volatile components, contamination from the sample itself cannot be ignored when considering vacuum faults. While the heavier contaminants will stay in the chamber, slowly increasing the astigmatism, the more volatile constituents, including water, will gather in the rotary pump oil until the backing pressure is adversely affected. For this reason the rotary pump should be left on gas ballast if there is a possibility of contaminants coming into the system or if the system is suspected of being contaminated. The flow of gas into the pump prevents condensation of contaminants in the oil, or helps to clean previous contamina-ation from it.

10.3. FAULTS WHICH RESULT IN AN OBVIOUS DISTORTED IMAGE OR IMAGE OF LOW QUALITY

10.3.1. Poor electron emission

The differences in electron emission for most metals are usually not so great as to make a serious difference in picture clarity. However, this is not the case with many other samples, especially those of low density; the poor electron emission and high

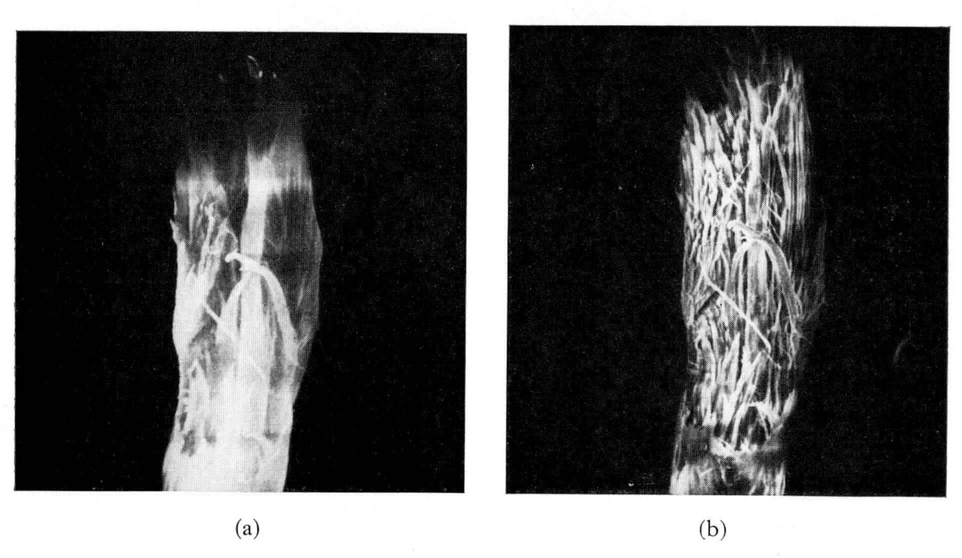

(a) (b)

FIG. 10.1. Fractured end of a cotton fibre taken at (a) 20 kV and (b) 5 kV, illustrating the loss of detail due to the extreme penetration of the beam at the higher accelerating potential.

beam penetration combine to give an image which is light and hazy with loss of detail (see Figs. 10.1, 4.5, 3.8, 7.1). The beam penetration can be brought down by reducing the voltage, and the electron emission raised by covering the specimen with a uniform layer of a good electron emitter, usually metal, so that what the beam "sees" is really a replica of the true surface, which lies below.

10.3.2. Charging

Charging can be defined as a reversible effect of the electron beam on the specimen which causes the picture to deteriorate. The processes which cause this deterioration are complex and not well understood, but may be controlled empirically quite successfully. In general, the secondary emission ratio for any particular material varies with the primary beam energy, as shown in Fig. 2.5. For beam voltages below the first

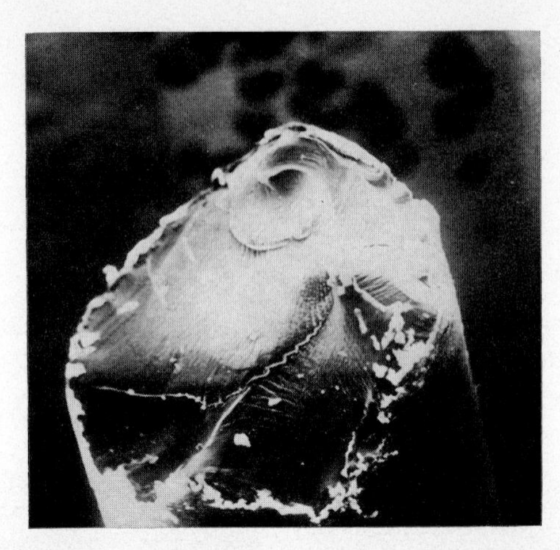

FIG. 10.2. Fractured nylon bristle illustrating charging as extremely bright areas.

cross-over point and above the second cross-over point there are more electrons arriving at the surface than being emitted from it and the surface should charge negatively. Between the cross-over points the surface should charge positively. If there is no path to earth the fields produced may distort the final image in a number of different ways. Strangely enough, although one might expect bright patches to indicate negative charging and dark patches positive charging, it is very usual to see both these effects on one specimen at the same time, showing that the charging pattern is a more complex one than the simple outline above. Indeed the patterns may be caused more by the deflection of the primary electrons to the collector (the analogue of a bright streak in optical observation) than to changes in emission. The sample may, in fact, be operating in part in the mirror mode described in Section 5.7 due to charging, rather than to deliberate connection to a high potential. The mechanism of charging has been described by VanVeld (1971) and Shaffner (1971).

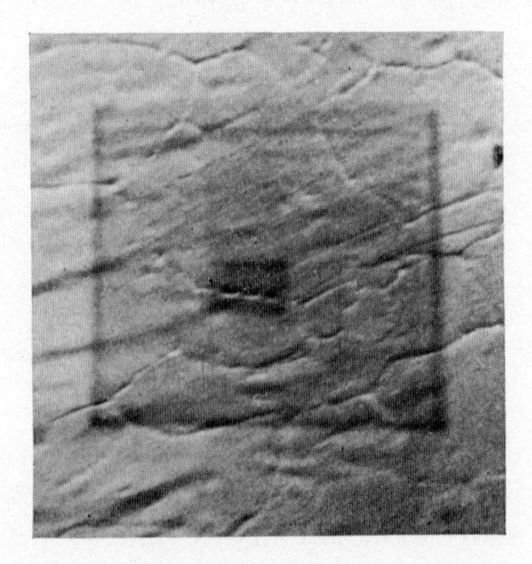

FIG. 10.3. Charging shown as dark areas corresponding to the areas scanned at higher magnifications.

FIG. 10.4. Tensile fracture of a polyethylene fibre showing charging as extreme contrast in the image.

Light or dark patterns, either static or moving, on the image of the specimen are indications of charging (Figs. 10.2, 10.3) as is extreme contrast in the image (Fig. 10.4), "banding" across part of the image (Fig. 10.5), or a distortion of the image of the specimen (Fig. 10.6) or the background (Fig. 10.7). The effects may be different on the same specimen at different magnifications although in general the charging gets worse as the magnification increases. With textile fabrics it is not unusual to get perfect pictures of individual fibres at high magnifications while suffering gross charging at the lower magnifications. Sometimes when the magnification is reduced a small square with white or black edges can be seen where the raster had been at a higher magnification, as in Fig. 10.3. If this disappears with time it can probably be counted as charging and suppressed by the usual methods; it can, however, be beam damage.

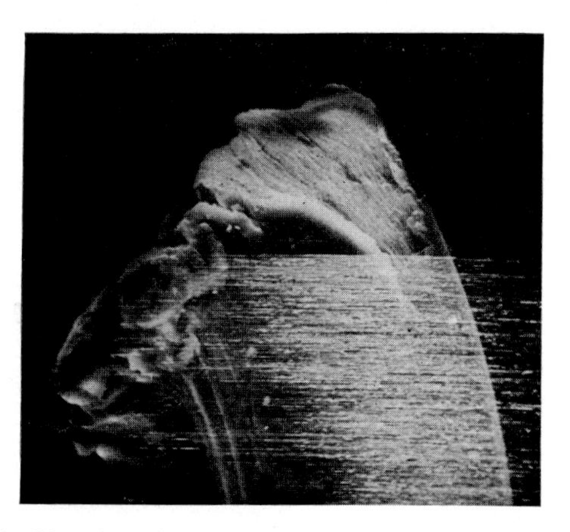

FIG. 10.5. Charging shown as bright banding across the image.

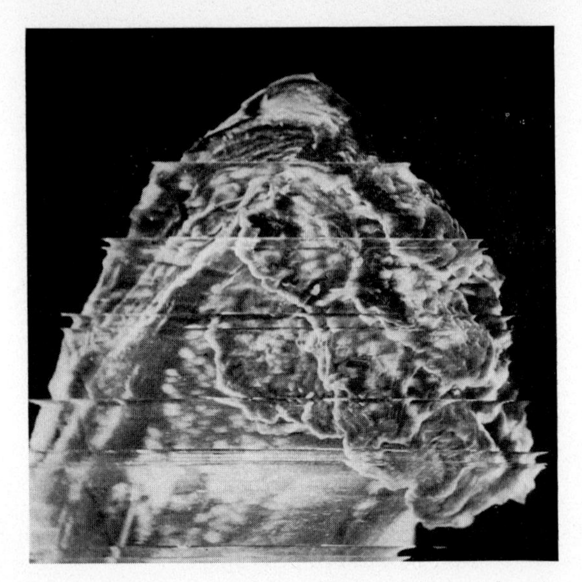

Fig. 10.6. An image which is distorted due to charging.

Charging may be controlled by changing the beam energy (in general this means lowering it) or by making a good conduction path to earth from all parts of the surface. This can be done either by coating with a continuous layer of conductive material more than 20 nm (200 Å) thick, or with a thin layer of an antistatic agent.

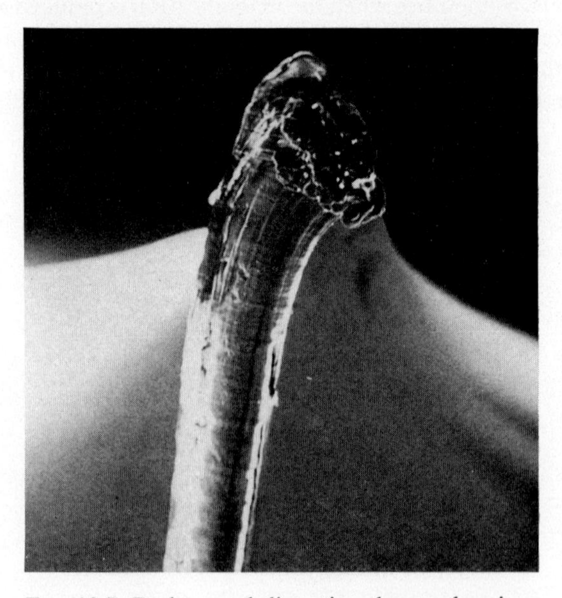

Fig. 10.7. Background distortion due to charging.

10.3.3. Beam damage

Beam damage may be defined as an irreversible effect of the electron beam on the specimen which alters its topography. It is commonly seen as a small light or dark square on the specimen corresponding to a raster at a higher magnification (Fig. 10.8) but it can also show as cracks (Fig. 10.9) or bubbles (Fig. 10.10) on the specimen. The nature of beam damage probably varies from specimen to specimen. It has been suggested that it is heat damage, irradiation damage, or that the beam acts as a source of

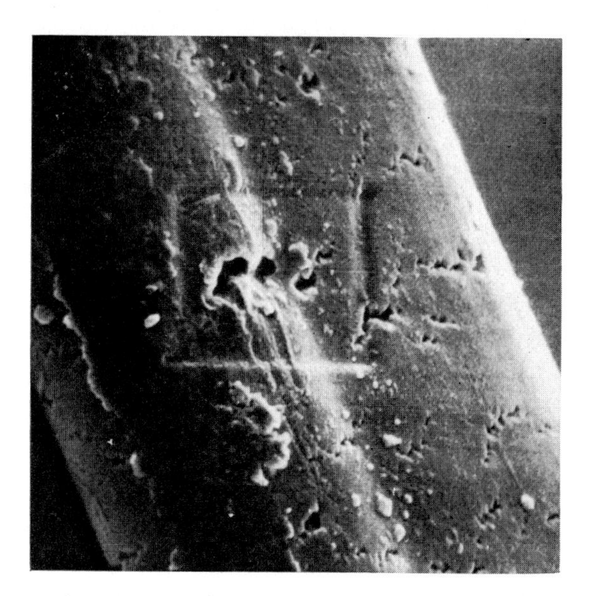

FIG. 10.8. Beam damage corresponding to the raster at a higher magnification.

energy to change polymer chain lengths, to produce monomers, or to cause the specimen and coating to interact. It is worth changing the coating material if there is any chance that the last could be correct. In general beam damage may be avoided by reducing the energy of the beam.

One effect that has been shown (Thornton, 1968) is that some oxide films on semiconductor devices can absorb a layer of the diffusion or rotary pump oil; the beam may then crack the layer. This may be overcome by using a cleaner (though not necessarily higher) vacuum system. Most of the trouble will come from the rotary pump so good fore-line cold traps can be added; in addition a heavier diffusion pump oil can be used with cold traps or an ion pump used instead of a diffusion pump. There are two notes of warning here: if silicon oil is used in the diffusion pump it will be much more difficult to clean out of the column when it finally becomes contaminated than if hydrocarbon oil is used; and if a magnetic ion pump is used it must be sited a little way from the instrument so that it does not affect the magnetic lenses.

Beam damage can usually be observed happening and it results from the same beam energy being dissipated over a smaller area and hence the greater the probability of

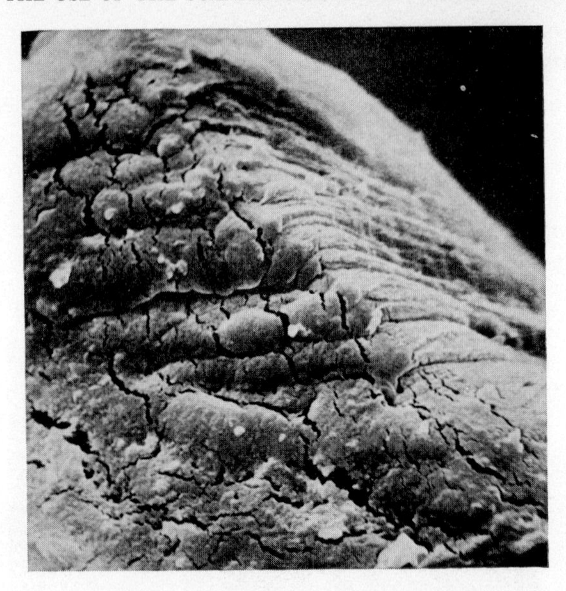

FIG. 10.9. Beam damage shown as cracking of the surface.

beam damage at a higher magnification. Within the raster square various types of damage may be seen, and the beam attacking some parts of the specimen preferentially. The first cure for beam damage is to reduce the energy of the beam by dropping the beam voltage or reducing the beam intensity by decreasing the final aperture size or increasing the lens currents. Damage can be reduced during photographic recording by using a faster scan if this is possible without introducing too much noise into the

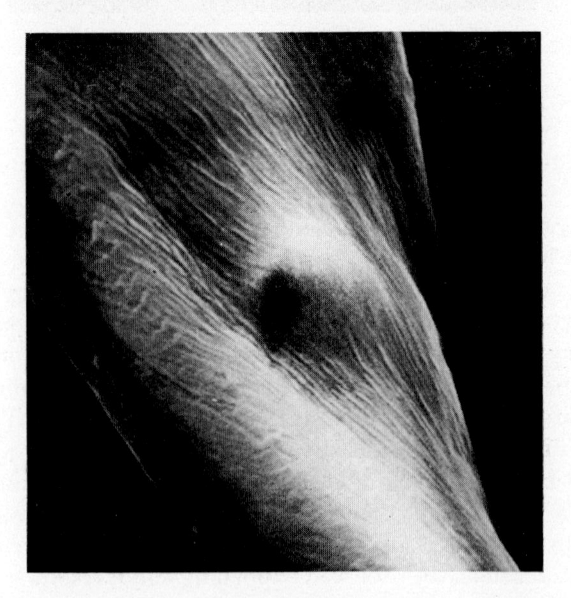

FIG. 10.10. Beam damage shown as a bubble forming on the specimen surface.

micrograph. It should be noticed here that the effect of increasing the lens currents will be to produce a noisy image on the display tube; however, a clearer micrograph than the one seen on the display screen may be produced by lengthening the line time on the record raster when taking photographs (see Figs. 10.11a, b). Other methods which may work for some samples are to decrease the thickness of the metal coating. Where beam damage is very extensive and not controlled by any of these methods, a move to scanning electron mirror microscopy might be considered. This is discussed more fully in Chapter 5.

10.3.4. Vacuum damage

Vacuum damage also is irreversible and a new specimen may have to be prepared for viewing once the damage is seen to be done. To determine vacuum damage it is usually necessary to have some idea of how the topography should look so that the damage, which often takes place before the image is seen on the screen, may be evaluated. Known sharp edges or protrusions should be checked for sharpness; and a flabby or wrinkled appearance in any part of a specimen that contains volatile liquids, especially

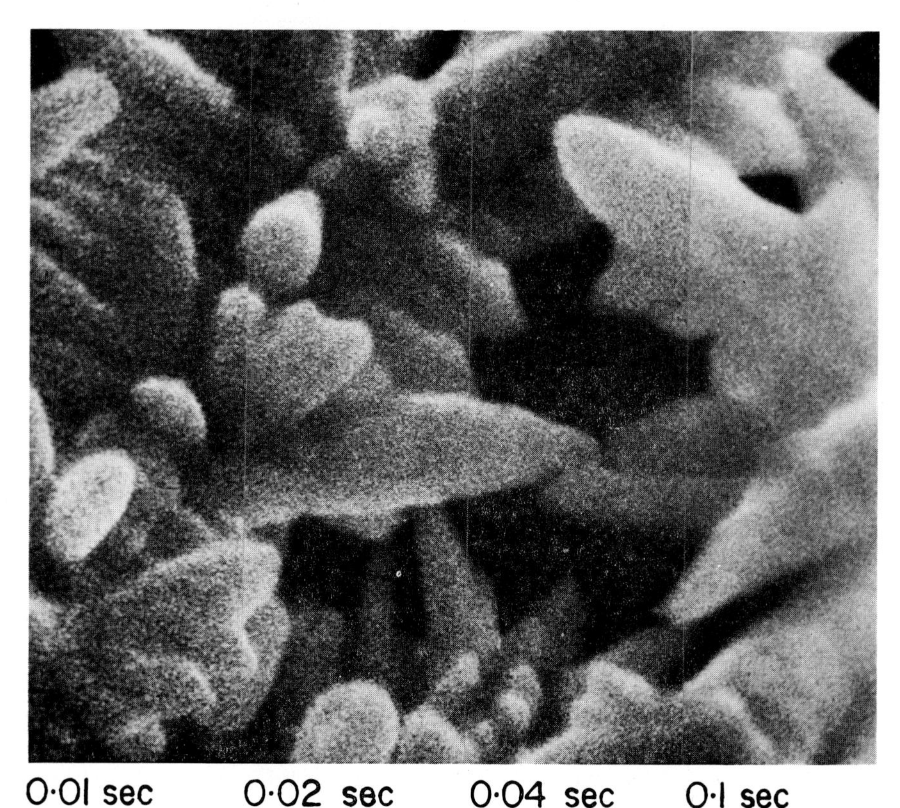

FIG. 10.11a. Illustration of noise reduction by using a slower line speed.

water, should arouse suspicions of vacuum damage at once. The only cure is a new specimen more carefully fixed or dried.

FIG. 10.11b. Correct elimination of noise by using a line speed of 0·2 second.

10.3.5. Poor image quality due to a magnetized specimen

If the picture becomes hazy and difficult to resolve at around 20,000 times instrument magnification it may well be assumed that this is the natural limit of resolution for that sample in the SEM, especially for biological or polymer samples. However, if this effect occurs at much lower magnifications, a sub-surface delay mechanism should be considered, and a method of blocking off the sub-surface to the beam sought; for instance, the sample could be coated with a metal, as if to prevent charging. This effect is sometimes shown by magnetic materials, especially steel, and can sometimes be cured by using a good conductive glue; the application of a non-magnetic metal coating will usually solve the problem by preventing the electron beam coming within the influence of the local fields. Magnetism is normally exhibited as gross astigmatism and may be reduced by the use of a long working distance if it is induced by the final lens.

10.3.6. Faults in the column and collector system

Most faults in the column which lead to distorted images are due to bad alignment of some components and to dirt. One of the most spectacular is due to a misposition of the scanning coils, which can happen on some instruments if air is admitted to the specimen chamber while the final aperture is still open to the column under vacuum. The image appears as if viewed through a hole, often on one side of the screen, or the image is grossly astigmatic. The fault is simply cured by repositioning the scanning coils. A similar appearance is given by misalignment of the final aperture; and if a first check does not show an obvious slip of the aperture, it is worth while looking to see if the aperture holder has been mishandled, so that it is bent or the positioning screws are loosened.

The most usual column fault, however, is dirt, mainly caused by cracked diffusion

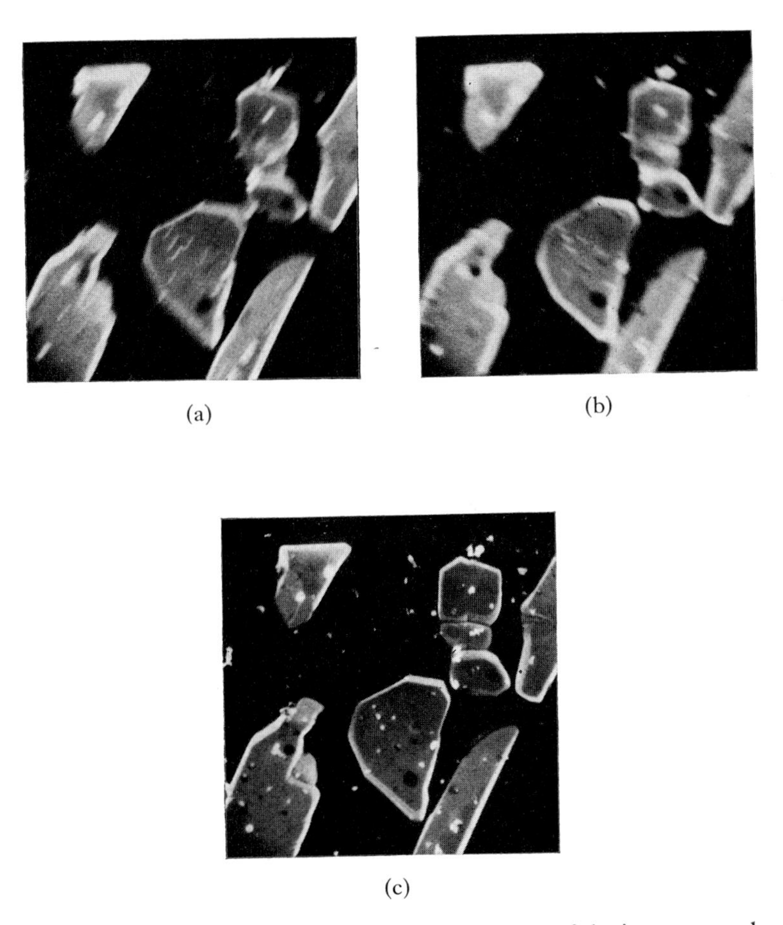

(a) (b)

(c)

FIG. 10.12. (a and b) Astigmatism shown as distortion of the image at angles to the scanning beam when focused above and below the focal point. (c) Focused image after astigmatic corrections are made.

and rotary pump oil, which causes trouble when it gathers around the electron optical axis. This may be reduced by using cold traps on the pumps. When the amount of dirt in the column is obviously affecting the image it shows itself as astigmatism. Loose dirt in the column also causes sudden image shifts. In an astigmatic image, details are elongated at an angle to the scanning axes when the picture is slightly off-focus; swinging onto the other side of the focus causes the elongation to swing through an opposite angle (see Fig. 10.12). This effect must always be carefully watched for, as it is easy to ignore in its early stages. The corrections for astigmatism are given in each manufacturer's handbook; in general the astigmatism will be greater at lower beam voltages. When the astigmatism cannot be corrected by the controls, the column should be dismantled and thoroughly cleaned, paying special attention to the spray apertures, final apertures and final aperture holder. The spray apertures should be cleaned according to the manufacturer's instructions or they may be "flashed" clean in a vacuum, by being heated electrically to cherry-red and left at that temperature for a few minutes. The last method does not always remove large amounts of burnt-on oil, and if the apertures are flashed for long periods they become cracked and contaminate very quickly when used in the instrument; however, it is a very gentle method for removing small amounts of dirt. The final apertures are usually cleaned by flashing, and the final aperture holder is cleaned with metal polish and washed with a volatile liquid. All these cleaning methods can be supplemented or replaced by ultrasonic cleaning, starting with a solution of a modern laboratory detergent, such as Quadralene or KBS 25. The components should be examined under a low-power optical microscope before reassembly to make sure that they are clean and scratch-free; apertures with bad scratches should be replaced.

A noisy picture is recognized by minute white specks moving over the whole area of the image or by a grainy appearance to the image (see Fig. 10.11). It can result from the following: (a) specimen charging, (b) poor electrical contact in the collector system, (c) reduced electron detector sensitivity, (d) scanning speed too fast, (e) excessive use of signal black level, (f) poor electron emission from the specimen, (g) incorrect specimen orientation in relation to the detector, (h) electron optical operating conditions incorrect, (i) poor electron-optical alignment and (j) poor filament alignment or saturation. If the image is noisy when the signal should be strong, that is when the lens currents are low and the final aperture is large, the filament should be checked to see that it is running on saturation current and the age of the filament should be considered. Some filament lifetimes are very regular and the age at which they become noisy can quickly be estimated; other filaments from different sources have more irregular lifetimes and may fail or become noisy at ages ranging from 12 hours to 60 hours. Filament alignment should also be checked.

If the noise increases over a period of time, but is not due to the filament, it is very likely to be caused by a stripping of the aluminium coating over the scintillator on the end of the light pipe which sometimes causes bright patches or bands on the image similar to those in Fig. 10.5. Check the condition of the coating by putting a strong light at the other end of the light pipe and examining the tip for discontinuities in the coating. Recoating of this tip is either done by the manufacturer or may be attempted by the operator. The old aluminium should be thoroughly polished off with a piece of silk and a light aluminium layer about 50–100 nm (500–1000 Å) vacuum-deposited on the scintillator while rotating the light pipe; the continuity of the layer should be checked as above before re-using.

Other causes of noisy images which can easily be checked by the operator are a hairy lead from the high-tension terminal to the scintillator, a non-perfect electrical contact at the scintillator, or a bad optical contact between the light guide and the photomultiplier. Failing these the fault is probably electronic in the amplification system. Sometimes apparent shifts in focus of the image are caused by not saturating the filament properly.

10.3.7. Electronic faults

Electronic faults are usually recognized by the non-functioning of the specific dials, or by the failure of the dials or the image to respond to the operation of controls. Faults in the scanning system may also show up by giving only a line, a moving spot or a reduced raster instead of a picture. A lack of stabilization of the power supply or mechanical vibration may give rise to a saw-tooth edge on the image of the specimen (see Figs. 10.13 and 10.14). The only electrical faults that can be cured by the average operator are blown fuses and dirty switches. It is, however, worthwhile writing down the exact history of the fault for the information of the maintenance engineer, especially for intermittent faults which may not show themselves during the engineer's visit. In general, it is wise to keep a detailed log-book of faults, maintenance and cleaning, so

FIG. 10.13. Mechanical vibration shown as a shift in the image.

that patterns and trends of faults may be noticed and the true cause of a recurring fault, and not just the symptoms, may finally be determined.

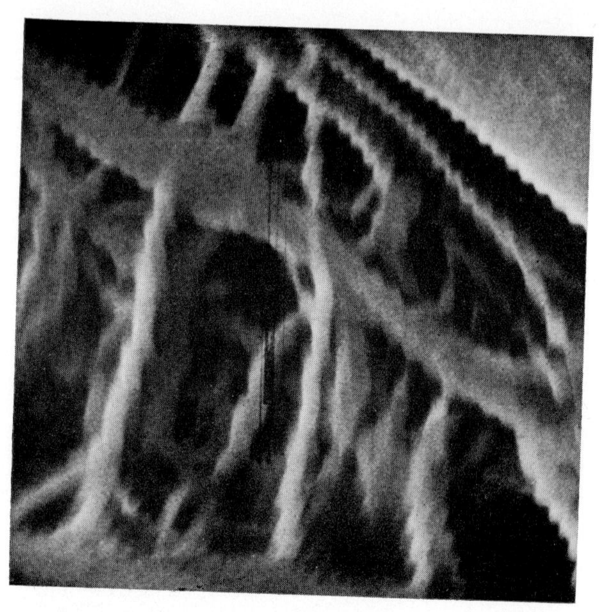

FIG. 10.14. Saw-tooth edges on the image due to an electronic fault.

10.4. FAULTS WHICH MAY NOT BE OBVIOUS ON THE FINAL PHOTOMICROGRAPH

In some ways, the most serious class of faults of all are those in which it is not obvious that anything is wrong; an operator may be lulled into a sense of false security by clear, sharp micrographs, which may nevertheless give completely false information as to the original topography of the sample. On the whole it is at the higher magnifications that one should be on the look out for these types of faults, although some occur at lower magnifications too.

10.4.1. Faults due to the condition and position of the specimen

One should always be watching for artefacts of preparation in the image, and where there is any doubt try similar samples prepared in different ways. Where cross-sections are examined the method of sectioning should be carefully reviewed; for instance, it has been shown with biological samples that tearing damages less of the structure than cutting or freeze fracturing. When samples have been prepared by spraying with anti-static, droplets and thin films of antistatic may be expected from time to time on the sample and must not be confused with the true detail of the sample. If any strange, unexpected structure appears, it should be considered that it might be dirt from the environment, or from the conducting paint used to produce a good conduction line to the stub; Aquadag in particular produces detailed and fanciful images which may be mistaken for true sample detail.

At high magnifications the screen may show a more diffuse image than the topography warrants due to a high beam spread in the sample. The type of material so affected usually tends to charge anyway, and may have a coating of metal vacuum deposited on it to inhibit charging. This will also have the effect of shielding the actual specimen surface from the beam, and so reducing the spreading. The coating itself, if it is too thick or non-uniform, may blot out details or introduce artefacts, and a nice compromise must be struck between the necessity to reduce charging and beam penetration and the necessity to keep the metal layer thin enough to contour clearly. The choice of a good contouring metal, such as silver, is indicated in this predicament.

Another compromise which arises is over the choice of beam voltage for the specimen. On the one hand, the greater the beam voltage the greater the theoretical resolution; on the other hand, the penetration of the beam rises as $KV^{1.7}$ where V is the beam voltage and K is a material constant. The greater the penetration, the greater is the spread of the beam; and also, since the information is coming from a sub-surface layer, much of the surface detail may be lost. This is particularly true of "soft" materials; Fig. 10.1 shows the effect with a broken cotton fibre. Other examples are shown in Figs. 2.19, 3.8, 7.1 and 4.5. It is wise, therefore, to check a range of voltages when attempting high magnification work, and to make your own decision as to which gives the most valid information for that particular sample.

If the actual surface detail is very smooth and rounded it may be difficult to distinguish on the display screen. If this is suspected make sure that the images at high magnification are studied at maximum contrast conditions and with the black level turned up. In reporting these micrographs, however, it must be clearly stated that the black level was boosted; otherwise a false impression of the depth of the details will be left.

There is one other condition of the sample which may lead to unsuspected faults and that is its position. Even at low magnifications with some samples one may see that the sample itself is affecting the primary beam and a distorted image of the background is given. Figures 1.6 and 10.7 show a fibre mounted upright on a round stub, the whole being tilted towards the collector. The stub appears "drawn up" around the fibre; the fibre is in fact acting like diverging lenses. The important question is whether the image of the fibre is itself distorted? It does not appear to be; but certainly micrographs like this should be studied only for the qualitative information they carry and a reservation as to possible distortion should be held. Similar results using a fine wire stretched horizontally a little below the final aperture have been obtained. Whenever a thin or pointed specimen is used it should be mounted so that the points have less chance of charging. Where this is impossible (as with fibre cross-sections) the possibility of distortion of the image should be kept in mind.

10.4.2. Faults due to instrument conditions

Another group of faults may be summarized as astigmatism, noise and low resolution; a small amount of astigmatism is particularly insiduous at high magnifications; since astigmatic images may be photographed as actual surface detail, it must be carefully watched out for, and, for high-magnification work, the instrument should be scrupulously cleaned beforehand so that the astigmatic effects are reduced and the resolution is increased. If high-magnification work is routine, the instrument is best cleaned once a week, especially if it is located near industrial pollution. If lower magnifications are all that is needed, then once a month is a convenient frequency of cleaning.

If the image is noisy it becomes difficult to focus clearly on small details, and therefore all precautions against noise should be used in high-magnification work. The filament should be saturated frequently, especially when new, and ideally it should be old enough to have stabilized properly and new enough not to be noisy. The collection system should be checked to see that it is perfect. The final photomicrograph should be taken using a slow frame speed on the record screen; this can reduce the noise dramatically and, provided that the image was focused correctly, show details which were never suspected on the display screen image.

For maximum resolution one should use a beam with the highest possible energy and the smallest possible spot. Under these conditions the energy of the beam can only be raised by increasing the beam voltage; but this increases the penetration and the information does not come from the surface, as discussed above, so that a compromise must be struck. For a small spot size, the condenser lenses can be strengthened, the final aperture narrowed and the working distance shortened (this has the effect of reducing the depth of focus but in high-magnification work that is not usually an important consideration). All these techniques unfortunately have the effect of decreasing the beam intensity so that the image becomes noisier and some kind of compromise must be struck again. To give sharper edges all machine vibration should be reduced to a minimum and one useful operation, if it can safely be carried out with the particular instrument, is to turn off the rotary pump during fine focusing and photographing. With some SEMs the pump can be left off for about 20 minutes without harming the diffusion pumps as long as the backing lines are shut.

In order to see some details on very smooth specimens the angle of tip to the beam must be considered; for a highly tipped specimen a sharp step of 20–30 nm (200–300 Å) may sometimes be seen. Tipping causes foreshortening and this should always be remembered when looking at any SEM image. The best way of getting the maximum safe information from a specimen is to take stereopairs, as outlined in Chapter 11.

REFERENCES

Shaffner, T. J. and VanVeld, R. D. (1971) *J. Phys. E: Sci. Inst.* **4**, 633.

Thornton, P. R. (1968) *Scanning Electron Microscopy—Application to Materials and Device Science*, Barnes & Noble, Inc., N.Y.

VanVeld, R. D. and Shaffner, T. J. (1971) *Proc. 4th Annual SEM Symposium*, IIT Research Institute, Chicago, Ill., p. 17.

CHAPTER 11

Dimensional Measurements

G. S. Lane

11.1. INTRODUCTION

The scanning electron microscope has many features which make it an ideal instrument for the examination of specimen topography at high resolutions. The following are particularly worthy of mention:

(a) In the normal operating mode the image contrast is largely representative of specimen topography.
(b) It accepts specimens in bulk so that artefacts associated with replication processes are avoided.
(c) Its depth of focus is very large.
(d) A wide range of adjustments are generally available so that the specimen can be viewed from many angles.

However, because of the large depth of focus, points separated by a large height difference can be in focus simultaneously. This can sometimes hinder interpretation by removing the suggestion of height differences afforded by variations in focus. Moreover, it is sometimes difficult to distinguish small protrusions from shallow depressions in a relatively flat surface. In both of these instances, stereographic techniques are invaluable. If two micrographs of the same area are taken from a slightly different viewpoint, the stereopair so formed can be examined in a special viewer to give a three-dimensional image. Even on a purely qualitative basis, a greatly increased insight can often be obtained in this way into the shape of the specimen surface. In addition, the signal-to-noise ratio is increased and details are often revealed which would otherwise have been hidden by noise.

11.2. THREE-DIMENSIONAL VIEWING

Normal visual perception of three dimensions depends on slightly different views being received by each eye. In a direction normal to the line joining the eyes the size of objects is the same to each, but in a direction perpendicular to this the apparent separation of two points is different. This difference, referred to as the "parallax", is proportional to the difference in the distances of the two points from the eyes. The parallax thus provides the information relating to the third dimension which the brain fuses into a three-dimensional image. It is important to note that parallax occurs only in a direction parallel to the line joining the two viewing positions, there being none perpendicular to this.

Similarly, the essential feature in the formation of a stereopair of electron micrographs is the requirement for two slightly different views of the specimen. These are normally obtained by tilting the specimen through a small angle between exposures. Angles between 5° and 20° can be used depending on the roughness of the specimen—the larger angles give greater magnification in the third dimension and are therefore used on relatively flat surfaces. Rotation about an axis not parallel to the electron beam can also be used to produce a stereopair, although this case is more difficult when a quantitative analysis is required. In the present treatment, the latter method will not be considered.

11.2.1. Unaided viewing

In order to obtain the three-dimensional image, each micrograph must be viewed by one eye only. Normally, of course, the lines of sight converge slightly and are approximately parallel only when a distant object is observed. The requirements for viewing of stereopairs, however, imply that the lines of sight are parallel and the eyes are focused on the micrographs, not infinity. This unnatural condition can be achieved with practice without external aids. The eyes are focused on a distant object and relaxed; the stereopair is brought into the field of view with the micrographs separated by the interocular distance; and the eyes are focused on them without converging. Alternatively the micrographs can be interchanged and the eyes crossed, the left looking at the right-hand micrograph and vice versa. Again with practice, three images are seen, the centre one being three-dimensional. Very short-sighted people can view stereopairs by placing the micrographs at their far points! In all cases the axis of tilt should be perpendicular to the line joining the eyes.

11.2.2. Optical stereoscopes

The simplest aid to viewing consists of a convex lens in front of each eye. If the micrographs are placed at the focal points of the lenses, their images will be at infinity. Such a device is made by Zeiss (the "pocket stereoscope") and has the advantages of cheapness and portability. However, the size of the micrographs accommodated is clearly small. This disadvantage can be overcome by arranging a series of mirrors between the micrograph and eye. This principle is used in "mirror stereoscopes" and instruments of this type are made by various manufacturers (Wild, Cassella, Hilger & Watts, Zeiss, etc.). Figure 11.1 shows a typical mirror stereoscope and a ray diagram illustrating the principle of operation. The light sources shown are part of the parallax measuring system (see Section 11.5.1). A detailed description of this stereoscope is given by Martin (1966).

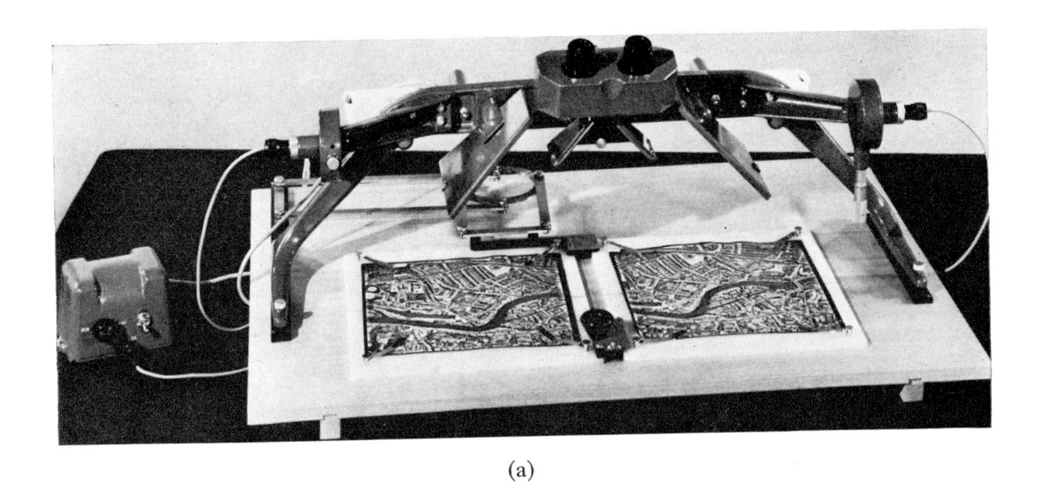

(a)

(b)

Fig. 11.1. Mirror stereoscope with parallax measuring system. (Courtesy of Rank Precision Industries, Hilger & Watts, Ltd.) (a) The mirror stereoscope. (b) Optical diagram showing parallax measuring device.

11.2.3. The viewing of stereopairs of electron micrographs

For correct viewing of electron micrograph pairs on photogrammetrical equipment such as the stereoscope, the following requirements must be complied with.

(a) Alignment of tilt axis. The parallax created by tilting a specimen in the electron microscope exists in directions perpendicular to the tilt axis. Comparison with the visual case discussed in Section 11.2 shows that the tilt axis must therefore be perpendicular to the line joining the eyes for correct three-dimensional viewing. Care must be taken to allow for beam rotation when locating the true tilt axis.

(b) On some specimens it is difficult to distinguish depressions from elevations. Indeed, if the micrographs are interchanged an inverse image is obtained which is seen without eyestrain if the surface relief is not too great. It is essential, therefore, to have a standard method of correctly orientating the micrographs. This can be achieved by using the following procedure:

(i) place one micrograph down with the tilt axis aligned as in (a);
(ii) knowing the direction of tilt, determine the direction a normal to the specimen surface would move on this micrograph during tilting;
(iii) place the second micrograph down similarly orientated to the first but displaced from it in the opposite direction to this;
(iv) align the micrographs under the viewer in the normal way while maintaining the spatial relationship determined in (iii).

(c) If eyestrain is to be avoided it is desirable to make the micrographs the same magnification (and photographic enlargement).

(d) For the best three-dimensional image, the separate micrographs should have similar contrast ranges, and be printed as diapositives for examination in transmitted light.

11.3. SIMPLE QUANTITATIVE MEASUREMENT TECHNIQUES

11.3.1. Lateral measurements using a single micrograph

Although heights can only be determined using a stereopair of micrographs, lateral dimensions are often all that is required. This is the case, for example, in particle-size determinations or area measurements. However, because of the variation of magnification with working distance, care must still be exercised on specimens which exhibit significant surface relief. The problem is illustrated in Fig. 11.2 which shows two different width features giving the same width image due to their height difference. Both can be in focus simultaneously because of the large depth of focus of the instrument. The effect is discussed by Christenhuss and Pfefferkorn (1968) who show that it is only serious at low magnifications, although every case must be treated on its merits. From a practical point of view, the danger can be avoided by focusing on each feature in turn at high magnification before returning to low magnification to determine its size. A stereopair can again be useful from a qualitative point of view in deciding whether or not a problem is likely to be encountered.

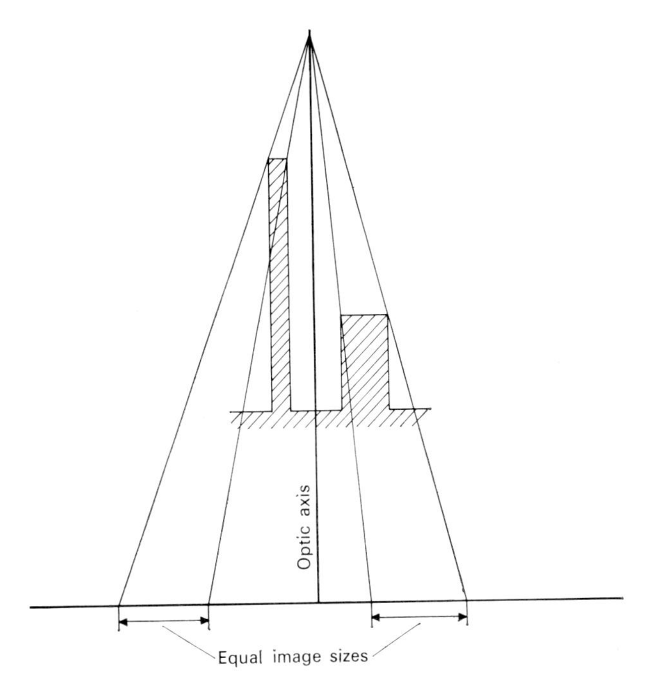

FIG. 11.2. Effect of height on lateral dimension measurement (exaggerated). (After Christenhuss and Pfefferkorn, 1968.)

TABLE 11.1. THE VARIATION IN MAGNIFICATION ACROSS A PLANE SPECIMEN AT 45° TO THE ELECTRON BEAM (STEREOSCAN Mk. IIA)

Nominal magnification	Angle scanned by beam	Change in rotation	Change in magnification
20	15°12′	7°42′	23%
50	6°14′	2°54′	8%
100	3°9′	1°30′	4%
200	1°13′	0°45′	1·8%
500	0°37′	0°20′	0·75%
1000	0°19′	0°11′	0·45%
2000	0°09′	0°05′	0·22%
5000	0°04′	0°02′	0·08%

Notes

1. The changes in magnification above refer to the differences between the top and bottom of the screen. The exact value depends on average working distance, which is taken to be that which gives the nominal magnification at the screen centre (11 millimetres on a Stereoscan. Mk. IIA). The total variation is expressed as a percentage of this magnification.
2. The changes in rotation refer to the magnitude of the beam rotation when the instrument is accurately focused at the top and then at the bottom of the screen.
3. The angle scanned by the beam is the total from extreme to extreme of the field width.

Even on plane specimens the magnification varies significantly across the field of view if the specimen is not normal to the beam. Boyde (1970a) has given approximate magnitudes of this effect for an operating angle of 45°. Table 11.1 gives the values together with the change in beam rotation which occurs when first the top and then the bottom of the micrographs is brought into focus. In theory, a variation in magnification also occurs at low magnifications for a plane specimen normal to the beam. However, even at magnifications as low as 10 times this only amounts to about 5%, and for all practical purposes can be ignored.

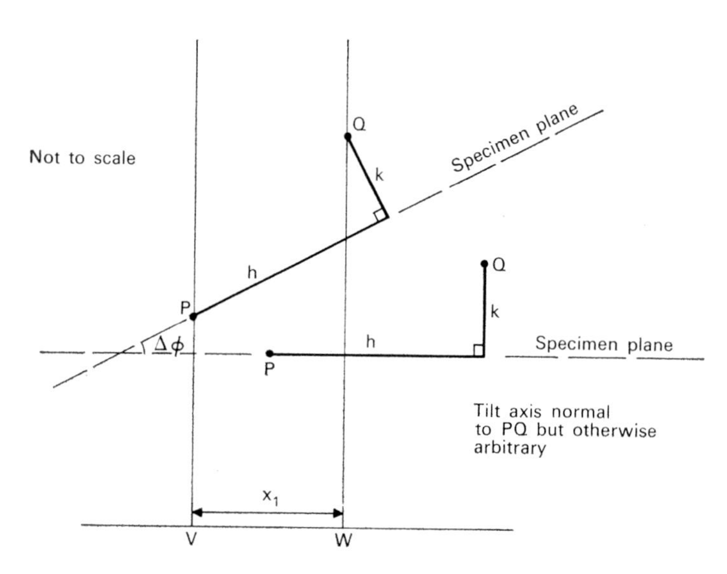

FIG. 11.3. The origin of parallax.

11.3.2. Measurement of height differences

In a similar way to the visual case described in Section 11.2, the parallax between two points on an electron micrograph stereopair is in certain instances proportional to the height difference between them. If it is assumed that the specimen starts normal to the beam and is rocked $\pm\ \Delta\phi$ about this position, then the height difference between two points P and Q can be determined from Fig. 11.3. Assuming the tilt axis is normal to PQ, the apparent separation of points P and Q when the specimen is tilted $+\Delta\phi$ is

$$x_1 = VW = M(h \cos \Delta\phi - k \sin \Delta\phi) \tag{11.1}$$

where (h, k) are the coordinates of Q with respect to P parallel and normal to the specimen surface; and M is the magnification. For the reverse tilt $-\Delta\phi$ we have similarly

$$x_2 = M(h \cos \Delta\phi + k \sin \Delta\phi).$$

Eliminating h and writing p = parallax = $x_2 - x_1$ we have

$$k = \frac{p}{2M \sin \Delta\phi}. \tag{11.2}$$

It should be emphasized that equation (11.2) only applies directly when the specimen is tilted $\pm\ \Delta\phi$ about a position normal to the beam, and when the magnification is such that the electron beam can be considered parallel, thereby producing an orthogonal projection system.

Tilting of the specimen away from the collector as implied by $-\ \Delta\phi$ is not a normal operation in scanning electron microscopy. It is possible up to 5° in the JEOL JSM-U3, but in the Cambridge "Stereoscan" and the MAC 700 and 400s it is not possible on the standard stages. Indeed, because of the reduction in signal which occurs, this mode of operation is not recommended. If the specimen is tilted from 0 to $2\Delta\phi$, the parallax observed is p' and can be determined from equation (11.1). Taking x to be a function of $\Delta\phi$

$$p' = x(0) - x(2\Delta\phi) = M\ \{h\,(1 - \cos 2\Delta\phi) + k \sin 2\Delta\phi\}.$$

The height k is not simply related to parallax; in fact

$$k = \frac{p'}{M \sin 2\Delta\phi} - \frac{h\,(1 - \cos 2\Delta\phi)}{\sin 2\Delta\phi}. \tag{11.3}$$

The second term in equation (11.3) is not negligible. For example, if $h = 2k$ and $2\Delta\phi = 5°$, this term is approximately 20% of the first. Simple parallax measurements are therefore not sufficient for the accurate determination of heights in this case. The value of h can be determined readily from $h = x(0)/M$, and substituting this in equation (11.3) gives

$$k = \frac{x(0)}{M \tan 2\Delta\phi} - \frac{x(2\,\Delta\phi)}{M \sin 2\Delta\phi}. \tag{11.4}$$

A similar equation is given by Kimoto and Suganuma (1968) for somewhat different geometry in connection with the JEOL instrument.

Application of the normal parallax equation (11.2) to the case when tilting is from 0 to $2\Delta\phi$ does not give the height measured normal to the specimen surface. If a plane is constructed through P at an angle $-\Delta\phi$ to the specimen plane (i.e. in the opposite direction to the tilt), then equation (11.2) will give the height k', in a direction normal to this plane. In terms of h and k,

$$k' = h \sin \Delta\phi + k \cos \Delta\phi. \tag{11.5}$$

Equations (11.3) and (11.4) cannot be applied directly to the normal operating mode of the SEM in which the operating angle ϕ is maintained around 30–45°, and stereo-pairs produced by tilting $\pm$ 5° or $\pm$ 10° about this position. Applying the simple parallax equation (11.2) to this case gives a height k'' measured normal to a plane at an angle $-\ \phi$ to the specimen surface where

$$k'' = h \sin \phi + k \cos \phi.$$

The relationship between k'' and the "true" height k is not simple, and it is possible that k'' will not be a quantity of interest. Under these conditions the full stereographic techniques described in the next section are required to determine both h and k, the separation of two points measured parallel and normal (respectively) to the specimen plane.

11.4. RIGOROUS TREATMENT

When the simple techniques described in the previous section are inadequate stereographic methods can be used to obtain a set of equations valid for any operating angle, any tilt, and without the restriction of orthogonal projection. The treatments of Garrod and Nankivell (1958) and Wells (1960) are rigorous but neither is directly applicable to the scanning microscope in its normal mode of operation. Both concentrate on the effects on non-orthogonal projection which is representative of the low-magnification case, and it is assumed that the specimen is tilted about a position normal to the beam. Two fundamental errors are introduced by the assumption of orthogonality, namely the "tilt" and "perspective" errors. The magnitude of these depends largely on the magnification, being worst at low magnifications. The former is due to the change in magnification which occurs on tilting when the tilt axis is remote from the area of interest. The latter is the effect mentioned in Section 11.3.1 and is due to the foreground of a micrograph being reproduced at higher magnification than the background, an effect which can be exaggerated by the operating angle in the microscope.

The work of Lane (1969) provides a theory easily applicable to the modern scanning electron microscope. The following items are taken into consideration:

(a) the wide range of magnifications (orthogonality is not assumed);
(b) the wide range of operating angles;
(c) the probability that the tilt axis is far removed from the field of view;
(d) the necessity for arbitrary adjustments of the x and y shift controls when taking the second micrograph of a stereopair (a consequence of (c)).

The theory can be divided into several cases, and these are described in the following sections.

11.4.1. The two-dimensional case

In order to avoid the assumption of orthogonality a geometric representation of the electronic method of image formation is assumed in which projection is from a focal point F onto a plane perpendicular to the optic axis. This representation, shown in Fig. 11.4, gives the correct relation between magnification and working distance, since magnification = image size/object size = $OV/SP = FO/FS$ = const/working distance at P.

Projection from F gives image points V, W of specimen points P, Q. If it can be assumed that the tilt axis is normal to the line PQ and that PQ intersects the optic axis, the geometry reduces to two dimensions as shown in Fig. 11.5. The position of this tilt axis is arbitrary, and the two specimen positions are denoted by suffixes 1 and 2. The objective is to determine the coordinates (h_0, k_0) of the point Q with respect to P in a system defined by the specimen stub and a normal to it. This will be referred to as the specimen coordinate system. As indicated above, full stereographic techniques are necessary for the determination of both h_0 and k_0 when the stereopair is formed by tilting about a non-zero operating angle, ϕ. Micrographs are thus taken at operating angles ϕ_1 and ϕ_2 where $\phi_1 = \phi - \Delta\phi$ and $\phi_2 = \phi + \Delta\phi$. From Fig. 11.5, for specimen position 1, $OW_1/OF = R_1Q_1/R_1F$, or

$$x_1/L = (h_0 \cos \phi_1 - k_0 \sin \phi_1 - d_1)/(w_1 - h_0 \sin \phi_1 - k_0 \cos \phi_1) \qquad (11.6)$$

where $L = OF =$ constant for a given magnification range, $w_1 = FS_1 =$ the working distance at P_1, and $x_1 = OW_1$.

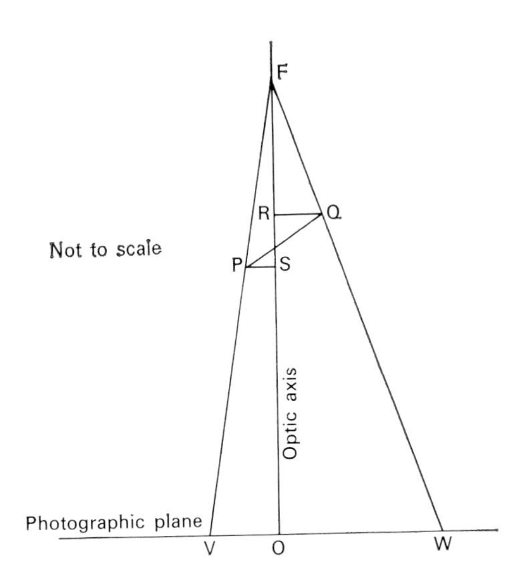

FIG. 11.4. Representation of image formation.

The quantity d_1 can be determined since $OV_1/L = d_1/w_1$ or

$$d_1 = w_1 \left(\frac{OV_1}{L}\right) = w_1 \left(\frac{x'_1}{L}\right). \qquad (11.7)$$

Analogous expressions to (11.5) and (11.6) are obtained for the second specimen position, and the simultaneous equations can be solved for h_0 and k_0 to yield

$$\left.\begin{aligned} h_0 &= \frac{\beta_2 w_2 (\alpha_1 \cos \phi_1 - \sin \phi_1) - \beta_1 w_1 (\alpha_2 \cos \phi_2 - \sin \phi_2)}{(1 + \alpha_1\alpha_2) \sin 2\Delta\phi + (\alpha_1 - \alpha_2) \cos 2\Delta\phi} \\ k_0 &= \frac{\beta_1 w_1 (\alpha_2 \sin \phi_2 + \cos \phi_2) - \beta_2 w_2 (\alpha_1 \sin \phi_1 + \cos \phi_1)}{(1 + \alpha_1\alpha_2) \sin 2\Delta\phi + (\alpha_1 - \alpha_2) \cos 2\Delta\phi} \end{aligned}\right\} \qquad (11.8)$$

where $\alpha_1 = x_1/L$,
$\beta_1 = (x_1 + x'_1)/L$

and similar relations for α_2, β_2.

Calculations based on equations (11.8) require two measurements off each micro-

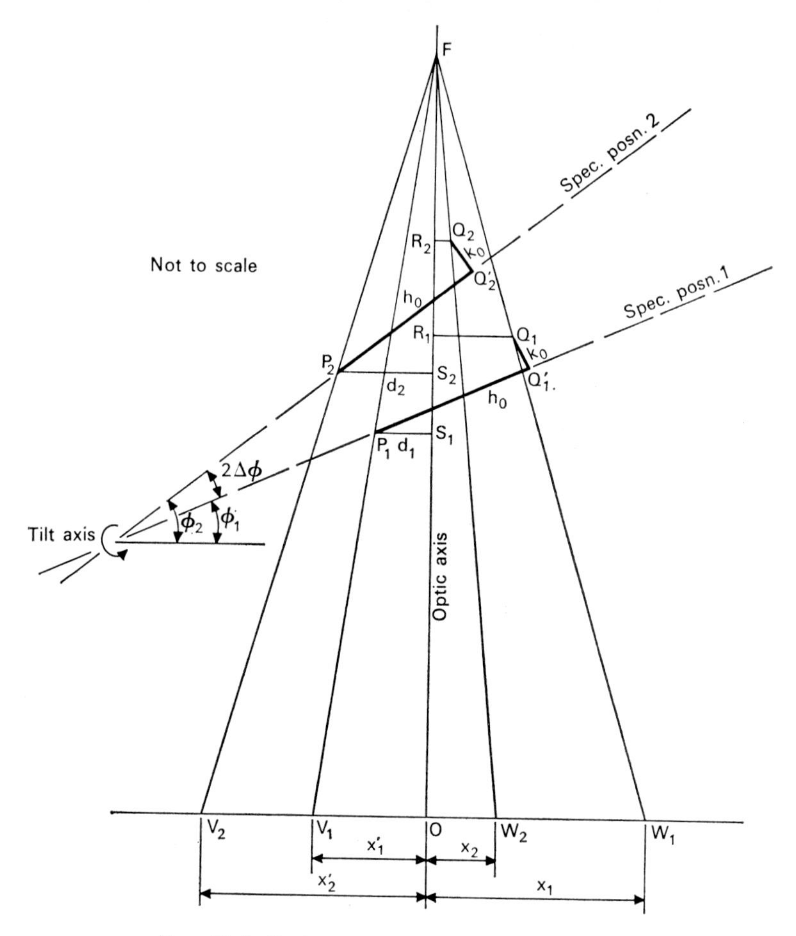

FIG. 11.5. Basic geometry in two dimensions.

graph, viz. x and x'. At high magnifications the equations can be simplified and the requirement reduced to one, viz. $x + x'$. If we write $y_{01} = x_1 + x'_1$ and $y_{02} = x_2 + x'_2$, then at magnifications of 1000× and above, equations (11.8) reduce to:

$$\left.\begin{aligned} h_0 &= \left(\frac{y_{01}}{M_1} - \frac{y_{02}}{M_2}\right)\frac{\sin\phi}{2\sin\Delta\phi} + \left(\frac{y_{01}}{M_1} + \frac{y_{02}}{M_2}\right)\frac{\cos\phi}{2\cos\Delta\phi} \\ k_0 &= \left(\frac{y_{01}}{M_1} - \frac{y_{02}}{M_2}\right)\frac{\cos\phi}{2\sin\Delta\phi} - \left(\frac{y_{01}}{M_1} + \frac{y_{02}}{M_2}\right)\frac{\sin\phi}{2\cos\Delta\phi}. \end{aligned}\right\} \quad (11.9)$$

These expressions give h_0 and k_0 in terms of the measured quantities y_{01}, y_{02}, M_1, M_2, ϕ, $\Delta\phi$.

11.4.2. The three-dimensional case

If the line PQ does not intersect the optic axis, the problem becomes three dimensional. The geometry for one specimen position is shown in Fig. 11.6. The quantities to be determined, h and k, can be related to h_0 and k_0 which are related to the measured

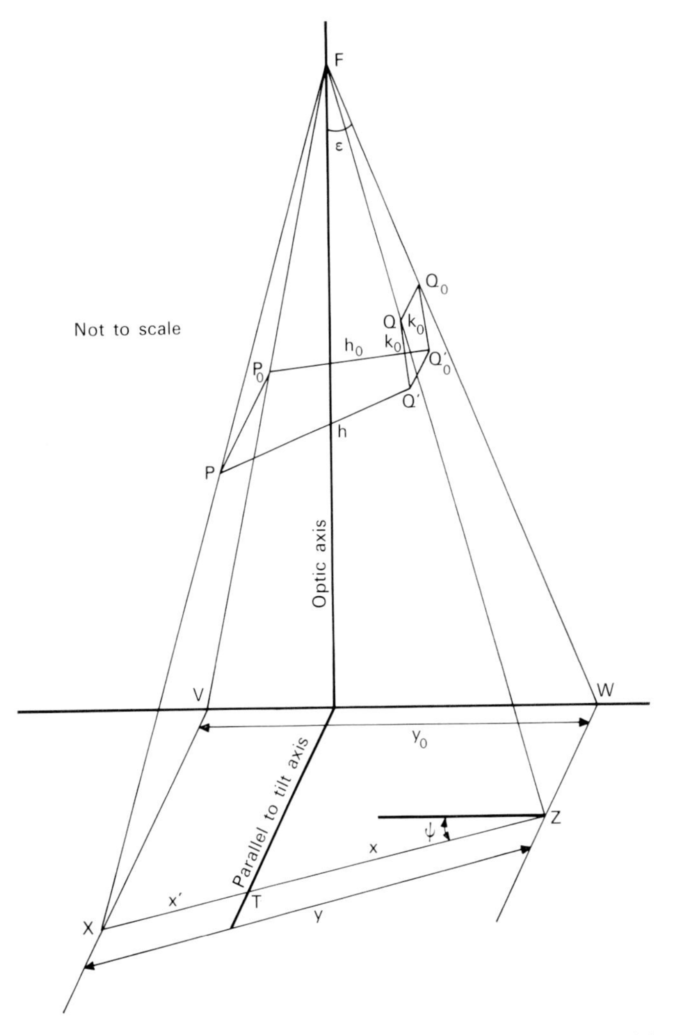

FIG. 11.6. Geometry in three dimensions for one specimen position.

quantities by the "two-dimensional" equations (11.9). The results are given in terms of the variables in equations (11.9), together with the angles ψ_1, ψ_2. Referring to Fig. 11.6, the images of P and Q are X and Z, and the line XZ makes angle $\pi/2 - \psi$ with a parallel to the tilt axis on the micrograph.

The value of k can be written explicitly, but the expression for h is cumbersome. Calculation is best carried out according to the following steps:

$$h_0 = \left(\frac{y'_1}{M_1} - \frac{y'_2}{M_2}\right)\frac{\sin\phi}{2\sin\Delta\phi} + \left(\frac{y'_1}{M_1} + \frac{y'_2}{M_2}\right)\frac{\cos\phi}{2\cos\Delta\phi}, \quad (11.10a)$$

$$k = k_0 = \left(\frac{y'_1}{M_1} - \frac{y'_2}{M_2}\right)\frac{\cos\phi}{2\sin\Delta\phi} - \left(\frac{y'_1}{M_1} + \frac{y'_2}{M_2}\right)\frac{\sin\phi}{2\cos\Delta\phi}. \quad (11.10b)$$

$$\theta = \tan^{-1}\{\tan\psi\cos(\phi + \tan^{-1}k_0/h_0)\} \quad (11.10c)$$

for (ψ_1, ϕ_1) or (ψ_2, ϕ_2),

$$h = \{(h_0^2 + k^2)/\cos^2\theta - k^2\}^{\frac{1}{2}} \quad (11.10d)$$

where $y'_1 = y_{01}\cos\psi_1$ and $y'_2 = y_{02}\cos\psi_2$.

In these expressions, θ is a constant angle and the relation (11.10c) holds for either ϕ_1, ψ_1 or ϕ_2, ψ_2. A useful check on the angle measurements is afforded by inserting these two pairs of values into (11.10c), which should give the same value of θ.

The equations (11.10) are valid for any values of the quantities ϕ_1, ϕ_2, ψ_1, ψ_2, k_0/h_0, k/h provided the magnifications are 1000 times or above.

11.4.3. The effects of measurement errors

Within their range of application, equations (11.10) correct automatically for tilt and perspective errors. However, the form of the equations is such that the effects on h and k of errors in a particular quantity can depend on the magnitude of the quantity itself or of the other measured parameters. For example, when ψ_1 and ψ_2 approach 90°, errors in these angles, and also errors in y_1 and y_2, become very significant. Again, differences in the errors in M_1 and M_2 are more significant than larger equal errors because of the occurrence of the term $(y'_1/M_1 - y'_2/M_2)$. This is particularly true if $(y'_1 - y'_2)$ is small. The effects of errors are therefore not easy to determine and each case should be treated separately. If the calculations are carried out using a computer, this can easily be done (see Section 11.5.2).

11.5. THE USE OF STEREOGRAPHIC EQUATIONS

11.5.1. Measurement of parameters

Calculations based on equations (11.10) require measurement of the parameters y_1, y_2, ψ_1, ψ_2, M_1, M_2, ϕ_1, ϕ_2. The first four are measured on the micrographs. For y_1 and y_2 a travelling microscope can be used, but if uncertainties of around ± 5% in h and k are acceptable, they can be measured using a ruler (read to ± 0·15 mm, say). However, measurements by any external medium make no allowance for possible non-linearities in the record CRT or the photographic processing. An excellent method of "internal calibration" in which a grid generated by the record CRT is superimposed on the micrograph has been suggested recently by Boyde (1970a). The image is recorded in the usual way, the camera shutter closed, and the photomultiplier switched off. Two patterns of, say, 100 lines horizontally and 100 lines vertically are then recorded in turn on the micrograph. The separation of points can now be measured using these grid lines accurate to the resolution limit of the CRT by interpolating to one-tenth of the grid spacing. Photographic enlargement up to a factor of 2 will facilitate this procedure. Measurements at any angle to the x-axis can be made by determining the separations (S_x, S_y) in the x and y directions. The distances required by equations (11.10) are then given by $\sqrt{(S_x^2 + S_y^2)}$ for each micrograph. A typical micrograph with the calibration grid superimposed is shown in Fig. 11.7; some non-linearity can be seen. This method of measuring off the micrographs is the most accurate and reliable yet devised.

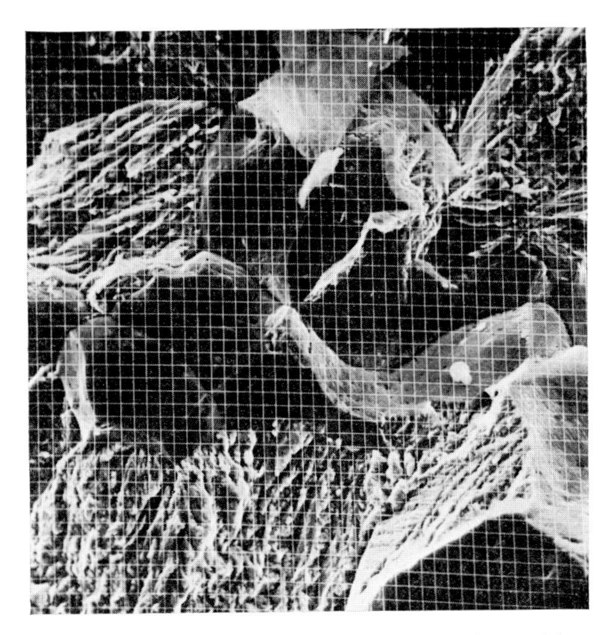

FIG. 11.7. Typical micrograph with 50 × 50 measuring grid superimposed (intergranular fracture 0·4% carbon steel, ×375).

The distance measurements cannot be made using the method of parallax determination used in aerial photogrammetry since this gives $y_1 - y_2$ whereas y_1 and y_2 are required explicitly. For this reason techniques based on the use of the parallax bar ("floating dot") and floating light spot (as in the Hilger & Watts stereoscope, Fig. 11.1) will not be described in detail. They can be used in the special case where the specimen is tilted $\pm\ \Delta\phi$ about a normal to the beam and for details the reader can refer to Martin (1966) or Boyde (1967). Determination of ψ_1, ψ_2 requires knowledge of the exact orientation of the tilt axis on the micrographs which varies according to the working distance because of spiralling of the electrons in the beam. This can be determined by moving the specimen mechanically in the direction of the tilt axis by operating the appropriate specimen stage shift control. Any given point then traces out a line parallel to the tilt axis on the screen. A useful calibration can be made of beam rotation against final condenser lens current. In the Cambridge Stereoscan the beam rotation can be automatically corrected for if the optional "Video Presentation Unit" is installed.

The second four parameters are instrument variables. M_1 and M_2 are the nominal magnifications of the two micrographs. These can be read off the meter directly or, more accurately, determined from a calibration against final condenser lens current. Although not essential for quantitative measurements, the condition $M_1 = M_2$ is desirable to make stereo viewing of the micrographs easier. On the Cambridge and MAC instruments this can be achieved by focusing for the second micrograph using the z-axis shift control. This is only possible in the JEOL-U3 at the large working distance setting. The Cambridge optional Video Presentation Unit is again useful since it gives (electronically) an infinitely variable magnification facility. The values of the tilt angles ϕ_1, ϕ_2 can be read directly off the appropriate stage control. However, care is required to avoid backlash. Measurements to $\pm \frac{1}{8}°$ (in a total tilt of 5°) are generally sufficient to give accuracies better than $\pm$ 5% in h and k. For high-accuracy work very careful calibration of the tilt mechanism is required, and optical lever methods have been described by Boyde (1970b) which allow $\Delta\phi$ to be determined to $\pm$ 3 minutes of arc, giving errors of $\pm$ 2% for a 5° tilt.

11.5.2. The use of the computer

Equations (11.10) are tedious to use manually, particularly if many sets of data are to be analysed. They do, however, lend themselves to computer calculation. A relatively simple programme will enable h and k to be calculated in the order suggested by the layout of the equations. Several valuable additional features can be included in the programme, for instance:

1. *Error calculations*

The effects of errors in the measured parameters can be determined by incrementing each in turn by an amount equal to the assessed error. Recalculation of h and k then gives the resulting errors unambiguously. It is strongly recommended that these calculations be carried out wherever possible since experience has shown significant variations in the errors determined in this way.

2. *Check functions*

As discussed in Section 11.4.2, values of the constant θ can be determined from the two pairs of values of ψ and ϕ. If the values are denoted θ_1 and θ_2, the programme should determine these and print an error message if, say, $|\theta_1 - \theta_2| > 1°$ when the angles ψ and ϕ have not been determined with sufficient accuracy.

3. *Various extra quantities can be calculated.* Examples are:

(a) the angle between the line PQ and the specimen stub $= \tan^{-1} k/h$;
(b) the absolute distance between P and Q, independent of coordinate system, $l = \sqrt{(h^2 + k^2)}$.

11.6. EXTENSION OF THE BASIC THEORY TO THE LOW MAGNIFICATION CASE

In Section 11.4.2 the three-dimensional stereographic theory was derived but only for high magnification (1000× and above) where, in effect, orthogonality can be assumed. The theory can be extended to remove this restriction (Lane, 1970). The method of calculation is similar to that of Section 11.4.2 except that the equations (11.7), which are without approximation, are used in place of the approximate equations (11.9). Determination of the angle θ is more complicated because of the non-orthogonality. It can be shown that

$$\tan \theta = \tan \psi \cos (\phi + \tan^{-1} k_0/h_0 - \epsilon)/\cos \epsilon \qquad (11.11)$$

where ϵ is an angle characterizing the non-orthogonality. In Fig. 11.6, ϵ is the angle between FQ_0 and the optic axis, and can be determined from $\tan \epsilon = OW/L$. Clearly, for the two specimen positions there are two values of ϵ, viz. ϵ_1 and ϵ_2.

At any working distance w, $L = Mw$ where M is the magnification and so a plot of w against $1/M$ will give a line with slope L for any microscope. The quantities ϵ_1 and ϵ_2 are thus fully determined. In a similar way to equation (11.10c) in Section 11.4.2, the present equation for θ is valid for either ψ_1, ϕ_1 and ϵ_1 or ψ_2, ϕ_2 and ϵ_2.

Determination of h and k can now be carried out as follows, using the notation of Section 11.4.2:

1. Measure x_1, x_2, x'_1, x'_2, ψ_1, ψ_2, ϕ_1, ϕ_2, M_1, M_2. See below.
2. Put $\alpha_1 = x_1/L$, $\alpha_2 = x_2/L$,
   $\beta_1 = (x_1 + x'_1)/L$, $\beta_2 = (x_2 + x'_2)/L$,
   $\epsilon_1 = \tan^{-1} (\alpha_1 \cos \psi_1)$,
   $\epsilon_2 = \tan^{-1} (\alpha_2 \cos \psi_2)$.
3. Calculate h_0 and k_0 according to equations (11.8).
4. Calculate θ_1 and θ_2 where
   $$\theta_{1,2} = \tan^{-1} \{\tan \psi_{1,2} \cos (\phi_{1,2} + \tan^{-1} k_0 h_0 - \epsilon_{1,2})/\cos \epsilon_{1,2}\}.$$
5. Check for $|\theta_1 - \theta_2| < 1°$.
6. Calculate $h = \{(h_0^2 + k_0^2)/\cos^2 \theta - k_0^2\}^{\frac{1}{2}}$.
   (for θ_1 or θ_2).
7. Since $k_0 = k$, the coordinates (h, k) have been calculated.

Clearly it is not reasonable to use these equations manually, but again a computer treatment based on the above sequence of operations is straightfoward. The measured parameters are three off each micrograph and four instrument variables. The determination of the quantities x requires special mention. A line parallel to the tilt axis is drawn on the micrographs through the origin, O (i.e. the centre of the micrograph where the "image" of the optic axis occurs). Referring to Fig. 11.6 this is OT, and we have

$$x' = XT$$

and

$$x = TZ,$$

X and Z being the images of the two points of interest P and Q. It can be seen that the quantity $XT \cos \psi = x' \cos \psi = OV$ is analogous to the "two-dimensional" OV in Fig. 11.5.

11.7. FURTHER APPLICATIONS OF STEREOGRAPHIC TECHNIQUES

The theories developed in Sections 11.4 and 11.6 enable the relative coordinates of any two points to be determined under any operating conditions encountered in the SEM. Extensions of the technique are now possible which allow the relationships between planes and lines to be investigated.

11.7.1. The equation of a plane

The equation of a plane in a coordinate system defined by the x, y and z-axes of the specimen stage can be determined from three points on the plane. If the three points are denoted A, B, C and B and C have coordinates (h_1, k_1), (h_2, k_2) with respect to A, then the x, y, z-coordinates of the points can be written:

$$\begin{aligned} A &= (x_0, y_0, z_0) \\ B &= (x_0 + h_1 \sin \psi_1, y_0 + h_1 \cos \psi_1, z_0 + k_1) \\ C &= (x_0 + h_2 \sin \psi_2, y_0 + h_2 \cos \psi_2, z_0 + k_2) \end{aligned}$$

where h, k, ψ have their customary meanings, and (x_0, y_0, z_0) are the coordinates of A. Taking the plane to be through the origin we can put $x_0 = y_0 = z_0 = 0$, and the equation of this plane is

$$\begin{vmatrix} x & y & z \\ h_1 \sin \psi_1 & h_1 \cos \psi_1 & k_1 \\ h_2 \sin \psi_2 & h_2 \cos \psi_2 & k_2 \end{vmatrix} = 0$$

or

$$Dx + Ey + Fz = 0$$

where
$$\begin{aligned} D &= h_1 k_2 \cos \psi_1 - h_2 k_1 \cos \psi_2, \\ E &= h_2 k_1 \sin \psi_2 - h_1 k_2 \sin \psi_1, \\ F &= h_1 h_2 (\sin \psi_1 \cos \psi_2 - \sin \psi_2 \cos \psi_1). \end{aligned}$$

Writing $G = (D^2 + E^2 + F^2)^{\frac{1}{2}}$ the direction cosines of a normal to the plane are $(D/G, E/G, F/G)$.

11.7.2. The angle between two planes

Applying the method outlined in Section 11.7.1 to two planes in turn gives the direction cosines of the normals from which the angle between the planes can be determined. If the direction cosines determined are (L_1, M_1, N_1) and (L_2, M_2, N_2) then the angle between the planes is $\cos^{-1} (L_1 L_2 + M_1 M_2 + N_1 N_2)$.

11.7.3. The angle between two lines

The determination of the angle between any two lines on a stereopair of micrographs can be made by performing three stereographic calculations. The angle, δ, between lines AB and AC (see Fig. 11.8) is given by the cosine formula as

$$\cos \delta = (l_1^2 + l_2^2 - l_3^2)/2l_1l_2$$

where

$$l_1 = |AB| = \sqrt{(h_1^2 + k_1^2)},$$
$$l_2 = |AC| = \sqrt{(h_2^2 + k_2^2)},$$
$$l_3 = |BC| = \sqrt{(h_3^2 + k_3^2)}.$$

The pairs of values (h_1, k_1), (h_2, k_2), (h_3, k_3) are the coordinates of B referred to A, C referred to A, and C referred to B, respectively. This method falls down if any of the lines is parallel to the tilt axis, but the specimen can normally be rotated to avoid this condition.

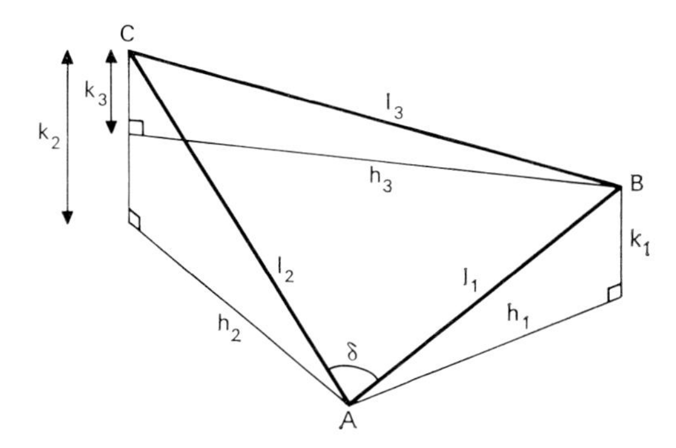

FIG. 11.8. Determination of the angle between lines.

11.8. MISCELLANEOUS MEASUREMENT TECHNIQUES

11.8.1. Contour mapping

In certain applications it may be very useful to have a contour map of a specimen and these have been prepared from stereopairs of transmission (Boyde, 1967) and scanning (Kimoto and Suganuma, 1968) electron micrographs. It would clearly be extremely laborious to determine contours on a point-to-point basis using the full stereographic methods, and simpler methods using adaptations of aerial photogrammetric techniques are required. Such techniques depend fundamentally on the proportionality between parallax and height, which only applies when a specimen is viewed normally in the microscope at high magnifications. At non-zero mean operating angles, a "height' is given by the parallax which refers to a plane at an angle to the specimen surface and is not simply related to the true height normal to the specimen surface (see Section 11.3.2).

If contour mapping is required, therefore, it is advisable to work at a mean operating angle as close to zero as possible, when the height determined refers to a plane close to the specimen plane and the effect of perspective error is least. If very high accuracy is not required, the problems associated with low magnifications may not arise. At a magnification of 100 times, for instance, non-orthogonality will introduce small height errors (0·02% to 0·1% depending on working distance) in the final contour map. This uncertainty is less than that introduced by perspective error, which can be estimated from the change in magnification with working distance. For instance, at 100 times, a height difference equal to 50% of the full field of view (not unreasonable in some applications) gives an error of up to 20% in dimensions measured at this height, with consequent uncertainty in the positioning of the contours. To compensate for the loss of signal which occurs when working at low operating angles, it may be useful to employ the Siemens pointed filament in place of the standard tungsten hairpin filament. This has been shown (Swift *et al.*, 1969) to give a much higher signal with consequent increase in signal-to-noise ratio.

Using a normal stereoscope and parallax bar, it is difficult to produce a contour map. Manual plotting of the parallax readings along a series of lines across the specimen can be used, but this is extremely time consuming. The Hilger & Watts stereoscope has a parallel linkage scanning mechanism and can be fitted with a drawing arm so that the points of equal parallax can be traced. The micrographs are scanned keeping the floating mark in contact with the ground, and the drawing arm traces out the corresponding contour. If access to more complex photogrammetrical equipment, such as an automatic plotting machine, is available, detailed maps can be produced readily. Kimoto and Suganuma (1968) show contour maps produced using the Wild Heerburg Ltd. Autograph A7. They start with the specimens at normal incidence and tilt 10° or 20° from this position, thereby complying with the conditions suggested above.

11.8.2. The production of three-dimensional models

As an alternative to contour mapping, Boyde (1967) has used the Hilger & Watts Stereosketch instrument (Fig. 11.9) to produce three-dimensional models directly from stereopairs. This instrument is basically a mirror stereoscope with semi-reflecting first mirrors, which permit a working surface to be seen at the same time as the three-dimensional image apparently "hanging in space". A suitable modelling material (e.g. plasticine) can then be moulded to fit the image while looking through the stereoscope. From this original model a permanent plastercast can be made. The effects of non-zero operating angles are less important with this technique since the model produced will exhibit the same tilt as the specimen, although the effects of perspective error will still be present.

11.8.3. The estimation of lengths and areas

If a specimen has a long twisted filament or fibre-type structure, the total length of these can be determined using the Stereosketch in a manner suggested by Boyde (1968). A long piece of flexible wire is bent into the shape of the feature and cut to length, while the three-dimensional image is viewed in the Stereosketch. The total length of the feature is then determined by straightening out the wire and measuring. At low magnification this procedure is open to some question due to the perspective error, but above about 100 times magnification the errors introduced would be less

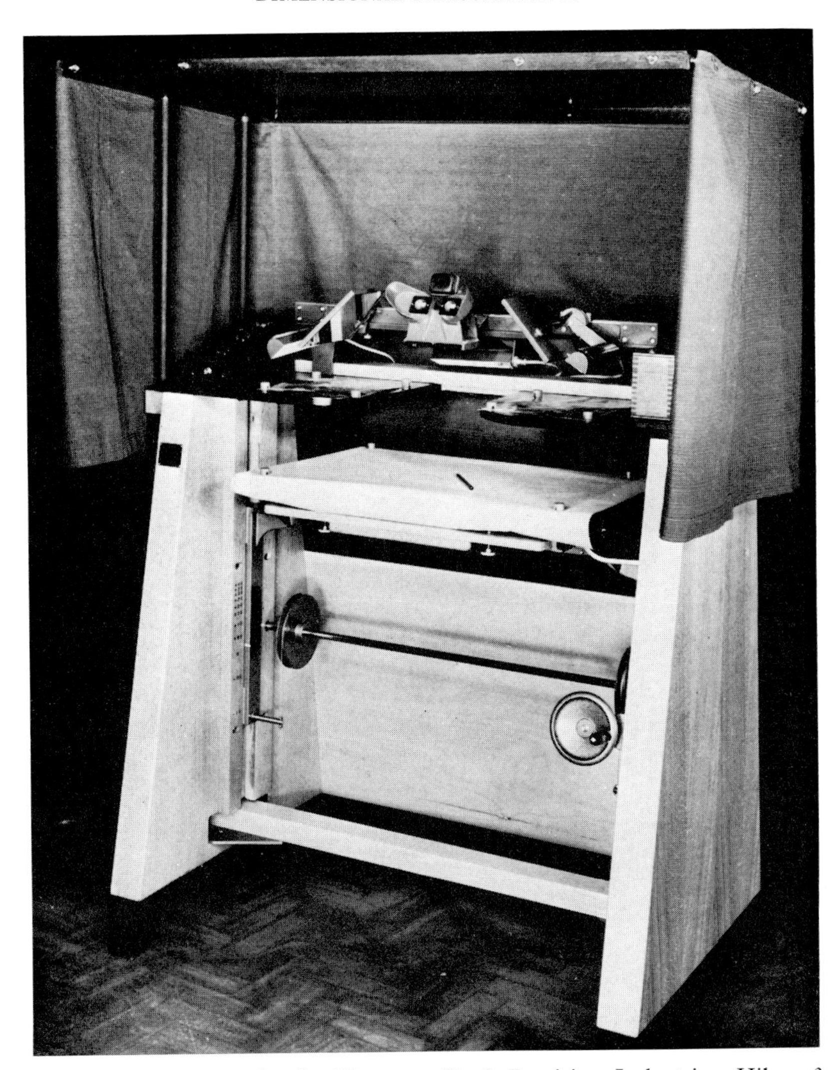

FIG. 11.9. The Stereosketch. (Courtesy Rank Precision Industries, Hilger & Watts, Ltd.)

than those involved in the experimental procedure. The size of simpler (straight) shapes can be estimated using a pair of dividers, with the same proviso. Direct measurement techniques such as these are extremely useful when the highest accuracy is not required.

On relatively flat specimens it is possible to estimate the areas of certain features by cutting out the image on a micrograph, and weighing. Boyde (1969) gives details of the method. Clearly the specimen and the region in question must be orientated normal to the beam for the true area to be given. If this is not the case the true area A_t is related to the apparent area A_a by

$$A_t = A_a/\cos\theta.$$

The angle, θ, between the normal to the plane and the optic axis, can be determined following Section 11.7.2 using stereographic techniques. Thus,

$$A_t = A_a/(L + M + N)$$

where L, M, N are the direction cosines of the area of interest.

Area measurement is a particular case of image analysis, and it seems likely that in the future more use will be made of the techniques of quantitative metallography in the analysis of all types of SEM images (topographical, compositional, or otherwise). At present, micrographs can be examined optically in quantitative metallography instruments and some image analysis performed but the scanning system of such instruments is ideally suited to the direct derivation of information from the SEM without the intermediate stage of a photograph.

Recently methods have in fact been described for the direct analysis of SEM signals along these lines. One approach (Dorfler and Russ, 1970) is to pass the input signal through discriminators and modulate it with a high frequency to obtain a digital signal which can be used to count the number of intercepts, spacing, and transverse lengths of each particle or phase. Logic units and counters are used to accumulate and display the basic parameters. White *et al.* (1970) convert the whole image into a series of binary picture points, again using discriminator and logic circuits. The computer-processed image so formed can then be analysed, and an example is given of the analysis of particle size distributions in particulate aluminas. This technique has a great deal in common with the analysis methods used in instruments for quantitative metallography such as the QTM (Metals Research Ltd., Cambridge).

ACKNOWLEDGEMENTS

Part of the work described above was carried out at the British Rail Research Department, Derby, and publication is with the permission of the Director of Research.

REFERENCES

BOYDE, A. (1967) *J. Roy. Microscop. Soc.* **86,** 359–70.

BOYDE, A. (1968) *Beitr. elektronenmicroskop Direktabb. Oberfl.* (*Münster*) **1,** 97–105.

BOYDE, A. (1969) *Z. Zellforsch.* **93,** 583–93.

BOYDE, A. (1970a) *Proc. 3rd Annual Scanning Electron Microscope Symposium* (*Chicago*), pp. 105–12.

BOYDE, A. (1970b) *Beitr. elektronenmicroskop Direktabb. Oberfl.* (*Münster*), to be published.

CHRISTENHUSS, R. and PFEFFERKORN, G. (1968) *Beitr. elektronenmicroskop Direktabb. Oberfl.* (*Münster*) **1,** 129.

DORFLER, G. and RUSS, J. C. (1970) *Proc. 3rd Annual Scanning Electron Microscope Symposium* (*Chicago*), pp. 65–72.

GARROD, R. I. and NANKIVELL, J. F. (1958) *Brit. J. Appl. Phys.* **9,** 214–18.

KIMOTO, S. and SUGANUMA, T. (1968) JEOL report No. SM-68018, Japanese Electron Optics Laboratory Co. Ltd.

LANE, G. S. (1969) *J. Sci. Instr.* (*J. Phys. E*) **2,** 565–9.

LANE, G. S. (1970) Unpublished work.

MARTIN, D. B. (1966) *2e Symposium International de Photo-interpretation* (*Paris*) III, 38–44.

SWIFT, J. A., BROWN, A. C. and SAXTON, C. A. (1969) *J. Sci. Instr.* (*J. Phys. E*) **2,** 744–6.

WELLS, O. C. (1960) *Brit. J. Appl. Phys.* **11,** 199–201.

WHITE, E. W., GÖRZ, H., JOHNSON, G. G. and MCMILLAN, R. E. (1970) *Proc. 3rd Annual Scanning Electron Microscope Symposium* (*Chicago*), pp. 57–64.

CHAPTER 12

Management of the SEM

A. J. Sherrin

12.1. INTRODUCTION

The scanning electron microscope is an investment of the order of £20,000 ($50,000) so it is essential that full value is obtained from the instrument. The evidence of the past few years has shown that well-managed installations can produce a stream of high-quality useful results. The foundation of the success of the SEM lies in its technical performance of being capable of producing unique images of surfaces. These images are produced over a wide range of magnification from samples which usually require the minimum preparation.

The widely varying specimens can be prepared and examined very quickly which results in a high work load. There are other modes of operation available, such as electron-probe X-ray microanalysis, which add to the SEM's versatility. If the instrument is well planned and managed then full use can be made of all the instrument's peculiar advantages.

Because the specimens are easy to prepare and the images are usually straightforward to interpret there is less need for background knowledge than there is in other forms of microscopy. It is not unusual for an operator to prepare and examine specimens as diverse as food and nuclear fuel without any detailed prior knowledge of these fields. In many ways it is a general instrument and can therefore provide a service to a variety of disciplines.

12.2. ORGANIZATION

12.2.1. Planning and policy

If the use of an SEM is well planned then many people can successfully obtain useful results from a single instrument. How an instrument will be used needs to be decided at the earliest possible stage. The terms of reference will include its purpose, its availability, the control and operation, the siting and the auxiliary services to be provided.

It is quite common to carry out a market survey of potential users when submitting a case for the purchase of an SEM and it is worthwhile determining at this stage the type of service that is required. This service will be different for each individual instrument as in some cases the instrument will serve a large area of the country or a university, while others will be tied to a very specific organization.

There are three basic ways of using the instrument: firstly, it can be used for "in house" work only; secondly, time can be rented on the instrument with an operator provided; or thirdly, the instrument can be rented without an operator. Some instruments are exclusively operated by full-time operators who take their instructions from the people wanting to make use of the instrument. Another method is to train anyone who wishes to use the instrument to operate it. It is quite usual for a combination of these systems to be used according to the needs of any given situation. It is, however, important to lay down the fundamental policy clearly so that priorities and procedures can be correctly understood. If the instrument is to be rented then a realistic charge should be made and the rules for rental clearly laid out. These will obviously vary from organization to organization and will depend on the individual service being provided.

12.2.2. Supervisor

The running of an SEM entails both technical and administrative problems involving a number of people, therefore it is essential that one person who is conversant with the microscope and its operation is made responsible. The duties involved would include responsibility for maintenance, training of operators, planning the work load and managing the administration. It is an interesting position with opportunities for experimentation and for providing a useful service for a wide variety of disciplines. Although not necessarily a full-time occupation it will take a considerable amount of time if done properly. If this duty is neglected, however, the efficiency of the system drops rapidly and the standards slip so that often poor or indifferent results are produced. It is only necessary to compare a poor micrograph with a good micrograph of the same specimen to realize how important it is to maintain high standards.

12.3. LOCATION AND FACILITIES

12.3.1. Accommodation and installation

The siting of the SEM should be planned well in advance of its installation. Details can be obtained from the instrument manufacturers so that the room can be prepared for installation. If this is carried out satisfactorily then the instrument can be operational within a matter of hours but if the room is not ready then the installation can be severely delayed.

The microscope should be situated so that it can perform satisfactorily and is convenient to use. The first consideration is to be able to meet the manufacturer's minimum specifications which lay down acceptable levels of ambient magnetic fields and vibration levels. Areas close to drop-forging equipment and high-frequency welding equipment are therefore not usually ideal.

An SEM is a large instrument which can be used to examine a large number of specimens and it therefore requires fairly substantial accommodation. The basic facilities necessary are:

A room for the SEM console.
A specimen preparation area.

Office space for records and administration.
Facilities for photographic processing.

12.3.2. SEM accommodation

The instrument console should be in a separate room which is of the minimum size suggested by the manufacturer. The room needs to have a solid floor capable of supporting the heavy console with sufficient space around the console to provide access for maintenance. Sufficient room is needed for the operator and at least two observers to work while seated in front of the instrument. A small bench will be needed to carry specimens, accessories and any equipment being currently used. There should be sufficient cupboard and drawer space to store accessories, spare parts and any other equipment. It should also be remembered that the SEM has high "visitor appeal" and on a number of occasions up to a dozen people will be squeezed in to view the instrument in operation.

The microscope will need an exclusive power supply and the laboratory will require an independent power supply. The instrument power supply is fed to the power packs which are often situated in an adjacent room to the instrument. The instrument manufacturers usually stipulate a minimum and a maximum distance between the power packs and the main console. The availability of suitable space for the power packs in relation to the instrument can be a critical aspect in selecting a suitable site for the instrument. The power packs and console are connected by a group of cables, including a high-voltage cable, which should be protected in a suitable conduit. This may require overhead trunking or service trenches for all or part of the route. The manufacturer usually requires to know the local mains voltage and frequency before the instrument is installed. Independent power points are required close to the instrument to supply service equipment during maintenance schedules.

A suitable cold-water feed and waste are required to provide cooling water for the diffusion pumps in the main console. The water should run to an open drain so that the rate of flow can be checked at any time. Although the water is usually drawn from the town supply, in hot countries or where mains pressure is low a refrigerated enclosed system can be operated.

It is advisable to have these supplies in position before the instrument arrives as it makes installation a straightforward operation.

Quite obviously the room will need to be dark and if there are windows these will need to be covered with darkroom quality blinds. The room should be well lit with the lights being controlled from the console as well as the normal wall switch. Many operators like a low-level illumination which can be provided by an angle poise lamp situated in a convenient place. As the room is enclosed the ambient temperature quickly rises because of the heat provided by the instrument. If the operator is to remain alert for long periods in a warm, dark room then the temperature and humidity will have to be controlled. Full air-conditioning is the most satisfactory solution to the problem as this also reduces the dust level in the atmosphere. The stable atmospheric conditions are beneficial to both the operator and the instrument as well as significantly reducing specimen contamination from airborne dust.

The door to the SEM room should not be lockable for safety reasons. There is always the possibility of an accident from an instrument which works at high voltages especially when there are adjacent running-water supplies. If an accident should occur then rapid access would be difficult if the door was locked. Suitably placed warning lights or notices are usually sufficient to keep the door shut when required. If there is an automatic fire-warning system in the building then it is advisable to have a sensing head over both the console and the power pack but *sprinkler systems should be avoided at all costs*. It is prudent to have access to a CO_2 type of fire extinguisher in case of emergencies.

12.3.3. Specimen preparation

Specimen preparation is not usually difficult except for some biological tissue which has to be prepared to withstand the vacuum of the instrument. However, some space adjacent to the main microscope room should be available so that specimens can be stored and prepared for examination. The most costly apparatus that is usually required is a vacuum evaporation unit which is used to deposit a thin conducting layer over the surface of non-conducting specimens. This unit should be a robust one and adapted for coating SEM specimens. The area should have some bench space, a laboratory sink and storage cupboards for any auxiliary apparatus not directly required in the microscope room.

If there is enough room available this area can conveniently be used to lay out micrographs side by side for the interpretation of results. This is a very useful facility which can also act as display area when demonstrations are arranged.

12.3.4. Office accommodation

Because the SEM is frequently used to a high capacity there is a considerable volume of filing and administration. It is not unusual to process over 10,000 negatives in a year, all of which have to be filed and cross-referenced. For each negative there will be at least one print and there will be many requests for reprints. This requires space and it takes up time so it should be decided at the outset whether the SEM laboratory is to be responsible for the filing of negatives or whether they should be dispatched to the sponsor of the work. This is a policy matter and can be most usefully resolved by deciding the ownership of the results; if these belong to the organization then the SEM laboratory can keep the central file but if the results belong to outside groups or individuals then they will be responsible for their own filing.

There is also a considerable amount of letter or report writing to be done in connection with various examinations. If the instrument is used by a number of people or time is rented then careful booking records will have to be kept and in some cases time records so that invoices can be made out. As the SEM operator works in a darkened room for long periods it is a welcome relief if the office accommodation has a pleasant outlook.

12.3.5. Photographic facilities

The resolution of the photographic recording screen is considerably better than that of the visual display so each specimen is usually photographed several times. An exhaustive study can use some tens of negatives on a single specimen. This large number of photographs require proper processing facilities. The ideal solution is a full-time photographic department who have sufficient capacity to deal with this volume of work. If this is not available then a proportion of the operator's time will be occupied in developing, printing and general photographic duties.

It is essential to have at least free access to a photographic darkroom for loading and unloading films. If, however, printing is to be carried out on any scale then a private darkroom will be required. Polaroid film, of course, avoids the problems of normal processing but there are certain problems with this material which will be discussed later.

It is advisable to have a lockable cupboard at some convenient place for the storage of photographic materials. This is not only a sensible precaution against loss but is usually a requirement of the Customs and Excise department for materials which have been exempt from purchase tax. The same cupboard can also be used for keeping any other precious materials such as gold/palladium wire.

12.4. OPERATION

12.4.1. Usage

The whole purpose of scanning electron microscopy is to produce an image or measured response from a specimen in the instrument. The chief factors controlling the quality of the work produced are the skill of the operator and the condition of the instrument. Maintaining the quality while obtaining a high work load from the instrument is the essence of the management of an SEM. The work must first be found, executed and finally the results dispatched.

Scanning electron microscopes all need "customers" who require the particular service that the instrument provides. There needs to be a sufficient work load to keep it employed and the original market survey is a good basis for promoting work in the early stages. The catchment area for potential users will be established by the policy for running the instrument. The technique of scanning microscopy is now sufficiently well known that potential users often promote their own subjects for examination. The supervisor ought also to be aware of the activities in the SEM field and keep potential "customers" in touch with their own specialities. Much of the work load is often promoted by the department or laboratory directly responsible for running it which is a satisfactory system as it provides the "bread and butter" work which keeps the instrument in constant operation.

The same principles are true if the bulk of the time on the instrument is rented except that there needs to be a formal sales programme. Few instruments are operated wholly on a rental-only system, therefore locally promoted investigations are still important.

12.4.2. Operators

Any operator of an SEM must be fully competent. It is only when the instrument is operated correctly that consistent high-quality results are produced. It is not a difficult instrument to operate and most people with a technical or scientific background can be successfully trained to use it. Training is not a difficult task but it needs a little time and preparation if it is to be done well. It should certainly be a rule governing any instrument that no person is allowed to operate it until they have been properly trained.

If the instrument is always worked by full-time operators then there must be enough of them to cope with the work load. It would need two operators to keep an instrument in full use provided that their photographic duties were at a minimum. It is important that all operators are kept up to date with the latest techniques and opportunities should be made for them to keep up with the subject.

12.4.3. Operator training

It is very tempting to reduce training to half an hour's demonstration of the controls and then allow the new operator loose on the instrument. It is certainly true that no one can be really competent until they have had several hours experience at the microscope. This experience is, however, more readily and surely acquired when it is backed by a sound practical knowledge of the way the instrument works.

An operator should be taught the basic principles on which the instrument works and then be shown how to operate the instrument to produce a good image and then photograph it. The instruction can be based on the simple block diagram of the instrument provided by the manufacturer which shows the construction of the electron optical column and the path of the accelerated primary beam. The collection of the secondaries and their conversion into an image on a cathode-ray tube should be described. The operator should then understand the purpose for each control and meter.

Once this is understood then the operator can begin to master the controls to produce first-class images and micrographs. This is a matter of instruction and practice at the instrument console. Each operator ought to be capable of producing high-quality images at all magnifications and be able to deal with difficult specimens. One exacting task is correcting for astigmatism when working at high magnification. The use of both 35-mm and Polaroid cameras should also be understood and practised. It cannot be stressed too strongly that the difference between a good and a bad micrograph of the same area of a specimen is so marked that it is always worth the effort of operating the instrument correctly.

All operators ought to be instructed in simple maintenance tasks such as filament changing. They should understand how to work the vacuum system manually so that they can deal with emergencies. The supervisor and another person ought to be able to deal with slightly more complex problems such as changing light guides.

12.4.4. Operating procedures

It is quite normal to examine twenty to thirty specimens in a day and this can result in over 100 micrographs being produced. The specimens are easily prepared and can be exhaustively examined in a time measured in minutes rather than hours. It is this high capacity which enables an instrument to be used for such a wide range of applications. Some instruments are used for both day and evening sessions so their output is considerable. It is therefore important to set out a suitable operating procedure to prevent any muddle or chaos.

There are three principle areas which need to be controlled. Firstly, there needs to be a system for allocating time on the instrument; secondly, there must be a system for recording specimens; thirdly, a system is needed to record and dispatch the final micrographs. Precisely how this control is carried out will vary according to circumstances but the responsibility for each area must be made clear.

12.4.5. Instrument allocation

This is a straightforward procedure but it needs to be carried out efficiently. It is obviously more significant when an instrument is being rented than if an instrument is run by two operators who directly control their own work programme. All that is required is a desk diary kept in a convenient place which is used exclusively for booking the SEM. The time required is shown against the name of the customer with any other relevant comments. From this operators' programmes can be planned for each week. In this way routine maintenance can be booked to run in smoothly with the instrument's programme.

If the operators work full time on the instrument it should be noted that their other duties away from the instrument take as much time as actually operating the instrument. It is therefore necessary to have the equivalent of two full-time operators if the SEM is to be kept fully employed.

12.4.6. Specimen processing

The number of specimens in an SEM laboratory can often run into hundreds at any one time. To prevent confusion each specimen should be recorded as soon as it is received making a note of date, specimen and the initiator. A sufficient number of columns should be available to record the progress of specimens until the results are finally dispatched. Many groups of specimens will be received which will appear to be identical and therefore great care should be taken to label each specimen individually. To prevent contamination and help identification it can be helpful to place each specimen in a separate envelope or plastic box. Plastic or cellophane envelopes are preferable as they are both transparent and also do not produce any contamination of the specimen surface. Boxes are useful for specimens which are easily damaged.

Specimens are usually placed on a specimen stub for examination in the SEM and each specimen can be identified by marking the base of the stub with an appropriate number. It is often desirable to preserve the specimens after they have been examined until the results have been studied in detail; therefore, the stubs are not reused very quickly. A large number of stubs are required and this may run into several hundred especially if there are a number which need to be adapted for a particular type of specimen.

Stubs with specimens can conveniently be supported on small blocks of plastic or metal with holes drilled into them to receive the shank of the stub. These blocks can be made to fit standard plastic boxes or a drawer of the small plastic racks which are currently available. Each block and drawer can be numbered so that any specimen may be easily retrieved by consulting the specimen book. This is particularly useful when specimens have to be re-examined after a lapse of a month or so. Finally, the stubs are not cheap and it is worth cleaning them for re-use when the specimens are no longer required.

12.4.7. Examination procedure

This is the most important part of the whole programme and all that has gone before and comes after is to make this process as efficient as possible. The real problems at this stage are to examine the specimen for the required detail and photograph the relevant areas. This means concentrating on the image and manipulating the controls to produce the most informative image. It is, however, essential to record all the details of each exposure as it is made. Most exposures last for at least 40 seconds which is usually sufficient time to write the necessary information. If recording the information is left until the specimen change then most of the details will be forgotten or worse still they will be confused.

The most satisfactory procedure is to fill a form in as the examination progresses. The detail on the form should be sufficient but not too great or this leads to apathetic use if there are columns which are rarely used or which are repetitive. A typical form is shown in Fig. 12.1 which has proved useful in practice.

Each form has a suitable heading for date, specimen, initiation, identification, etc. The form is then subdivided into forty spaces which is the usual number of exposures obtained from the standard 35-mm cassette when it has been self-loaded. As each exposure is made the details are filled in with any special comments which may be relevant. These can include such details as stereo-pairs or any particular points relating to the specimen. A sheet is thus completed for each cassette and the negative numbers on the film are recorded after the film has been processed.

The details on each particular form are likely to vary but in principle they are the same. When the instrument is "on rental" then each customer will have separate forms even if a film is only half exposed. It is normal practice for the customer eventually to retain both the film and the details on the form. In addition rental periods should be carefully controlled with starting and finishing times being recorded as well as any substantial periods of interruption. It may be useful to have duplicate carbon copies of these details which should be signed at the end of the period by both the operator and the customer. A similar procedure should be adopted when the instrument is leased for a period with the operator being provided by the customer.

DATE	SUBJECTS	INITIATOR	DEPARTMENT

No.	NEG. No.	STUB	SPECIMEN	COMMENTS	MAG.	FILE No.
1						
2						
3						
4						
5						
6						
7						
8						
9						
10						
11						
12						
13						
14						
15						
16						
17						
18						
19						
20						
21						
22						
23						
24						
25						
26						
27						
28						
29						
30						
31						
32						
33						
34						
35						
36						
37						
38						
39						
40						

Fig. 12.1.

12.4.8. Filing and dispatching

If the original procedure forms are carefully written then they can provide the basis for the master file. When each negative has been printed it can be given a unique number which is recorded in the file against the negative. These forms can then be placed in loose-leaf binders to be kept as a permanent record. The 35-mm negatives can be cut into strips of six and stored in one of the commercial binders available for filing 35-mm film. These binders contain pockets for each strip of film and pages for recording the details of the film. Prints can also be kept if desired but they require much more space. At this stage it can be useful to record cross-references on a card-index system which can relate to subject-matter or initiators of the work. All this is necessary because repeat prints are often required months or years later and there has to be some system for retrieving the right negatives.

If this type of procedure is followed the recording of specimens, micrographs and filing take place as part of the normal process of instrument operation. When the prints are labelled they can be dispatched to the initiator of the work with a note giving any necessary information. If a good system is worked out then it prevents those dreadful periods when all the work has to stop whilst everyone catches up on the filing and correspondence.

12.5. MAINTENANCE

12.5.1. SEM

An SEM represents a £20,000 investment and this capital asset should be protected by keeping it in good running order. It is not unreasonable to expect to spend $2\frac{1}{2}$–$3\frac{1}{2}$% of this capital value on annual maintenance. (This is a figure well within the level of expenditure that the average motorist incurs in servicing his car.) It just is not possible to maintain an instrument of this type on a shoe-string budget and expect to obtain good results. This attitude towards maintenance is worth establishing when the instrument is purchased and should not be delayed until the guarantee period has expired. It is difficult to obtain an annual grant of £500 or over if it has not been budgeted for in advance. There are two methods for obtaining maintenance service—either by contract maintenance from the instrument manufacturer or by using suitably qualified laboratory staff and relying on the manufacturer to provide a back-up service when required. Whichever system is used it is preferable to work on a system of regular inspection rather than waiting for problems or breakdowns to occur.

The contract maintenance system usually provides for servicing at regular intervals plus an emergency service which is all supplied for a fixed annual sum. Spare parts are often included in this type of agreement and this acts as a useful insurance policy against the failure of expensive components. The advantages of this system are that the engineers have experience of a number of instruments and the regular inspections often lead to faults being anticipated before they critically affect the performance.

In some cases a suitably qualified electronic engineer on the staff may be available to undertake the maintenance of the instrument. If this is so then the SEM should be made part of his duties and a regular servicing schedule made out. On some occasions the manufacturer may be called in for particularly difficult repairs and this can usually be arranged. The advantages of this system are that the engineer is "on site" and the instrument is regularly serviced by one man. Although this system appears to be cheaper the staff member may be more usefully employed on other tasks and also a larger reserve of spare parts has to be carried.

All operators ought to be capable of carrying out routine maintenance and should also be capable of interpreting simple telephoned instructions from an engineer.

It is useful for the supervisor to keep in close contact with the maintenance engineer so that he is always aware of the condition of the instrument. This is sometimes a useful feed-back system for discovering faulty operational procedures which need to be corrected from time to time.

12.5.2. Ancillary apparatus

Although it is not as critical the ancillary apparatus should be kept in good order. In particular the vacuum evaporation unit needs regular attention if it is to perform satisfactorily. This is a straightforward job of cleaning the chamber and changing the oil in the pumps but it is a long, messy procedure. There are vacuum specialists who will undertake this type of maintenance job.

All optical instruments such as enlargers and binocular microscopes need to be regularly cleaned and inspected. The lens surfaces are particularly sensitive to dirt and damage. If an air-conditioning plant is installed the filters need to be changed regularly and the plant serviced annually.

12.6. RENTAL SERVICE

When providing a rental service it is important to remember that it is a commercial operation and the customer is looking for value for money. Instruments are usually charged on a time basis with a minimum period of hire. The rate will depend entirely upon the circumstances under which the hiring takes place. A research institute may charge less to its members who have already subscribed to the capital cost of the instrument than a commercial laboratory which is working for a profit.

Because time is limited the instrument tends to be operated more intensely during rental periods and a few points are worth noting. The customer should arrive, preferably on time, knowing clearly what programme he wishes to follow. Specimen preparation should be kept to a minimum by providing specimens which only require to be suitably mounted and perhaps coated. It is worth having more specimens than time will allow because sometimes a line of investigation proves to be fruitless and then an alternative line can be pursued.

A good operator should understand the requirements of the customer and help him by explaining the images on the SEM.

12.7. PHOTOGRAPHY

One of the cornerstones of microscopy is the production of a visual image. Scanning microscopy is no exception and it is usually true that more detail can be seen on the final micrograph than on the visual screen. Great care should therefore be taken over the photography and it is valuable to have the advice of a professional technical photographer. It is ideal if the bulk of the photographic processing can be carried out by an established photographic department because they already possess the necessary equipment and skills for dealing with large volumes of photographic material.

Both Polaroid and 35-mm film are usually standard for SEM work although there is no fundamental reason why other formats should not be used. Polaroid, of course, has the advantage of producing quick results and reducing the cost of processing to practically nothing. However, from experience it seems to be more difficult to produce high-quality results from Polaroid than from 35-mm film. Polaroid exposures often have to be repeated and this adds to the expense of an already expensive material. It is a tedious process to produce duplicate prints unless the large format P/N 55 package with a redeemable negative is used. 35-mm film, however, is much cheaper and easier to use for long runs and it is not difficult to process. All exposures should be carefully controlled to produce uniform results as this aids the subsequent processing of the film. The development should follow a precise procedure so that each film is correctly processed. Ideally the operator should quickly scan the wet film to check that there are no mistakes which need to be rectified. This is particularly important on rental service when the specimens are often available for a strictly limited period.

However, printing 35-mm films takes a long time and it requires a fully equipped photographic processing unit. There are, however, some automatic printers which semi-automate the processing stages and produce satisfactory prints. It is worth noting that most SEM negatives are square whereas the usual commercial printing papers are oblong (5×4: 10×8) and so if the full negative is printed there is a waste of up to 20% of photosensitive paper. However, photographic companies will supply printing paper cut to any specified size at no extra cost provided a sufficient amount is ordered. Printing is not an expensive process provided it is organized correctly and it is possible to produce prints at a cost of a few pence each.

The final print is the evidence which the SEM finally provides and it is worth taking considerable trouble to make sure that these are always of high quality.

12.8. VISITORS

The SEM has great "visitor appeal" and this should be catered for within the normal work of the laboratory. It is useful to have some sets of the more interesting micrographs ready for display and to keep suitable specimens for demonstration purposes. If these are arranged as standard items then a worthwhile demonstration can be mounted at short notice and with the minimum delay to the running of the department.

12.9. CONCLUSION

The SEM can produce a whole wealth of information if it is properly run. These thoughts on managing an SEM are a collection of my own experiences and those of other people with whom I have discussed the subject at various times.

My thanks are to the Directors of Dunlop Ltd. for allowing me to contribute this chapter.

CHAPTER 13

The Future of Scanning Electron Microscopy

J. W. S. HEARLE and D. C. NORTHROP

13.1. THE PATTERN OF DEVELOPMENT

When this book was first planned the scanning electron microscope was passing out of a brief infancy, characterized by a concentration on taking pictures of an ever-increasing number of experimental specimens, and by a rather small choice of commercially available instruments. There was developing a much more serious intention to use the instrument systematically in research, and it was being suggested that the mere recording of topographical detail might be supplemented by a range of specialized techniques. The pace of progress is such that attachments to commercial instruments now allow many of these techniques to be used, and they have been described in the preceding chapters of this book.

Inevitably this has produced a two-pronged development amongst commercial instruments. The research instruments, on which we shall concentrate in this chapter, are becoming more complex and more expensive to allow the full use of all the possible modes of operation. There is also a number of cheaper and less versatile instruments suitable for teaching and for routine inspection and testing in those areas of work where scanning electron microscopy has become established as an indispensable routine tool.

The SEM has become, in itself, a miniature laboratory. From the general appearance of the sample—presented in various ways, in reflection or transmission, at various angles, and at accelerating potentials of 1 to 50 kV, or even from an auxiliary optical microscope view—it is possible to switch to a study of its electrical characteristics, to the identification of elements by the X-ray or Auger electron emission, to a study of cathodoluminescence effects, to an investigation of the crystallography by electron diffraction or electron channelling lines, to the performance of dynamic experiments and the recording of moving pictures, to etching of the sample by ion bombardment, to processing by the electron beam, and so on. In the most advanced commercial instruments, the specimens themselves may be as large as 10 cm or so in size, and the guaranteed resolution is 150 Å (15 nm), though under optimum conditions it may be much better.

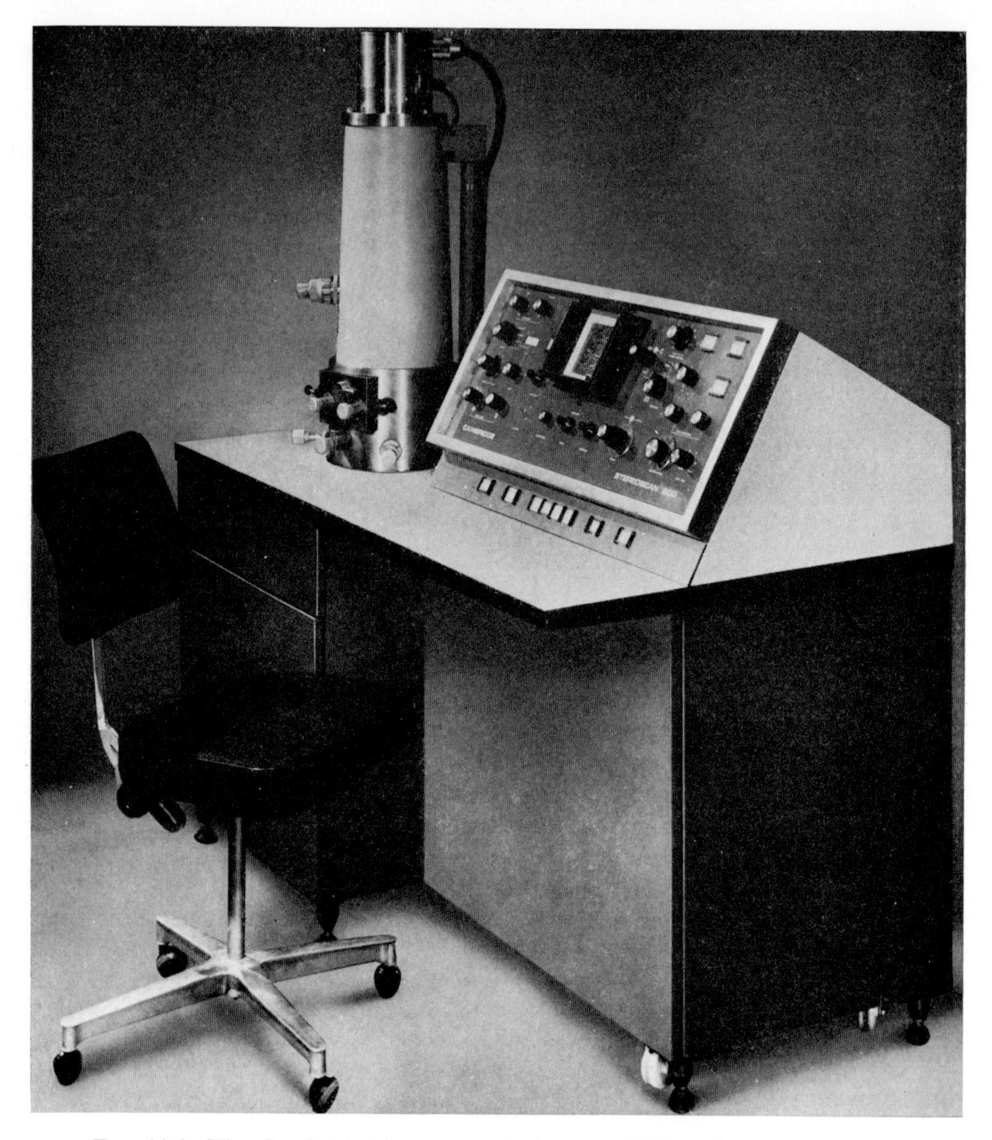

FIG. 13.1. The Cambridge Stereoscan 600 a new high-performance compact SEM which is easy to operate and ideal for routine sample checking. (Courtesy of Cambridge Scientific Instruments Ltd.)

Apart from these advances in facilities there are, of course, continuing improvements (and maybe some reverses) in the engineering design of the microscopes—in particular the use of solid-state (transistor) circuits is simplifying the electronics, and some instruments have ion-pumps to give a better and cleaner vacuum. The Cambridge Stereoscan 600, shown in Fig. 13.1, is an example of the simpler and cheaper microscopes; others are the JEOL JSM-61 and the Perkin–Elmer SSM-2A. They contrast with expensive, advanced microscopes such as the AMR-900, the Stereoscan S4, and the JEOL JSM-U3.

In future, we are thus likely to see these two main families of commercial scanning electron microscopes developing side by side. The first, marketing at under £10,000, would be general-purpose instruments, designed for economy and ease of use. The second, marketing at over £20,000, would be the advanced instruments, intended for the most demanding research studies and probably specially adapted to the needs of a particular user.

Instruments which are cheaper still can be made if the performance requirements are reduced: for example, the VG Miniscan has electron optics giving a spot size with a resolution of only 1000 Å. As another possible economy measure for some owners of electron microscopes, Philips have announced an SEM attachment for their high-resolution transmission electron microscope EM 300.

Going beyond the advanced commercial instruments, we find some even more specialized research scanning electron microscopes: super-instruments either in their resolution or their accelerating voltages.

In the remainder of this chapter, we shall describe briefly some developments which the scanning electron microscopist should watch for and give some additional information on the latest techniques, which have not been fully incorporated into the general pattern of usage described in previous chapters.

13.2. INSTRUMENT DEVELOPMENTS

13.2.1. High-resolution versions of current instruments

In a number of research laboratories, scanning electron microscopes basically similar in design to commercial instruments have been operating so as to give improved resolution; some of their features are now appearing in commercial instruments. O. C. Wells, speaking at the 1970 Chicago Symposium, mentioned three particular features in his own microscope. The first was a modified design of electron collector, which omitted the grid, placed an improved scintillator carefully in position, and had a straight light pipe. The second was the use of a "brighter" electron gun of which further details have been given by Broers (1970). The filament is made of lanthanum hexaboride with a tip 1 μm in diameter instead of the usual tungsten filament, and makes it possible to get a spot of 25 Å (2·5 nm) or less. The third feature, required for guns of this type, is the use of a higher vacuum. In addition it must be remembered that improved resolution will not be achieved unless the instrument is free from vibration, stray fields, lens asymmetries or supply variations.

As a summary of the position, and a reminder that the real limits are often in the specimen, we may quote Broers (1970): "Little further reduction in probe diameter below 25 Å (2·5 nm) for surface microscopy and 10 Å (1 nm) for transmission microscopy can be expected as the brightness of electron guns is increased. Higher beam currents will result, however, and these should allow improved signal/noise ratios in the microscope images as well as shorter exposure times. Electron penetration effects rather than electron optical limits will in most cases set the limit for surface microscopy." A lanthanum hexaboride gun with an ion-pump system has been introduced commercially in the Ultrascan SM 3.

13.2.2. Crewe's microscopes

Much greater changes have been pioneered by A. V. Crewe at the University of Chicago. He uses a tungsten field emission gun. A good general account of this development is given by Crewe (1971) in *Scientific American*. The field emission source has a brightness over a thousand times greater than a conventional source though it also requires a vacuum 10,000 times higher than is needed in other electron microscopes: current high-vacuum technology allows this to be done.

Crewe's aim in this work was to make a high-resolution scanning microscope which would exceed the best direct electron microscope in resolving power when used in transmission, and so enable single atoms to be detected. If this could be achieved it would be of enormous benefit in the characterization of complex natural organic molecules.

Crewe points out that the similarity of the two kinds of microscope—the principle of reciprocity discussed in Chapter 1—might plausibly suggest that no more could be achieved with the scanning electron microscope than with the direct transmission microscope. What makes the difference is the fact that in the scanning microscope all the electron optics is in the incident beam, which has a very narrow spread of energy, and all the scattered electrons are available for collection and characterization. By an energy analyser, Crewe separately collects the elastically scattered (N_e) and inelastically scattered electrons (N_c) and then makes use of the fact that:

$$N_e = 46{\cdot}5\,(Nn/\sigma\, V)\, Z^{\frac{4}{3}},$$

$$N_i = 868\,(Nn/\sigma\, V)\, Z^{\frac{1}{3}},$$

where N is the total number of incident electrons, n is the number of atoms in the beam, σ is the cross-sectional area of the beam, V is the accelerating voltage, and Z is the atomic number of the element.

It is then possible to process the signal to form an image from the ratio of the collected signals (N_e/N_i): this will depend only on Z and will give pure atomic number contrast. In this way Crewe has located individual uranium and thorium atoms attached to organic molecules.

Another advance came almost by chance. With the field emission source a 100 Å (10 nm) spot could be produced with a rather simple electron lens only 4 cm from the gun. This has made it possible to make an SEM which is smaller and simpler than other instruments. A commercial microscope, the Cwikscan/100, which utilizes these principles has now been introduced. Apart from simplicity and mobility the major advantage claimed is the fact that the great brightness enables high-quality and high-resolution pictures to be obtained at fast (TV-rate) scan speeds. This gives a flicker-free picture and would be useful for dynamic studies. It remains to be seen how microscopes of this type, with an extra-high vacuum, will perform in general use.

13.2.3. A high-voltage SEM

Another advanced development is the high-voltage SEM described by Cowley *et al.* (1970). This operates with an electron acceleration potential of 600 kV. Among the particular applications described are the observation of diffraction patterns and dislocation studies. However, the most interesting possibility is what has been called "al fresco" electron microscopy, with the specimen out of the vacuum. In order to do this in direct transmission electron microscopes, the specimen has to be put in a special cell with two electron transparent windows. But in a scanning microscope, because there are no electron optics beyond the specimen, all that is needed is one window for the scanning electrons to emerge from the microscope. The specimen and detector can be placed in the outside atmosphere as illustrated in Fig. 13.2.

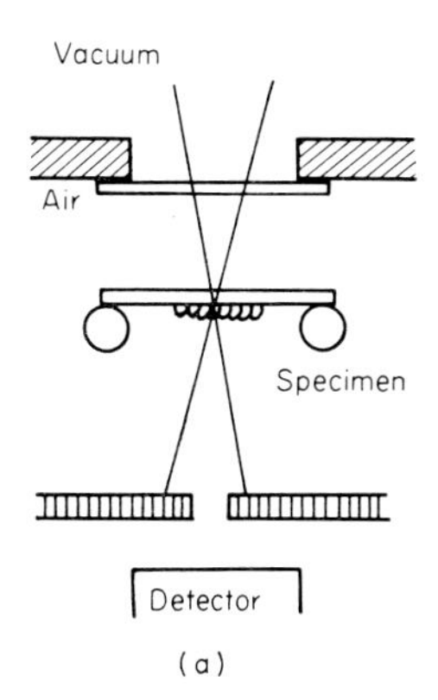

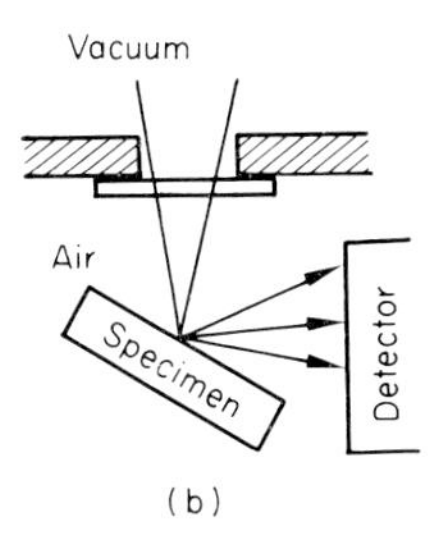

Fig. 13.2. Diagrams illustrating arrangements for "al fresco" electron microscopy of (a) transmission specimens, (b) reflection specimens (after Cowley *et al.*, 1970).

13.3. TECHNIQUES

13.3.1. Dynamic viewing

Recently, high-speed scan systems have been introduced. These are compatible with commercial television and so enable a picture to be seen on the screen of an ordinary TV monitor. This gives an obvious advantage in picture quality and enables dynamic events to be viewed in real time. An enlarged view of a watch movement makes an impressive demonstration in an exhibition.

However, it must be remembered that it was not just a desire to be different that originally led to the use of slow scan speeds: these are essential if a high enough signal-to-noise ratio is to be achieved, particularly in difficult specimens. The new facility of TV scanning will be useful in some circumstances but not in all.

For dynamic studies, which can be themselves slowed down, a time-lapse ciné-photography technique, as described by Hearle *et al.* (1971), is often preferable.

13.3.2. Environmental stages

The use of a cold stage is described in Section 9.3.4 and hot stages are also easily made. What is less obvious is that the vacuum requirement in the immediate environment of the specimen need not be so stringent. Cross and Cross (1968) described studies of evaporating ice, in which the water vapour from the specimen was continually being removed. Other specimens, such as hair, can be examined by putting

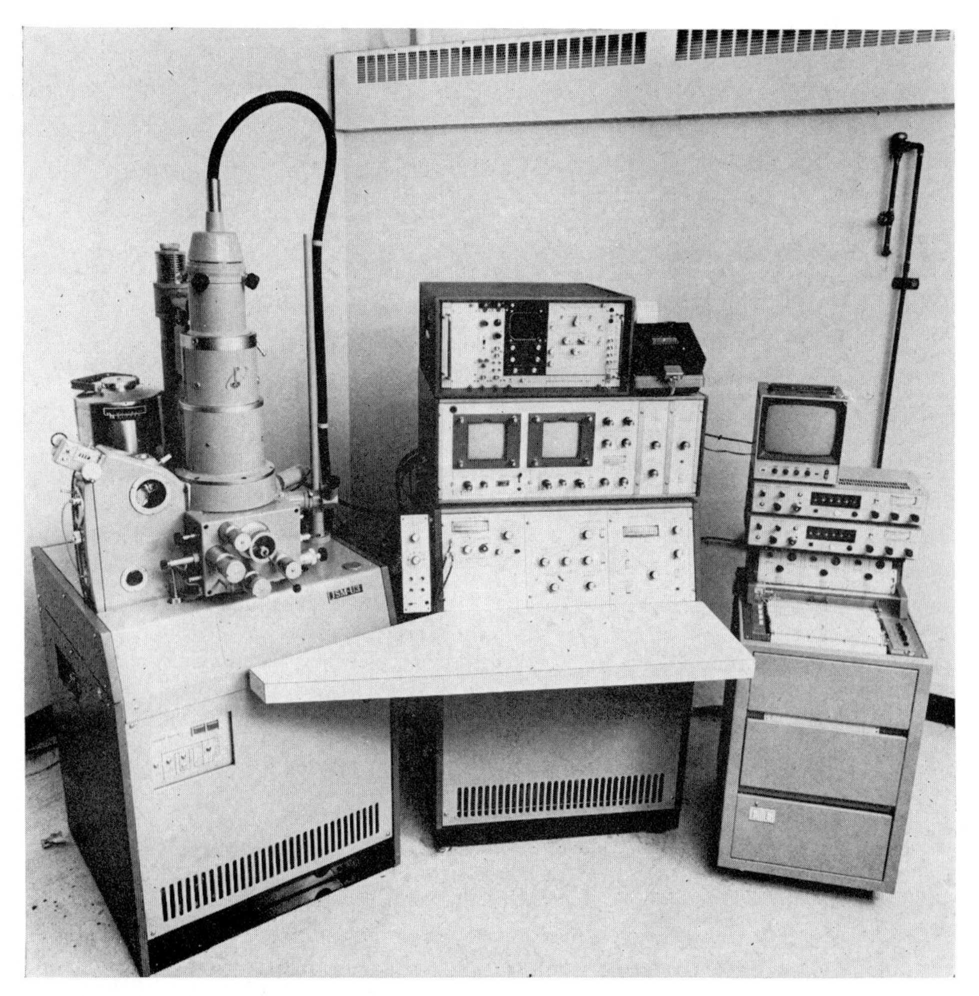

FIG. 13.3 Nuclear diodes solid-state X-ray energy analyser system (non-dispersive type) shown interfaced to JEOLCO Model JSM-U3 SEM. (Courtesy of A.E.P. International Ltd., Staines, Middlesex.)

them in the microscope wet and allowing the moisture to evaporate slowly during 20 minutes or so of examination.

In another stage, described by Lane (1970), an enclosed reservoir with small annular openings is located below the specimen which thus remains in an atmosphere of moderate humidity.

13.3.3. Element identification

The availability of X-ray analysis devices for attachment to scanning electron microscopes for identification of elements has now become commonplace. A review of the methods is given by Sutfin and Ogilvie (1970). In the dispersive system a crystal diffracts the X-rays so as to form a spectrum which can be analysed; in the non-dispersive system, the energy of the photons falling on the detector is used as the means of discrimination. For most SEM applications the latter has most advantages.

As an example, Fig. 13.3 shows a nuclear diodes solid-state (non-dispersive) X-ray energy analyser connected to a JEOLCO scanning electron microscope. A typical application is illustrated in Fig. 13.4 where a collection of particles can be separately identified according to whether they contain aluminium, silicon or zinc. Another method of use, the element analysis at a single point in the specimen, was illustrated in Figs. 7.20 and 7.21.

The chief limitation of X-ray analysis is that it cannot be applied to the lighter elements. For these elements, Auger electron analysis, as described by MacDonald *et al.* (1970), can be used. An ion-pumped vacuum is needed for this application, and it must be noted that what is detected is the element composition in a surface layer only a few angstroms deep (as compared to the micron depth of X-ray analysis): surface contamination is therefore a particular hazard.

13.3.4. Electron channelling

Electron channelling is a phenomenon which can be observed only in high-quality single crystal specimens. It arises because in such crystals the electron density is non-uniform on an atomic scale, and there are therefore certain crystallographic directions in which electron–electron scattering is comparatively low. These are the "open" directions which can be easily identified by inspection of a crystal model. It therefore follows that if the primary electron beam in an SEM is incident on the specimen in a channelling direction the mean range of the primary electrons will be greater than that for a non-channelling direction. This in turn means that the resulting secondary emission from the surface is reduced because the secondaries must escape from defects in the crystal. By a similar argument it follows that energetic electrons at a given depth below the specimen surface have the greatest chance of escaping if they are travelling along a channelling direction. These are the two basic mechanisms by which light and dark bands are produced when the surface of a uniformly flat single crystal is scanned by the electron beam.

There are two ways in which electron-channelling patterns may be produced in an SEM. The first, which involves no modification to the instrument, consists of scanning the beam over the specimen surface at low magnification so that the angle of incidence

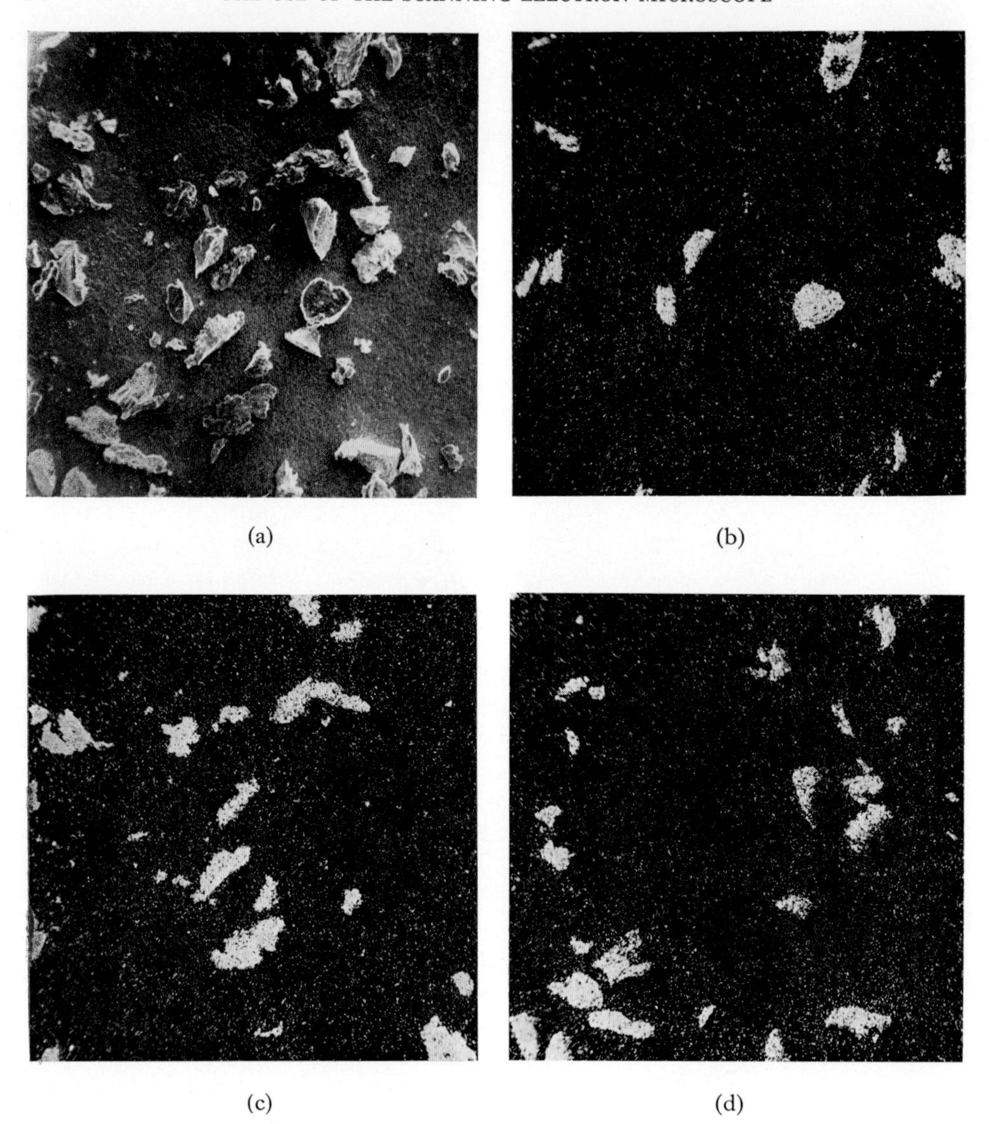

(a) (b)

(c) (d)

FIG. 13.4. (a) Normal SEM image, (b) Al X-ray image, (c) Si X-ray image, (d) Zn X-ray image. (Courtesy of A.E.P. International Ltd., Staines, Middlesex.)

on the surface is a function of the scan coordinates. The secondary electron emission depends on the angle of incidence. The disadvantages of this method are, firstly, that it requires a fairly large specimen ($\sim$ 5 mm) and low magnification ($< 100\times$) in order to obtain large variations in angle of incidence, and, secondly, that each angle of incidence is investigated at a different point on the crystal surface.

A more advanced method which overcomes these objections uses a modified scanning system in which the electron beam is held on a particular point on the surface and is made to "rock" over a wide range of angles of incidence by two equal and opposite linear displacements of the beam. This allows channelling patterns to be obtained from very small single crystals, in theory no bigger than the beam diameter, and for constant surface conditions.

Channelling patterns bear a strong resemblance to Kikuchi electron diffraction patterns and can be used to obtain the same kinds of information about the specimen. Firstly, and most obviously, the patterns can be used to determine crystal structure and orientation and unit cell size, and in principle this could be used to identify small crystallites. Secondly, the data can be used to provide information about point defects and other departures from perfect crystallinity. Crystal imperfections of any kind will clearly reduce the probability of electron channelling and will therefore change the intensity distribution and may also distort the channelling pattern. The detailed interpretation of these changes is not a simple matter, and much work needs to be done before channelling patterns can be fully utilized.

Readers interested in electron channelling are advised to consult the *Proceedings of the 4th Scanning Electron Microscope Symposium*, 1971, and the review article by Booker (1970).

13.3.5. Stroboscopic techniques

Stroboscopic techniques are commonly used to examine high-frequency periodic phenomena under flash illumination where the flash duration is much shorter than the period of the phenomenon and the flash frequency is very slightly less than the frequency of the phenomena. In this way the apparent frequency of the phenomenon is slowed down to the difference between its real frequency and the flash frequency.

The possibilities of using an SEM in this way are strictly limited because the flash frequency corresponds, in this case, to the frame frequency. This is such a low frequency that there is no need to use stroboscopic techniques for phenomena which are of even lower frequency. It must, however, be borne in mind that the electron beam is capable of modulation at very high frequencies, and this leaves open the possibility of stroboscopic work where the beam modulation defines the "flash" duration and frequency.

One possible application is being attempted by Gopinath (1971) who aims to make observations of high field domains in Gunn oscillators using microwave modulation of the electron beam without scanning the beam on the specimen surface. The idea is to measure the field as a function of time, at a point on the surface of the device, and then to move the beam and repeat the measurement, so piecing together a complete picture of the field distribution and its variation with time.

At lower frequencies it may be possible to use beam modulation and a single-line scan in order to obtain a complete picture.

13.4. SIGNAL PROCESSING

13.4.1. The range of the SEM

It has been made clear in many places in this book that the fact that the information from the object is transformed into an electron signal offers many possibilities. Various modes of display have already been mentioned.

The signal can also be manipulated in various ways. If it is differentiated, regions of rapid change will be sharply shown up—this can often make a picture clearer, though care must be taken in interpretation. Image processors have therefore become available as SEM attachments.

Where quantitative measurements have to be made of numbers of particular particles, areas, and so on, image analysis techniques are available. For Auger electron analysis, and other purposes, MacDonald's instrument at North-American Rockwell is connected to a computer. And we have mentioned how Crewe takes the ratio of two signals, from electrons of different energy, to get a contrast proportional to atomic number.

13.4.2. The isolation of potential contrast from surface topology

Oatley (1969) has described a beam-modulation technique which allows the surface topology contrast to be suppressed, thereby allowing a much more precise measurement

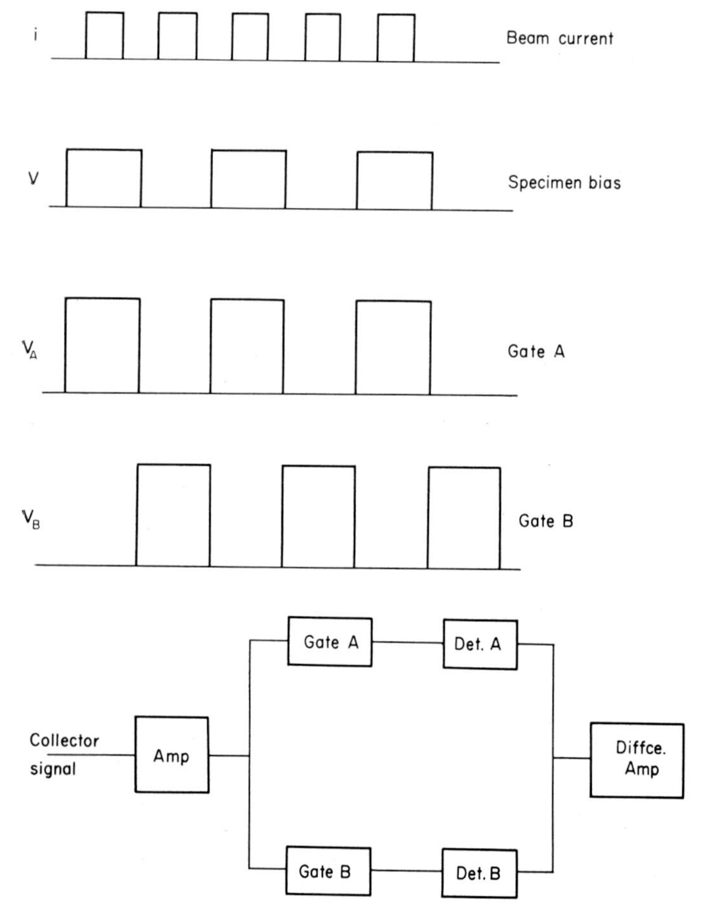

FIG. 13.5. Circuit schematic for the isolation of potential contrast.

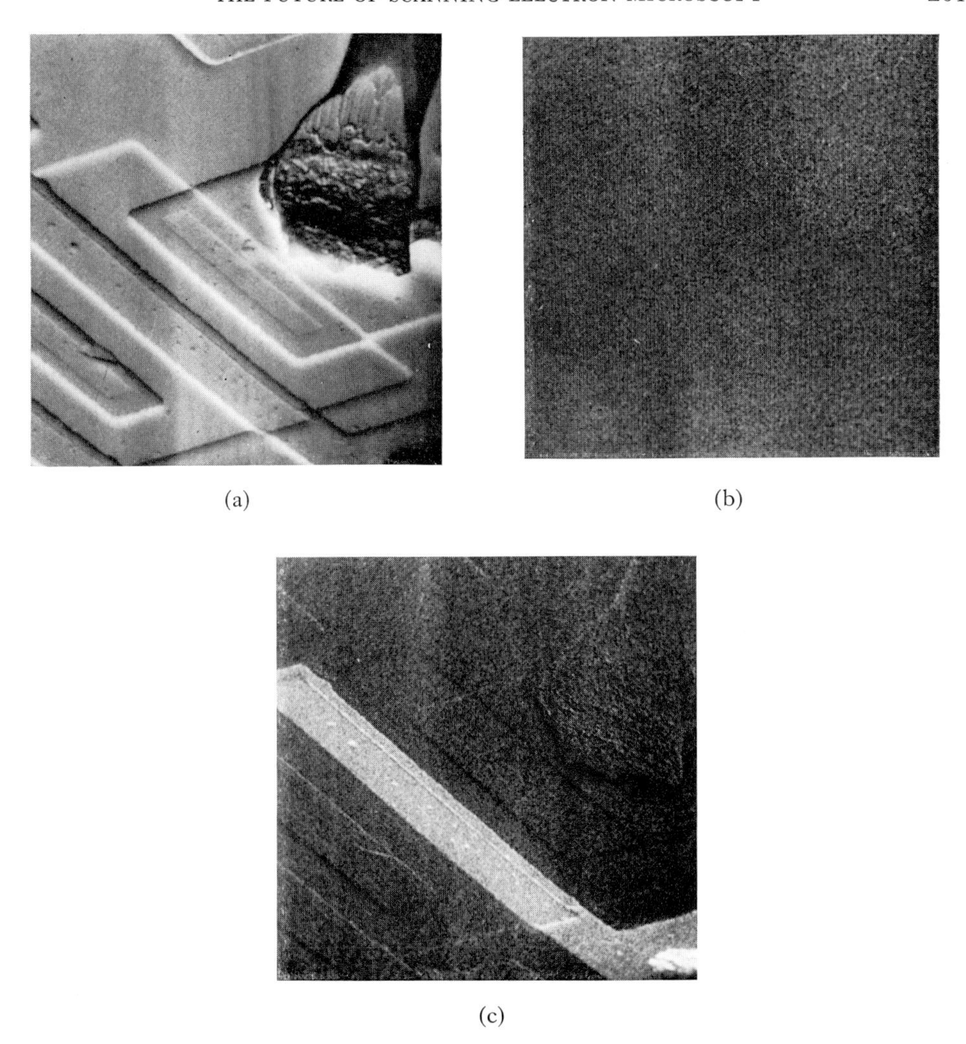

(a) (b)

(c)

FIG. 13.6. Micrographs of an electronic shift register. (a) Display from one channel only. (b) Both channels with specimen bias. (c) Both channels with bias applied to specimen. (We are indebted to Professor C. W. Oatley who supplied the original micrographs for this illustration.)

of the potential distribution in the specimen. The method is indicated by the circuit and waveform diagrams in Fig. 13.5. The beam current must be chopped by a square wave voltage applied to deflection plates mounted between the electron gun and the first lens (100 V peak to peak is suggested). The frequency of the square wave must be high enough not to cause loss of resolution in the resulting picture, and 200 kHz is adequately high for a frame scan time of 30 seconds. The specimen bias is modulated at half this frequency and the resulting signal to the electron collector in the microscope is then fed to two detector circuits via gate circuits A and B. The signal fed to channel A is that due to surface topography and potential distribution whilst that fed to channel B

is due to surface topography alone. Thus the difference signal A–B is the potential contrast alone. Figure 13.6 illustrates clearly the value of this technique, by which potential differences as low as 0·1 V should be detectable.

13.5. CONCLUSIONS

13.5.1. Sources of information

In a book on a new subject like scanning electron microscopy, it is finally important to list the sources of the most recent information on new developments. For this reason the names of eleven manufacturers of scanning electron microscopes and the manufacturers of auxiliary equipment and materials are listed at the end of the chapter.

Papers which include the results of SEM studies are now widely dispersed in the scientific literature; and more particular information on the technique itself will be found in *Journal of Scientific Instruments* (*Journal of Physics E*), *Journal of Microscopy*, *Review of Scientific Instruments*, and in conference proceedings such as the Proceedings of the 25th anniversary meeting of the Electron Microscopy and Analysis Group of the Institute of Physics published as a book under the same title in 1971. However, the most important sources are the proceedings of the annual scanning electron microscopy symposium organized annually since 1968 by Dr. Om Johari at IIT Research Institute, Chicago. Apart from the conference papers, this includes a bibliography started by Miss. V. Johnson and now being kept up by Dr. O. C. Wells of IBM Watson Research Centre, Yorktown Heights, N.Y. Authors of papers are encouraged to keep Dr. Wells informed so that the bibliography can be kept up to date.

13.5.2. Review of 1971 Chicago Symposium

A good way to describe the changing state of the art of scanning electron microscopy as this book goes to press is to review the papers presented at the 1971 Symposium held at IITRI. A classification of the papers is given in Table 13.1. The dominant fields of application are metals, circuits and devices, and biological materials, with some more exotic specimens such as lunar dust. Polymers and fibres are surprisingly under-represented, possibly because of a symposium on scanning electron microscopy of polymers and coatings held in the autumn of 1970 (Princen, 1971).

The general background of the work described in most of the papers is included elsewhere in this book, but a few new developments and highlights are worthy of special mention. The two instrument developments involve the use of additional electron lenses. In the first (Yew, p. 33), an additional lens is added in order to correct the focus as the beam moves across a specimen. This cannot, of course, improve the depth of focus in a complicated specimen, but it can counter the effects of change in level from one side to the other in a specimen which is tilted towards the collector. Dynamic focusing, as this is called, is available on the ETEC SEM. In the second (Dinnis, p. 41) a deflection system (an electron prism rather than an electron lens) is placed between the lenses and the specimen. This enables stereo-pairs to be obtained by changing the direction of "observation" without moving the specimen (cf. Fig. 1.11)

TABLE 13.1. CLASSIFICATION OF PAPERS FROM THE 1971 IITRI SYMPOSIUM
(Numbers are page references in symposium proceedings—Johari and Corvin, 1971.)

Topic	*Papers*
General methodology	1, 9, 17, 521, 529
Instrument developments	33, 41
Contamination in SEM	505
Signal processing	393
Special techniques:	
transmission SEM	27
X-ray emission	65, 73, 81, 537
Auger	89
cathodoluminescence	401, 409
electric signals	443, 457
mirror	433
back-scattered electrons	145
electron channelling	465, 473, 481, 489, 497
dynamic studies	97, 113
micro-manipulation	105
signal synchronization	451
device fabrication	417
Preparation of biological specimens	225, 249, 257, 265
Image examination methods	49, 57
Applications	
Metal morphology, etc.	163, 193, 201, 217, 471
Metal deformation and fracture	97, 113, 121, 129, 137, 143
Analysis	73, 169, 177
Inorganic non-metals, soil, etc.	49, 153, 209, 385, 409
Circuits and semiconductors	81, 145, 425, 433, 441, 349, 457, 517
Fibres	185
Palaeobiology	233
Microorganisms	241, 273, 321, 511
Biological tissues, etc.	251, 281, 297, 305, 313, 337 345, 353, 361, 369, 377
Biological cells	249, 265, 289, 329
Forensic studies (special workshop)	537, 545, 553, 561, 569, 577, 585
Bibliography	587

and is also useful in aligning the beam direction for the observations of channelling patterns.

Boyde (p. 1) gave the keynote paper on interpretation of scanning electron micrographs. His views on several topics are strong and controversial, but worth noting:

On three-dimensional analysis

"Interpretational errors . . . which stem from the analysis having been stinted by the availability of only two-dimensional pictures."

"It is impossible to believe that we can analyse complicated surface morphology without the aid of stereoscopic vision and geometrical analysis and reconstruction."

On distortion due to tilt of the specimen

"It is common practice to put magnification markers on single SEM images . . . this can only serve to perpetuate errors of concept."

"The best practice would be to give the nominal width of the field of view of the image . . . at a height halfway through the depth of field."

On distortion due to the tube

"CRT's of most currently available commercial SEM's do not have a raster presenting straight lines, evenly spaced and orthogonally orientated."

On etching

"The interpretation of the morphology of etched surfaces requires thinking back to what was present in the layers which have been removed rather than what is present in the layers which remain."

On derivative processing

"Derivative signal processing makes it possible to study fine detail in dark holes or bright crests. Although it is often possible to see the same information in an 'unprocessed' normal video image after the event, it is our experience that detail was not seen, and would not have been seen without this signal processing."

On colour coding

"Although the 'colour' itself can have no real meaning . . . the use of colour to combine different sorts of image is generally helpful."

Charging is a nuisance in scanning electron microscopy and the effects are often not well understood. Van Veld and Shaffner (p. 17) give a very useful account of the way in which bright and dark spots result from the deflection of the beam to a region where many electrons emerge and reach the collector or the reverse. In the analogy with conventional viewing these are "sources of light or darkness which are reflected in the charged specimen". It is this rather than any other effect which probably gives rise to most troubles due to static; and careful attention needs to be paid to the effective electric currents in the specimen and its surroundings. In order to avoid charging troubles the value of the time constant CR for the system must be small enough.

Among techniques of novel interest we find the use of phase-contrast methods and other contrast techniques in transmission micrography (Zeitler, p. 25); the sensitivity of back-scattered electron images to chemical composition (Price and Johnson, p. 145); and the study of cathodoluminescence spectra (Muir *et al.*, p. 401). In image analysis there were reports on the use of optical diffraction techniques to pick up periodicities (Tovey, p. 49) and high-resolution enhancement by *a posteriori* holographic image processing (Stroke *et al.*, p. 57). The latter technique demonstrates the latent information present in a scanning electron signal and shows that in conditions where the best resolution obtained by ordinary methods would be 200 Å, the holographically sharpened images show a resolution better than 100 Å.

REFERENCES

BOOKER, G. R. (1970) Modern Diffraction and Imaging Techniques in Material Science, North Holland Publishing Co., pp. 613–53.

BROERS, A. N. (1970) *Proc. 3rd Annual SEM Symposium*, IIT Research Institute, Chicago, Ill. (ed. O. Johari), p. 1.

COWLEY, J. M., SMITH, D. J. and SUSSEX, G. A. (1970) *Proc. 3rd Annual SEM Symposium*, IIT Research Institute, Chicago, Ill. (ed. O. Johari), p. 9.

CREWE, A. V. (1971) *Scientific American*, **224**, No. 4, 26.

CROSS, J. D. and CROSS, P. M. (1968) *J. Sci. Instr.* (*J. Phys. E*), series 2, **1**, 1123.

GOPINATH, R. (1971) private communication, University College North Wales, Bangor.

HEARLE, J. W. S., CLARKE, D. J., LOMAS, B., REEVES, D. A. and SPARROW, J. T. (1971) Proc. 25th Anniversary Meeting of EMAG, Institute of Physics, London, p. 210.

JOHARI, O. and CORVIN, IRENE (eds.) (1971) *Scanning Electron Microscopy*, IIT Research Institute, Chicago, Ill.

LANE, W.C. (1970) *Proc. 3rd Annual SEM Symposium*, IIT Research Institute, Chicago, Ill. (ed. O. Johari), p. 41.

MACDONALD, N. C., MARCUS, H. L. and PALMBURG, P. W. (1970) *Proc. 3rd Annual SEM Symposium*, IIT Research Institute, Chicago, Ill. (ed. O. Johari), p. 25.

OATLEY, C. W. (1969) *J. Sci. Instr.* (*J. Phys. E*), series 2, **2**, 742.

PRINCEN, L. H. (ed.) (1971) *Scanning Electron Microscopy of Polymers and Coatings*, Interscience Publishers, New York.

SUTFIN, L. V. and OGILVIE, R. E. (1970) *Proc. 3rd Annual SEM Symposium*, IIT Research Institute, Chicago, Ill. (ed. O. Johari), p. 17.

LIST OF MANUFACTURERS AND SUPPLIERS

The following is a list of instrument manufacturers and suppliers of some of the consumable and non-consumable items which will be needed for operation of a SEM laboratory. This list is by no means complete and is only given as a guide to help new users of the SEM.

Instrument manufacturers

Cambridge Scientific Instruments Ltd.,
Chesterton Road,
Cambridge, CB4 3AW,
England

Kent Cambridge Scientific Inc.,
8020 Austin Avenue,
Morton Grove,
Illinois 60053,
U.S.A.

JEOLCO (U.K.) Ltd.,
JEOLCO House,
Grove Park,
Edgeware Road,
Colindale,
London, N.W. 9.

JEOLCO (U.S.A.) Inc.,
477 Riverside Avenue,
Medford,
Massachusetts 02155,
U.S.A.

CAMECA,
103 bd St-denis,
92 Courbevoie,
France.

CAMECA Instruments Inc.,
101 Executive Boulevard,
Elmsford,
New York 10523,
U.S.A.

ETEC Corporation,
2284 Old Middlefield Way,
Mountain View,
California 94040
U.S.A.

Materials Analysis Company,
1060 East Meadow Circle,
Palo Alto,
California 94303,
U.S.A.

Perkin-Elmer,
Box 10920,
Palto Alto,
California 94303,
U.S.A.

Philips Electronic Instruments,
750 South Fulton Avenue,
Mount Vernon,
New York 10550,
U.S.A.

Applied Research Laboratories,
P.O. Box 129,
Sunland,
California 91040,
U.S.A.

Vacuum Generators Ltd.,
Charlwoods Road,
East Grinstead,
Sussex.

Coates & Welter Instrument Corporation,
2191 Ronald Street,
Santa Clara,
California 95050,
U.S.A.

Ultrascan Company,
18530 South Mills Parkway,
Cleveland,
Ohio 44128,
U.S.A.

Vacuum coating equipment

Edwards High Vacuum Ltd.,
Manor Royal,
Crawley,
Sussex.

Edwards High Vacuum Inc.,
3279 Grand Island Boulevard,
Grand Island,
New York 14072,
U.S.A.

Balzers High Vacuum Ltd.,
Northbridge Road,
Berkhamsted,
Herts.
Balzers Hochvakuum GmbH,
6, Frankfurt 90,
Postfach 900144,
West Germany.
N.G.N. Ltd.,
Kirk Road,
Accrington,
Lancs.
High Voltage Engineering Corp.,
South Bedford Street,
Burlington,
Massachusetts 01803
U.S.A.

Evaporation sources
Tungstan Manufacturing Co., Ltd.,
Fishergate Works,
Portslade,
Brighton,
Sussex.
C. W. French Division,
EBTEC Corp.,
5 Shawsheen Avenue,
Bedford,
Massachusetts 01730,
U.S.A.

Coating materials
Johnson Matthey Metals Ltd.,
81 Hatton Gardens,
London, E.C.1.
Balzers High Vacuum Ltd.,
(see address above)
Edwards High Vacuum Ltd.,
(see address above)
Degussa,
6450 Hanau,
Postfach 622,
West Germany.
C. W. French Division,
EBTEC Corp.,
(see address above)

Rotary coating devices
Materials Science North West Ltd.,
29 Lowther Street,
Kendal,
Westmorland.
Industrial Development,
Dean Street,
Bangor,
North Wales.
EBTEC Corporation,
(see address above)
Applied Research and Engineering Ltd.,
Parsons Estate,
Washington,
County Durham.

Apertures and filaments
Siemens (U.K.) Ltd.,
Great West Road,
Brentford,
Middlesex.
Siemens A.G.
WWM-Berlin,
Abtlg. Elektronen-Mikroskopie,
1 Berlin.
EBTEC Corporation,
(see address above)

Mountants
Degussa,
6, Frankfurt,
Postfach 3993,
West Germany.
Fischer & Fischer,
758 Bühl (Baden)
Postfach 1440,
West Germany.
Aquadag, Silverdag
Acheson Colloids Ltd.,
Prince Rock,
Plymouth,
Devon.
Durofix
The Rawlplug Co., Ltd.,
London Road,
Kingston,
Surrey.
Twinstik
Evode Ltd.,
Stafford,
Staffs.

Antistatic spray
Duron
John Sheard Ltd.,
P.O. Box 309,
Bradford,
Yorks.
Hansawerke Lurmann,
Schütte & Co.,
28, Bremen-Hemelingen,
West Germany.
Soromin TX 1360
BASF,
67, Ludwigshafen,
West Germany.

Cleaning materials

Inhibisol
Penetone Co., Ltd.,
Cramlington,
Northumberland.

Lintless tissues
General Paper & Box Co.,
Treforest Industrial Estate,
Glamorgan,
South Wales.

Quadraline
Quad Inst. Cleaning,
Quad Chem. Ltd.,
Liversage Street,
Derby.

Ultrasonic cleaners
Kerry's (Ultrasonics Ltd.),
Chester Hall Lane,
Basildon,
Essex.
MEL Equipment Co. Ltd.,
Salfords,
Redhill,
Surrey.

Additional microscopy accessories

Ernest F. Fulham, Inc.,
P.O. Box 444,
Schenectady,
New York 12301.

Polaron Instruments Ltd.,
4 Shakespeare Road,
Finchley,
London, N.3.

Wironit,
Karl Hammacher, GmbH,
565 Solingen-Aufderhöhe,
West Germany.

Special stages

Applied Research and Engineering Ltd.,
Parsons Estate,
Washington,
County Durham.

X-ray equipment

Nuclear Diodes Inc.,
P.O. Box 135,
Prairie View,
Illinois 60069.

A.E.P. International Ltd.,
14 High Street,
Staines,
Middlesex.

Ortec,
100 Midland Road,
Oak Ridge,
Tennessee 37830.

Ortec GmbH,
8, Munchen 13,
Frankfurter Ring 81,
West Germany.

Author Index

Subject Index